Low-Rank Coal Applications in Agriculture

Low-Rank Coal Applications in Agriculture

Humic Analyses, Products, and Performance

L. Edwin Liem

WILEY

This edition first published 2021

Registered Offices
John Wiley & Sons, Inc., 111 River Street, Hoboken, NJ 07030, USA
John Wiley & Sons Ltd, The Atrium, Southern Gate, Chichester, West Sussex, PO19 8SQ, UK

Editorial Office
The Atrium, Southern Gate, Chichester, West Sussex, PO19 8SQ, UK

For details of our global editorial offices, customer services, and more information about Wiley products visit us at www.wiley.com.

Wiley also publishes its books in a variety of electronic formats and by print-on-demand. Some content that appears in standard print versions of this book may not be available in other formats.

Library of Congress Cataloging-in-Publication Data

Name: Liem, L. Edwin, author.
Title: Low-rank coal applications in agriculture: humic analyses, products, and performance / L. Edwin Liem.
Description: First edition. | Hoboken, NJ: Wiley, 2021. | Includes bibliographical references and index.
Identifiers: LCCN 2020030609 (print) | LCCN 2020030610 (ebook) | ISBN 9781119734758 (hardback) | ISBN 9781119734819 (adobe pdf) | ISBN 9781119734727 (epub)
Subjects: LCSH: Soils–Humic acid content. | Coal. | Soil biochemistry.
Classification: LCC S592.8 .L54 2021 (print) | LCC S592.8 (ebook) | DDC 631.4–dc23
LC record available at https://lccn.loc.gov/2020030609
LC ebook record available at https://lccn.loc.gov/2020030610

Cover Design: Wiley
Cover Image: © Rustic/Shutterstock, © Christian85/Shutterstock

Set in 9.5/12.5pt STIXTwoText by SPi Global, Pondicherry, India
Printed and bound by CPI Group (UK) Ltd, Croydon, CR0 4YY

C9781119734758_170321

Contents

List of Figures

List of Tables

Units

#	number
$	dollar
%	percent
°	degree
cfu	colony forming unit
cm	centimeter
cmol	centimole
dm	decimeter
g	gram
ha	hectare
kcal	kilocalorie
kg	kilogram
kJ	kilojoule
kPa	kilopascal
l	liter
lm	lumen
m	meter
meq	milliequivalent
mg	milligram
ml	milliliter
mm	millimeter

mM	millimole/l
mmol	millimole
mole	6.02×10^{23} atoms
mpn	most probable number
mS	millisiemens
nm	nanometer
ppb	part per billion
ppm	part per million
ppt	part per trillion
t	tonne (= 1000 kg)
µg	microgram
µm	micrometer

1

Introduction

Coal is the most abundant type of fossil fuel, accounting for 64% of globally recoverable resources in the world, compared to oil (19%) and natural gas (17%). Coal is traditionally used in the energy sector, generating about 40% of the world's electricity. The demand for coal is expected to increase by over 60% from 2006 to 2030, of which developing countries will account for over 90%. Coal-fired electricity generation is still a major energy source in North America and Australia. In addition, coal is also used in the metallurgical sector, where 70% of the world's steel industry depends on it. The challenge here has been in how to maximize productivity, reduce energy consumption, and drastically reduce carbon dioxide (CO_2) emissions (Osborne and Gupta 2013). Efforts have been made to offer a more efficient and cleaner use of coal, including its use in the production of electricity, steel and its associated products, and energy-related chemicals, as well as increasing the use of coal byproducts (Osborne et al. 2013). While significant progress has been made, it is still worth seeking new environmentally friendly and efficient usages of coal beyond these well-known applications. Such usages are not set to immediately replace the existing applications of coal, however, but instead provide alternatives so that whenever the opportunity arises, new market sectors can be generated and expanded as quickly and efficiently as possible.

Among different coal materials, low-rank coal has been one of the most discussed and has seen its popularity grow in recent years. This is in part because of its unique yet challenging nature, and because of its versatile applications. Low-rank coal contains lower calorific values than those of anthracite, bituminous, and subbituminous. However, it is rich in organic matter (OM) and humic substances (HS, or "humic") that have been proven to be beneficial for several different purposes. Unfortunately, these purposes have so far been largely unrecognized or misunderstood. This is chiefly because they fall outside of the scope of the usages that people are more familiar with when it comes to coal materials. In the

Low-Rank Coal Applications in Agriculture: Humic Analyses, Products, and Performance, First Edition. L. Edwin Liem.

province of Alberta, for example, low-rank coal has been recommended as a source of calories for electricity generation (Alberta Energy 2017).

The main purpose of this book is to show that low-rank coal can be commercialized and marketed in sectors beyond those in which it is currently employed, such as electricity generation. Consequently, the new usages of this type of coal should be championed as a more efficient and cleaner way forward. As summarized below, at least four alternative usages have been identified, while others are possibly to follow in the near future.

The first alternative usage of low-rank coal is in oil/gas drilling. In this sector, drilling fluid is required to protect the stability of the wellbore from fractures, pores, and other openings. It is needed to transport cuttings for their separation at the surface, suspend solids within the fluid, and maintain stability of uncased sections of the borehole. In the water-based drilling fluid system, colloidal materials (i.e. clay) are added to the fluid to reduce water loss through the porous media of the formation. However, this may result in an increase in the drilling fluid viscosity, reducing its rate of penetration. This can be compensated for by adding low-rank coal to deflocculate the fluid, reducing both its viscosity and also reducing water loss. In the oil-based drilling fluid system, low-rank coal is added as an emulsifier, resulting in smaller droplets of a high film strength. A reduction of water loss is achieved with a minimal increase in viscosity (Dearing et al. 2004 and Caenn et al. 2017). Low-rank coal has been used more commonly in the water-based system as a thinning/dispersant and secondary filtration control agent. At high temperatures, this material performs better than that of lignosulfonates (Canamara-United Supply Ltd. 2001).

The second alternative usage of low-rank coal is in mining remediation. This sector deals with the presence of inorganic and organic contaminants that lead to a decline in soil quality. There are two in-situ treatment methods available for the contaminated soil: bioremediation and phytoremediation. Bioremediation relies on microbial metabolism of contaminants, while phytoremediation relies on uptake and processing by plants (Fangueiro et al. 2018). Low-rank coal has been found to be effective when added as a soil amendment, enhancing the growth of soil organisms and accelerating the breakdown of organic contaminants (Liem et al. 2003a). It has been found to be effective in increasing plant cover grown in mining tailings ponds after several years of treatment (Szczerski et al. 2013). In the laboratory, low-rank coal has been reported to improve the chemical and physical properties of soil (Bekele et al. 2013; Liem et al. 2003a and 2003b).

The third alternative usage of low-rank coal is in foundry. This sector concerns metal casting, which includes melting, molding, casting, solidification control, and other post-casting processing operations. Greensand (clay and water) is a molding material commonly used in the industry, in which impressions of the molten material can be formed (Campbell 2015). Between uses, water is added to the sand and later mulled (mixed and smashed) in order to regenerate it. During this step, when low-rank coal is added to the blend, it has been found to reduce clay viscosity, as well as improving mulling efficiency, mold permeability, water retention, and shakeout efficiency. This material does not affect clay durability, and in fact lowers foundry emissions by reducing the amount of smoke in the cooling line (Van Leirsburg 1999).

The fourth alternative usage of low-rank coal is in agriculture. This usage is what this book is all about. In this context, agriculture is defined as any practices related to the improvement of soil, crops, and livestock productivity. The emphasis is on the content of HS within the material, in which commercial products can be manufactured, i.e. soil amendments and feed supplements. HS are active ingredients that have been scientifically proven to enhance the productivity of soil, crops, and livestock. The main objective of this book is to show that not only does low-rank coal benefit agriculture, but also that avoiding its use in electricity generation will result in the reduction of CO_2 emissions. This can be seen as a win-win scenario with respect to both the coal and agricultural sectors, thus improving environmental credentials of the coal industry.

Before going into the detail of this proposal, it is essential to understand the properties of different coal materials. Those who are not familiar with the sector will most likely associate low-rank coal with other coal materials (especially with their potential hazards) and therefore question the feasibility of its having a green use. Chapter 2 discusses the unique physical and chemical properties of low-rank coal. In this book, low-rank coal is defined as a type of coal with lower calorific values (19 300 kJ/kg or less) compared to those of anthracite, bituminous, and subbituminous (with calorific values of 19 300 to 34 900 kJ/kg). This material falls under the category of lignite (including its weathered/oxidized form, i.e. leonardite), which is in the bottom rank among all coal materials. The main information in this book comes from a concern with low-rank coal from Alberta (also identified as humalite), which is in the author's line of work. It contains only small impurities and is therefore a good-quality material. This material has been found to be non-hazardous with respect to its contents as well as its handling.

While low-rank coal contains lower calorific values, it is rich in OM and HS. These are the determining parameters with respect to the benefit of this material in agriculture. Since HS are complex in nature and have unique physical and chemical properties, the various methods for their analysis must be understood and compared with respect to their accuracy and practicality. The colorimetric and gravimetric methods are the most popular in the commercial community due their (relative) simplicity and accuracy, while others are more popular within a theoretical or research context. Different low-rank coal materials are analyzed for their humic components and compared to one another. The main focus of Chapter 3 is on original work completed using both the colorimetric and gravimetric methods, which resulted in the modification of the International Organization for Standardization (ISO) 19822 method. This chapter also presents work on the comparison of various commercial OM products available on the market. Products to be evaluated include dry and liquid humic products, biochar, calcium lignosulfonate, soybean extract, peat moss, and molasses. These are humic and non-humic products of similar properties, which are commonly mistaken for one another.

Chapter 4 summarizes the benefits of HS in agriculture and includes information on their background science. Due to the vast collection of information and existing data on soil, crops, and livestock available from public domains, a literature review is provided. Firstly, saline and compacted soils have commonly been problematic in commercial fields in that they lower productivity. The use of HS has been found to improve the quality of such soil. Secondly, how to improve crop yields or maintain them at reduced fertilizer inputs is of interest. And thirdly, this chapter looks at the production of livestock and focuses on the growth, feed efficiency, and health of farmed animals. Efforts to reduce the emission of gaseous byproducts from livestock production are also discussed. The disadvantages of HS are also considered, i.e. that an increased application of the product does not necessarily produce a better result. This chapter ends with detailed discussions on the registration of commercial humic products with various regulatory bodies, including the US Department of Agriculture-National Organic Program (USDA-NOP), the Canadian Food Inspection Agency (CFIA), local states in the US, and other organizations. Topics of interest include status, efficacy, safety, and the claim of the registered products.

Chapter 5 is about the characteristics of Alberta low-rank coal (humalite) and the commercial products manufactured from this material. Historical data obtained from the mine site are discussed first. This discussion includes information concerning on the quality and safety of humalite. Next, commercial humic products are discussed in great detail, particularly with respect to the properties of dry, liquid, pelletized, and dry soluble products. These products can either be standalone (containing only HS) or can be packaged in combination with nutrients for soil, crops, and livestock applications. Understanding their characteristics is important (each are quite different) in order to make the appropriate decisions regarding product handling (including transfer, reprocessing, repackaging, storage, and field application) at the client or sub-contractor's site. This includes the selection of the right equipment types and materials to ensure their proper functioning, the integrity of the product, as well as the practicality and safety of the whole work.

Chapter 6 evaluates the performance of commercial humic products manufactured from Alberta low-rank coal (humalite) in agriculture. This evaluation is based on completed work and original data. Their compatibility with common nutrients is firstly evaluated. This information is critical to observing whether they can be blended with certain fertilizers, so that when applied on the field they will not create any problems with field applicators. Next, their efficacy as proven in finished projects is presented, which is the most important information contained in this book. Two product efficacy projects to improve saline and compacted soils, and nineteen projects to improve yields of rice, wheat, canola, lentil, barley, lettuce, chrysanthemum, tea, tomato, coffee, watermelon, corn, soybean, and shrub have been completed in outdoor fields. Three laboratory projects have been completed on the growth of lettuce, emergence of lentil and pea seeds, and growth of a beneficial type of fungus. One product efficacy project on livestock has been completed in the laboratory and in a test pen. Parameters to be evaluated include feed digestion efficiency and methane (CH_4) gas emission. Projects with significant statistical results or the ones showing critical information are presented in sub-sections. Not every project delivers complete data or shows significant statistical results, as they have been completed within practical and operational constraints during field work. The results are included in this chapter as they exhibit strong trends with interesting consequences. This chapter also discusses the economy of humic products for the crop application, including a cost–benefit study.

This book has been written within a practical context of low-rank coal applications in agriculture. The information and data provided, as well as their evaluation and interpretation, are therefore presented accordingly. Table 1.1 shows the format of this book, including the content and main source of information for each chapter, as well as its target readers. A literature review is included to show that the science behind the use of low-rank coal and HS in agriculture is valid. Examples of completed work and original data are presented to show the real-life applications and benefits of this material, as well as to show that its commercial products are marketable and indeed useful for the agricultural community. This book aims to accommodate the interests of various readers who are interested in this subject matter from a practical point of view. Each chapter has been written to include the necessary information for it to be read independently of the others, allowing readers to prioritize their main topic of interest.

It is to be acknowledged that many early parts of the work on low-rank coal's characterization, development of the most basic dry and liquid products, and the initial product efficacy projects in the laboratory and test plots in the early 2000s have been completed by the Department of Soil Science and the Department of Biological Sciences at the University of Manitoba, Winnipeg, Manitoba, and the Eastern Cereal and Oilseed Research Centre of Agriculture & Agri-Food Canada, Ottawa, Ontario. Some of the work has been contributed by Dr. Morris Schnitzer of Agriculture & Agri-Food Canada, a well-known Canadian scientist in soil organic chemistry and HS. All information is widely available in public domains.

Table 1.1 Book format.

Chapter and content	Main source of information	Target readers
1 General information Book format	Chapters 2 to 6	All
2 Material quality Material safety Documentation	Original data	Field practitioners End users/marketers Operators Regulators Researchers/academics
3 Analytical methods Product comparison	Original data	End users/marketers Researchers/academics Field practitioners Operators Regulators
4 Benefits and disadvantages Efficacy results for soil/crops Efficacy results for livestock Product registration	Literature	Field practitioners End users/marketers Researchers/academics Regulators Operators
5 Raw material Commercial products Product handling	Original data	End users/marketers Operators Field practitioners Regulators Researchers/academics
6 Product applicability Efficacy projects for soil/crops Efficacy project for livestock Cost–benefit study	Original data	End users/marketers Field practitioners Regulators Researchers/academics Operators
7 General information Summary	Chapters 2 to 6	All

All of the excellent collaboration with coworkers, contractors, and consultants during mine site, laboratory, and field-related activities is greatly appreciated. Product efficacy projects have been completed in collaboration with Battle River Research Group (Forrestburg, Alberta); Agriculture & Agri-Food Canada (Lethbridge, Alberta); Chinook Applied Research Association (Oyen, Alberta); Northern Alberta Institute and Technology (Edmonton, Alberta); Alberta

Innovates-Technology Futures (Edmonton, Alberta); 20/20 Seed Labs Inc. (Nisku, Alberta); fertilizer manufacturers, distributors, and commercial farms in North America (Sturgeon Valley Fertilizers Ltd., J.R. Simplot Co., Melmont Bean & Seeds Inc., Bates Farm Inc., AU Farms Inc., and Tranquility Agricultural Farm Inc.); as well as Cantho University, Cuu Long Rice Research Institute, Ho Chi Minh City, and the Department of Agricultural and Rural Development in Vietnam.

Last but not least, thanks are due to Scientific Research & Experimental Development of Canada Revenue Agency, Alberta Treasury Board and Finance, Industrial Research Assistant Program of National Research Council Canada, and the Canadian Agricultural Adaption Program-Agriculture & Food Council of Alberta for providing financial support for work related to the relevant developments in technology, including the formulation of new products and their field evaluation.

Reference

Alberta Energy (2017). Coal types. http://www.alberta.ca/coal-types.asp (accessed 13 August 2020).

Bekele, A., Roy, J.L., and Young, M.A. (2013). Use of biochar and oxidized lignite for reconstructing a functioning topsoil: plant growth response and soil nutrient concentrations. *Soil Science* 178(7): 344–358.

Caenn, R., Darley, H.C.H., and Gray, G.R. (2017). Introduction to drilling fluids. In: *Composition and Properties of Drilling and Completion Fluids*, 7th ed. (ed. R. Caenn), 1–34. Amsterdam, Holland: Gulf Professional Publishing is an imprint of Elsevier.

Campbell, J. (2015). *Complete Casting Handbook: Metal Casting Processes, Metallurgy, Techniques and Design*, 2nd ed., 797–818. Waltham, MA: Butterworth-Heinemann is an imprint of Elsevier.

Canamara-United Supply Ltd. (2001). *Humalite: A Fluid-Loss and Rheology Control Additive for Water-Based Drilling Fluids* 1–16. Calgary, AB.

Dearing, H., Simpson, J., Huddle, D., and Daniel, S. (2004). Flexible drilling fluid formulation and application. *AADE 2004 Drilling Fluid Conference* 1–11. Houston, TX: American Association of Drilling Engineers.

Fangueiro, D., Kidd, P.S., Alvarenga, P., et al. (2018). Strategies for soil protection and remediation. In: *Soil Pollution from Monitoring to Remediation* (eds. A.C. Duarte, A. Cachada, and T. Rocha-Santos), 251–281. Amsterdam, Holland: Academic Press is an imprint of Elsevier.

Liem, L.E., Facey, R.M., Carrigan, I., et al. (2003a). Enhanced hydrocarbon bioremediation through the addition of humic substances. *Mining and the*

Environment III Conference 1–8. Sudbury, ON: Canadian Land Reclamation Association/Centre for Environmental Monitoring.

Liem, L.E., Facey, R.M, and Fischer, M. (2003b). Enhanced hydrocarbon bioremediation through the addition of humic substances part II: improved land treatment and static-pile composting. *ESAA Remediation Technologies Symposium* 1–13. Banff, AB: Environmental Services Association of Alberta.

Osborne, D.G. and Gupta, S.K. (2013). Industrial uses of coal. In: *The Coal Handbook Towards Cleaner Production Vol. 1: Coal Production* (ed. D. Osborne), 3–30. Philadelphia, PA: Wood Haven Publishing.

Osborne, D.G., Sharples, M., Lien, L., et al. (2013). Future directions toward more efficient and cleaner use of coal. In: *The Coal Handbook Towards Cleaner Production Vol. 2: Coal Utilisation* (ed. D. Osborne), 497–528. Philadelphia, PA: Wood Haven Publishing.

Szczerski, C., Naguit, C., Markham, J., et al. (2013). Short- and long-term effects of modified humic substances on soil evolution and plant growth in gold mine tailings. *Water, Air, and Soil Pollution* 224(3): 1–14.

Van Leirsburg, M. (1999). Leonardite & its effects on green sand molding part 2: applications in green sand systems. *Ductile Iron News* 2: 9–11.

2

Low-Rank Coal

2.1 Overview

Coal is defined as an organic sedimentary rock containing varying amounts of carbon (C), hydrogen (H), oxygen (O), nitrogen (N), and sulfur (S), as well as other trace elements (ASTM D121-15 2015). It is formed by the decomposition and alteration of vegetation through the process of compaction, resulting in temperature and pressure change. The original sources of coal are a variety of ancient plant forms and their woody precursors. The biomass of the dead plants has been deposited in anaerobic and aquatic conditions with a low O level, thus preventing its degradation and the release of carbon dioxide (CO_2). The different presentations of coal, based on the coalification process, include wood, peat, lignite, subbituminous, bituminous, and anthracite (see Table 2.1). The order of listing in the table is determined by the increase in pressure and temperature at which the substance is formed (Speight 2015).

According to Bowen and Irwin (2008), Alberta Energy (2017), US Energy Information Administration (2018), and ASTM D388-18 (2018), coal has been ranked into four groupings: anthracite, bituminous, subbituminous, and lignite. The rank is determined based on the calorific value, moisture, degree of transformation into C, and composition. The last two characteristics are usually represented by volatile matter (VM) and fixed carbon (FC). In Alberta, lignite and subbituminous are considered to be lower-ranked coal materials, used only for the generation of electricity. Bituminous and anthracite are higher-ranked coal materials, used for thermal and metallurgical purposes, as well as household cooking and heating fuel (Alberta Energy 2017).

In addition to these four types of coal, there is a material not included within this ranking system. This material is identified as leonardite, which is defined as the weathered or oxidized form of lignite (Fowkes and Frost 1960, Kohanowski 1970,

Low-Rank Coal Applications in Agriculture: Humic Analyses, Products, and Performance, First Edition. L. Edwin Liem.

Table 2.1 Coalification process.

Component	Coal rank	Pressure	Temperature
Wood	–	Low	Low
Peat	–	↓	↓
Lignite	4		
Subbituminous	3	High	High
Bituminous	2		
Anthracite	1		

Sources: Speight (2015), Bowen and Irwin (2008), Alberta Energy (2017), US Energy Information Administration (2018), and ASTM D388-18 (2018).

and Siemers and Wadell 1977). Another material, originating from Alberta, is identified as humalite. It is defined as the weathered product of subbituminous coal (i.e. low-rank coal) and carbonaceous shale materials. Leonardite and humalite are similar naturally occurring materials, formed by the natural weathering or oxidation of their parent materials (Hoffman et al. 1993). They have higher contents of O (meaning lower calorific values) than those of their parent materials. Leonardite and Alberta low-rank coal are therefore expected to be ranked within the lignite group (rank 4).

Table 2.2 compares VM, FC, moisture, and calorific values of different coal materials. Literature data show that higher calorific coal is small in VM and moisture, but high in FC, while lower calorific coal is the opposite. The table shows that anthracite, having a calorific value of up to 34 900 kJ/kg, sits at the top (containing less than 10% VM, at least 85% FC, and less than 15% moisture), followed by bituminous and subbituminous. With a calorific value as low as 9300 kJ/kg, lignite sits at the bottom of the table (containing at least 40% VM, at most 35% FC, and up to 60% moisture). They can be compared to data from Alberta, i.e. Luscar coal and its low-rank coal. Looking at these four parameters, the closest rank for Luscar coal (with calorific values of 21 200 to 25 100 kJ/kg) is subbituminous, while low-rank coal from Alberta (with calorific values of 13 700, 17 400, and 12 800 kJ/kg and with a mean value of 14 600 kJ/kg) is in the middle range of lignite. VM, FC, moisture, and calorific value have been analyzed according to ASTM D2013/D2013M-18 (2013), ASTM D3175-18 (2017), ASTM D3172-13 (2013), ASTM D3302/D3302M-17 (2017), and ASTM D5865-13 (2013); or their earlier versions.

ASTM D388-18 (2018) adds subgroups for each rank of coal. In the case of lignite, it is separated into lignite A and lignite B, each with calorific values of 14 700 to 19 300 and less than 14 700 kJ/kg, respectively. Alberta low-rank coal is therefore ranked between lignite A and lignite B. Arnold (2013) provides coal

Table 2.2 Coal ranks.

Literature data				
Type (rank)	**VM (%)**	**FC (%)**	**Moisture (%)**	**Calorific value (kJ/kg)**
Anthracite (1)	< 10	85–98	< 15	30 200–34 900
Bituminous (2)	10–31	45–86	2–15	24 400–34 900
Subbituminous (3)	35–50	35–45	10–45	19 300–30 200
Lignite (4)	40–50	25–35	30–60	9300–19 300
Data from Alberta				
Type (rank)	**VM (%)**	**FC (%)**	**Moisture (%)**	**Calorific value (kJ/kg)**
Luscar coal (3) based on 3 data	31.6–40.5	43.1–50.6	8.2–19.2	21 200–25 100
Alberta low-rank coal (4) based on 3 data	47.3–48.5	30.8–42.9	28.9–34.3	12 800–17 400 (mean = 14 600)

Sources: Bowen and Irwin (2008), Center for Climate and Energy Solutions (2017), ASTM D388-18 (2018), Arnold (2013), and Canadian Humalite International Inc. (2020a); *note:* caloric values converted from imperial to metric units and rounded to closest 100.

categories based on the petrographic composition (the amount of vitrinite). The main parameter is the mean random reflectance range (R_0). Lignite and subbituminous are categorized as low-rank coal materials, consisting of low-rank C (lignite C) with < 0.4 R_0 (moisture > 35%), low-rank B (lignite B) with < 0.4 R_0 (moisture ≤ 35%), and low-rank A (subbituminous) with R_0 from 0.4 to < 0.5 (see Table 2.3). Note: see further discussion in this section, that with respect to its application in agriculture, subbituminous is not considered low-rank coal. Alberta low-rank coal (R_0 value = 0.48, Canadian Humalite International Inc. 2020a) is categorized as low-rank A. This analysis has been completed according to ASTM D2797/D2797M-11a (2011) and ASTM D2798-11a (2011). R_0 indicates how mature the material is with respect to the coalification process. For example, peat has a value of approximately 0.2, compared to approximately 5.0 for anthracite. Coal and dispersed organic matter (OM) with R_0 values of 0.55 or less are considered to be low rank (Mukhopadhyay 1994).

With respect to its formation, it is hypothesized that coal is best formed during a long period of organic accumulation over a large flat region, followed by a repeated sequence of rapid soil or sand inundations (Speight 2015). As shown in Table 2.4, major coal depositions have been recorded during the Cretaceous, Permian, and upper-Carboniferous periods (65 to 144 and 248 to 320 million years ago). Minor depositions have been reported 2 to 65, 144 to 213, and 320 to 408

Table 2.3 R_0 of low-rank coal.

Rank	R_0	Moisture
Low-rank A (subbituminous)	0.4–< 0.5	N/A
Low-rank B (lignite B)	< 0.4	≤ 35%
Low-rank C (lignite C)		> 35%

Source: Arnold (2013); note: R_0 for Alberta low-rank coal = 0.48 (Canadian Humalite International Inc. 2020a), N/A = not available.

Table 2.4 Coal deposition.

Era	Period	Million years ago	Coal deposition
Cenozoic	Quaternary	< 2	None
	Tertiary	2–65	Minor
Mesozoic	Cretaceous	65–144	Major
	Jurassic	144–213	Minor
	Triassic	213–248	Minimal
Paleozoic	Permian	248–286	Major
	Upper-Carboniferous	286–320	Major
	Lower-Carboniferous	320–360	Minor
	Devonian	360–408	Minor
	Silurian	408–438	Unknown
	Ordovician	438–505	None
	Cambrian	505–590	None

Source: Bend (1992).

million years ago during the Tertiary, Jurassic, lower-Carbonaceous, and Devonian periods (Bend 1992).

Anthracite in Pennsylvania is estimated to have been formed during the upper-Carboniferous period (299 to 318 million years ago) and Alberta low-rank coal and subbituminous are estimated to have been formed during the Cretaceous period (65 to 145 million years ago), together with lignite from Colorado, New Mexico, and Utah. Lignite from Wyoming, Montana, North Dakota, and Texas is estimated to have been formed during the early Tertiary period (35 to 65 million years ago) (McCabe et al. 1988 and Palmer 2012). Alberta low-rank coal is therefore a younger material than that in Pennsylvania and is closer to lignite from other parts of North America.

Within the limited scope of this book, only coal materials with calorific values of 19 300 kJ/kg or less are considered suitable for agricultural applications, leaving out rank 1 (anthracite), rank 2 (bituminous), and rank 3 (subbituminous). Low-rank coal is therefore limited to rank 4 (lignite) only, including its oxidized or weathered form (leonardite), as well as Alberta low-rank coal (see Table 2.2 as the reference). The definition of low-rank coal here is therefore different from that of Alberta Energy (2017), in which subbituminous is considered low-rank coal due to its use in the generation of electricity.

Further on, this material has been assigned a Chemical Abstract Service (CAS) Registry Number of 129521-66-0, that of "lignite" (American Chemical Society 2020). Harmonized Codes (HCs) of 2701.19.00.50 (known as "coal; briquettes, ovoids and similar solid fuels manufactured from coal: coal, whether or not pulverized, but not agglomerated: other") and 2702.10.00.00 (known as "lignite, whether or not agglomerated, excluding jets: lignite, whether or not pulverized, but not agglomerated"), assigned by US International Trade Commission (2019), can be used to represent low-rank coal.

2.2 Material Quality

Material quality is determined by the amounts of C, H, O, N, S, and ash in coal, as analyzed using the ultimate analysis according to ASTM D2013/D2013M-18 (2013), ASTM D5373-16 (2016), ASTM D2492-02(2012) (2012), ASTM D3174-12 (2012), ASTM D3176-15 (2015), and O by difference; or their earlier versions. The sum of elemental C, H, O, N, and S represents OM, while ash represents the inorganic matter. Therefore, OM + ash + moisture = 100% of the composition. Ash contains mainly aluminum (Al) and silica (Si) that form sand and clay (the main impurities within the material) as well as other trace elements, including metals. Since OM contains mainly C compounds, it can also be quantified as the amount removed at a very high temperature, during the loss on ignition process (ASTM D7348-13 2013). The lower the ash content, the better the quality of the material, which applies for both high- and low-rank coal materials.

The following are results for Alberta low-rank coal and other coal materials from North America. Alberta low-rank coal samples have been collected from a mine site in Alberta, according to ASTM D2234/D2234M-17 (2017), ASTM D5192-09(2015) (2015), and ASTM D7430-18a (2018); or their earlier versions. Other samples have been received from their places of origin and analyzed accordingly (Canadian Humalite International Inc. 2020a). Results for Kootenay coal, Luscar coal, and Coalspur coal have been collected from MacDonald et al. (1989) and Langenberg et al. (1989).

Table 2.5 Coal materials from Alberta.

Coal material	C %	H %	O %	N %	S %	Ash %	OM %
Kootenay coal	64.7–88.7 (78.6)	3.3–4.0 (3.7)	0.8–6.7 (2.3)	0.8–1.4 (1.1)	0.1–6.5 (0.5)	13.8	86.2
Luscar coal	52.2–87.2 (76.8)	2.7–4.6 (4.0)	2.4–10.9 (4.6)	0.9–1.6 (1.2)	0.1–0.6 (0.3)	13.1	86.9
Coalspur coal	44.0–78.7 (68.7)	3.1–5.4 (4.5)	7.4–17.6 (14.3)	0.3–1.3 (0.9)	0.1–2.8 (0.3)	11.3	88.7
Low-rank coal	40.0–59.0 (49.9)	2.2–3.5 (2.9)	14.8–40.3 (24.9)	0.9–1.8 (1.3)	0.5–0.9 (0.6)	20.4	79.6

Sources: MacDonald et al. (1989), Langenberg et al. (1989), and Canadian Humalite International Inc. (2020a); *note:* mean values are in brackets.

Table 2.5 compares four different coal materials mined in Alberta: Kootenay coal (based on 8 data), Luscar coal (based on 104 data), Coalspur coal (based on 49 data), and Alberta low-rank coal (based on 26 data). The table shows minimum, maximum, and mean values (in brackets), based on the dry matter analysis (excluding moisture). Compositions of the elemental C and O clearly show that Alberta low-rank coal (containing at most 59.0% C and at least 14.8% O) is well positioned below other coal materials, with Kootenay coal on the top (containing at least 64.7% C and at most 6.7% O), followed by Luscar coal (containing at least 52.2% C and at most 10.9% O), and then Coalspur coal (containing at least 44.0% C and at most 17.6% O). In general, coal materials from Alberta are of good quality, as indicated by their low ash values (with mean values of 11.3 to 13.8% for coal materials, and 20.4% for Alberta low-rank coal). In this context, OM can simply be calculated as the sum of C, H, O, N, and S, ranging from 79.6 to 88.7%.

Table 2.6 shows C, H, O, N, S, ash, and OM of lignite materials mined from five different mine sites in the states of North Dakota (based on two data), Wyoming (based on one datum), and New Mexico (based on two data). The table shows minimum, maximum, and mean values (in brackets). Note: "lignite" in this case may refer to its trade name and/or its actual rank. New Mexico has the lowest C and O (up to 43.1% and 19.6%, respectively), North Dakota has the highest S (at least 2.2%), while Wyoming has the least ash (17.1%). Calorific values for lignite from North Dakota, Wyoming, and New Mexico are 14 500, 16 400, and 13 800 kJ/kg, respectively (each is based on one datum). Looking at their calorific values, these materials are all in the middle range of lignite. This table shows that they are similar to Alberta low-rank coal with a calorific value of 14 600 kJ/kg, containing 49.9% C, 2.9% H, 24.9% O, 1.3% N, 0.6% S, and 20.4% ash (all mean values). These analyses have been completed according to ASTM D2013/D2013M-18 (2013), ASTM D5373-16 (2016), ASTM D2492-02(2012) (2012), ASTM D3174-12 (2012),

Table 2.6 Lignite materials from the US.

Lignite material	C %	H %	O %	N %	S %	Ash %	OM %	Calorific value kJ/kg
North Dakota	48.0–53.0	3.1–3.2	20.6–23.2	0.6–1.0	2.2–2.5	20.2–22.5	77.5–79.8	14 500
Wyoming	52.3	3.4	25.5	1.2	0.5	17.1	82.9	16 400
New Mexico	40.9–43.1	2.5–3.2	16.5–19.6	0.8–0.9	0.4–0.5	35.7–36.1	63.9–64.3	13 800

Source: Canadian Humalite International Inc. (2020a).

ASTM D3176-15 (2015), and ASTM D5865-13 (2013); or their earlier versions. O has been calculated by difference and OM = 100% – ash.

It is important to note that Tables 2.5 and 2.6 also indicate that the lower the calorific value, the (slightly) higher the ash content within the material. This implies that high calorific coal contains less ash than its counterpart, i.e. 11.3 to 13.8% for Kootenay, Luscar, and Coalspur coal materials, compared to 17.1 to 36.1% for Alberta low-rank coal and lignite from North Dakota, Wyoming, and New Mexico. This indicates that high calorific coal contains more OM than that of its counterpart at 86.2 to 88.7% compared to 63.9 to 82.9%. However, most of the OM is not in a useful form for agriculture (see Chapters 3, 4, and 6).

Table 2.7 shows Hardgrove Grindability Index (HGI) mean and standard deviation values of selected coal materials. This parameter measures how easy it is for the material to be pulverized into a certain mesh size; it is originally intended for coal used for the firing of electricity station boilers (Agus and Waters 1971 and Csoke et al. 2003).

Agus and Waters (1971) organize material grindability into six groups: very hard (40 or less), hard (40 to 60), medium (60 to 80), soft (80 to 100), very soft (100 to 120), and extremely soft (120 and higher). Table 2.7 shows Alberta low-rank coal to have the largest index at 111 (analyzed according to ASTM D5003-19 2019). It sits well below Luscar coal (mean value of 64), bituminous (mean value of 55), and anthracite (mean value of 46), and slightly below subbituminous (mean value of 110). Alberta low-rank coal can therefore be classified as very soft. Lignite is not available in this table, which is likely positioned at a similar spot as that of Alberta low-rank coal.

The fundamental difference between high- and low-rank coal materials can be understood in terms of the proportions of C, H, and O in each. High-rank coal is very high in C but very low in O (indicating a low level of oxidation), with a high calorific value. Low-rank coal is low in C and high in O, with a low calorific value. H is generally low for each of them. Tables 2.5 and 2.6 indicate that Alberta

Table 2.7 HGI of coal materials.

Coal material	# of data	HGI mean (SD)	Ash	Calorific value
Anthracite	117	46 (72.9)	Low	High
Bituminous	65	55 (21.3)	↓	↑
Luscar coal	2	64 (N/A)		
Subbituminous	16	110 (20.9)	High	Low
Alberta low-rank coal	1	111 (N/A)		

Sources: Gillmore et al. (1956), Agus and Waters (1971), and Canadian Humalite International Inc. (2020a); note: SD = standard deviation, N/A = not available.

low-rank coal and lignite from North Dakota, Wyoming, and New Mexico all are different from high calorific coal materials. This can easily be visualized by plotting ratios of H over C (10 × H:C on the vertical axis) vs. O over C (on the horizontal axis), known as the modified Van Krevelen diagram (Van Krevelen 1950).

Figure 2.1 shows that Kootenay, Luscar, and Coalspur coal materials are closer to the high calorific zone. Each dot represents a mean value of 8 data for Kootenay

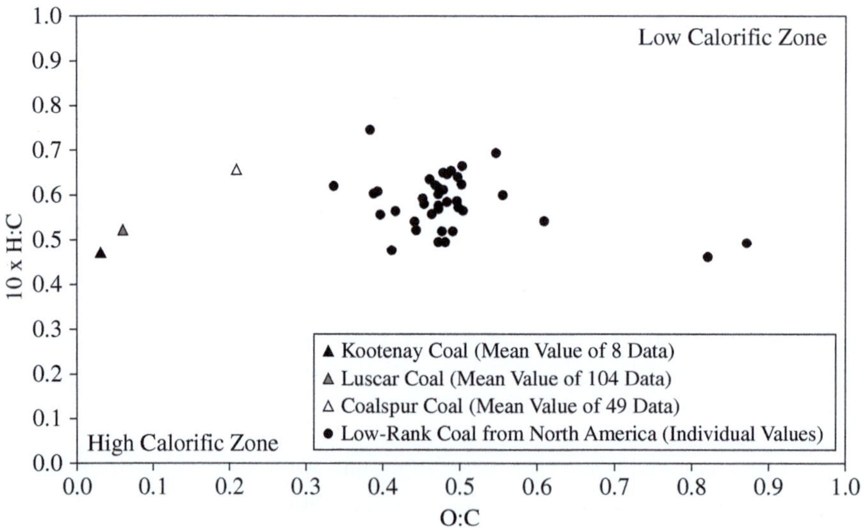

Figure 2.1 Modified Van Krevelen diagram.

coal, 104 data for Luscar coal, and 49 data for Coalspur coal. The figure shows that Alberta low-rank coal and lignite from North Dakota, Wyoming, and New Mexico are closer to the low calorific zone. Each dot represents an individual value of 33 data from Alberta, 2 data from North Dakota, 1 datum from Wyoming, and 2 data from New Mexico (Canadian Humalite International Inc. 2020a).

2.3 Material Safety

In addition to their quality, low-rank coal materials should also be evaluated with respect to their safety during handling and field application. All analyses have been completed for low-rank coal samples collected from a mine site in Alberta according to ASTM D2234/D2234M-17 (2017), ASTM D5192-09 (2015) (2015), and ASTM D7430-18a (2018), as well as from commercial facilities for the manufacture of humic products using Alberta low-rank coal as the raw material, according to Puskar (1991). The following results are therefore applicable for Alberta low-rank coal only.

During the combustion testing, the material is pulverized into dust (preferably at the lowest moisture) to maximize exposure of each particulate under a controlled source of energy to reach its combustion stage. Therefore, it provides the worst-case scenario since the material usually comes in the form of chunks and coarse particulates, or at worst, powder with some moisture in it. Parameters to be evaluated include the minimum ignition temperature (MIT), minimum ignition energy (MIE), minimum explosible concentration (MEC), maximum pressure (Pmax, how much pressure is created if the material combusts), and explosion severity (Kst, how quick or bad the explosion is if the material combusts).

Table 2.8 shows that particulate moisture and its sizing are critical. The smaller and drier the particulates, the easier it is for the material to combust. Between 60 and 80 g/m^3 dust in air of 3% moisture and 31 μm mean particulate sizing has been reported to trigger combustion at a temperature of 270 °C with an MIE of over 1000 mJ. Alberta low-rank coal has a Kst of 135. If combustion indeed happens, it creates a Pmax of 730 kPa in the surrounding environment. As a comparison, wood flour and sulfur have much lower MIE values of 10 and < 1 mJ, respectively. Typical coal materials have Pmax values in the range of 500 to 1600 kPa (Dalhousie University 2002 and Fike Corporation 2015). Occupational Safety and Health Administration (2009) classes dust with Kst of < 200 as "St 1," of a "weak explosion" characteristic. Also grouped under this class are powdered milk, sugar, charcoal, zinc, and sulfur. This indicates that Alberta low-rank coal is generally non-hazardous and is in the same class as that of other materials commonly used in our day-to-day life. These combustion parameters have been

Table 2.8 Combustion parameters.

Test #	Particulate moisture %	Mean particulate sizing μm	Dust in air g/m^3	MEC g/m^3	MIT °C	Pmax kPa	Kst
1	3	31	≤ 2000	60–80	270	730	135
	0	48	250	250	550	480	138
2	37	48	250	No explosion			
	25	48	≤ 2000	450	590	40	11
	30	48	≤ 2000	1000	590	340	20

Sources: Fike Corporation (2015) and Dalhousie University (2002); note: each test completed using a 5000 mJ ignition source and > 1000 mJ MIE.

analyzed according to ASTM D2013/D2013M-18 (2013), ASTM D5709-09(2015) (2015), ASTM D3302/D3302M-17 (2017), ASTM E1226-12a (2012), ASTM E1491-06(2012) (2012), ASTM E2019-03(2013) (2013), and ASTM E1515-14 (2014); or their earlier versions.

Respiratory hazard is related to the presence of free Si within a material. This is a component of Si which is not chemically bound with any other elements in the form of silicates. Free Si includes amorphous and crystalline Si. Crystalline Si consists of respirable quartz, cristobalite, and trymidite Si, which has been the component to cause most concern due to its carcinogenic potential. Any material containing over 0.1% crystalline Si must follow the Occupational Safety and Health Administration (OSHA)'s regulation to meet federal guidelines on worker training and hazard communication (US Bureau of Mines 1992). Within the crystalline Si group, respirable quartz specifically has been regulated by the Government of Alberta (2020) and Occupational Safety and Health Administration (2019). Si analyses have been completed according to the National Institute for Occupational Safety and Health (NIOSH) Method No. 7500 (2003).

Table 2.9 shows that Alberta low-rank coal contains small amounts of free Si (up to 1.7%), up to 3.4% of respirable quartz Si, and up to 3.4% crystalline Si. Their mean values are 1.3, 1.7, and 1.7% respectively; in brackets are the standard deviation values. This indicates that this material must follow the OSHA's regulation. Within the same sample, crystalline Si cannot actually be larger than free Si. Discrepancies here are caused by analyses that have been completed on different samples. Minimum reporting limits (MRLs) for respirable cristobalite and trymidite Si are 0.25 and 1.50%, respectively.

The enforcement of regulations around respirable quartz, respirable cristobalite, and crystalline Si has been determined with respect to the workers'

Table 2.9 Si in material.

Parameter	# of data	Value %	Mean (SD) %	Limit %
Respirable quartz Si	5	0.5–3.4	1.7 (0.99)	N/A
Respirable cristobalite Si	5	< MRL	N/A (N/A)	N/A
Respirable trymidite Si	5	< MRL	N/A (N/A)	N/A
Crystalline Si[a]	5	0.5–3.4	1.7 (0.99)	0.1[b]
Free Si[a]	3	0.9–1.7	1.3 (0.40)	N/A

Sources: Canadian Humalite International Inc. (2020a);
note:
[a] discrepancies between crystalline and free Si due to analyses completed on different samples,
[b] US Bureau of Mines (1992; beyond this limit, material to follow OSHA's regulation),
SD = standard deviation, MRL = minimum reporting limit, N/A = not available.

exposure during material handling, not with respect to the composition of the material itself. The relevant limits for each material here are 25, 25, and 50 $\mu g/m^3$ air, respectively, averaged over an 8-hour shift (Government of Alberta 2020 and National Safety Council 2016). According to the State of California (2018), the permissible exposure limit for crystalline Si is 50 $\mu g/m^3$ of air, as an 8-hour time weighted average. The action level for crystalline Si is 25 $\mu g/m^3$ of air, as an 8-hour time weighted average.

An air monitoring study can be completed for a specific commercial facility. Samples are to be obtained from different areas within the facility, including processing plants, storage sections, offices, and laboratories. Each is taken from an air sampler attached to an individual during an 8-hour working shift, and analyzed for respirable quartz, respirable cristobalite, and respirable trymidite Si. Ideally, more than one sample should be taken from each area. The results are then averaged and compared with the limits set by the relevant regulatory body (see Table 2.10). For work related to low-rank coal, results for respirable cristobalite and respirable trymidite Si are typically below their minimum reporting limits. Therefore, respirable quartz Si, in this case, is identical to crystalline Si. In most cases, the results have been within the limits (Canadian Humalite International Inc. 2020a). Still, taking proper precautions during material handling is critical, that has to be understood and practiced as discussed in Section 2.4.

For selected metals taken from Alberta low-rank coal, each sample is acidified prior to analyses to extract all the elements, including those locked within the material. Their contents are obtained using the Inductively Coupled Plasma (ICP) analysis. Table 2.11 shows minimum, maximum, mean, and standard deviation

Table 2.10 Si in air.

Component	Limit	Unit	Source
Respirable cristobalite Si	25[a]	µg/m^3	Government of Alberta (2020)
Respirable quartz Si	25[a]	µg/m^3	Government of Alberta (2020)
Respirable trymidite Si	N/A	µg/m^3	Government of Alberta (2020)
	25[b]	µg/m^3	State of California (2018)
Crystalline Si	50[c]	µg/m^3	State of California (2018)
	50[c]	µg/m^3	National Safety Council (2016)
	50[a]	µg/m^3	Government of Alberta (2020)

Note:
[a] occupational exposure limit,
[b] action level,
[c] permissible exposure limit.

values of selected metals contained within Alberta low-rank coal. The table shows that this material contains Al (at 2.1%), sodium (Na, at 0.3%), arsenic (As, at 7 ppm), cadmium (Cd, at less than 1 ppm), chromium (Cr, at 4 ppm), mercury (Hg, 0.06 ppm), nickel (Ni, 9 ppm), and lead (Pb, at 5 ppm) (all mean values). All analyses have been completed according to ASTM D2013/D2013M-18 (2013), ASTM D6357-11 (2011), and ASTM D6722 (2011); or their earlier versions.

Table 2.11 also includes guidelines and limits for agricultural soil and livestock applications (Canadian Council of Ministries of the Environment 2002a, Canadian Food Inspection Agency 2017a, National Research Council 2005, and Canadian Food Inspection Agency 2017b). The most stringent limitations are those that apply to livestock feed and agricultural soil applications. For the livestock feed application, the limitations are 1000 ppm for Al, 8 ppm of As, 0.2 ppm of Cd, 0.2 ppm of Hg, and 8 ppm of Pb. For the agricultural soil application, the limitations are 64 ppm of Cr and 36 ppm of Ni. The table shows that Alberta low-rank coal, based on its mean values, will likely meet all established guidelines and limits when applied at typical rates. For example, recommended rates of this material for livestock have been <4% of total feed by weight (see Section 6.6.12), thus meeting its Al guideline at 1000 ppm. This indicates that this material is non-hazardous and that it can be used in agricultural soil and livestock feed applications. The low level of Na indicates that it is from a freshwater origin.

Considering that a typical mine site is quite remote from any human and animal activities, the presence of pathogens in low-rank coal is quite unlikely. As shown in Table 2.12, Alberta low-rank coal contains negligible amounts of *E. coli* (below detection limit), fecal coliform (below detection limit), and *Salmonella*

Table 2.11 Selected metals.

Parameter	# of data	Value	Mean (SD)	Unit
Al	54	0.4–6.1	2.1 (1.59)	%
Na	54	< 0.1–1.4	0.3 (0.29)	%
As	463	1–28	7 (3.0)	ppm
Cd	51	≤1	< 1 (N/A)	ppm
Cr	34	1–15	4 (3.1)	ppm
Hg	47	≤ 0.2	0.06 (0.003)	ppm
Ni	36	1–15	9 (6.2)	ppm
Pb	51	1–34	5 (2.5)	ppm

	Agricultural soil		Livestock	
Parameter	Guideline[a] ppm	Limit[b] kg/ha	Guideline[c] ppm	Limit[e] ppm
Al	N/A	N/A	1000	N/A
Na	N/A	N/A	N/A[d]	N/A
As	12	15	30	8
Cd	1.4	4	10	0.2
Cr	64	210	100	N/A
Hg	6.6	1	0.2	N/A
Ni	45	36	100	N/A
Pb	70	100	10	8

Source: Canadian Humalite International Inc. (2020a);
note:
[a] soil quality guidelines for agriculture (Canadian Council of Ministries of the Environment 2002a),
[b] maximum acceptable cumulative additions to soil over 45 years (Canadian Food Inspection Agency 2017a),
[c] maximum tolerable levels for poultry, swine, and cattle feeds, use the smallest values (National Research Council 2005),
[d] 1.7% as NaCl,
[e] current action levels for livestock feeds (Canadian Food Inspection Agency 2017b),
SD = standard deviation, N/A = not available.

(negative). All analyses have been completed according to Christensen et al. (2002) and Reid (2009). Limits have been set for fecal coliform and *Salmonella* for the agricultural soil application at 1000 mpn/g and "not detected," respectively

Table 2.12 Pathogens.

Pathogen	# of data	Value	Limit for soil[a]	Limit for feed[b]	Unit
E. coli	2	BDL	N/A	N/A	mpn/g
Fecal coliform	5	BDL	1000	N/A	mpn/g
Salmonella	6	Negative	N/D	N/D	N/A

Source: Canadian Humalite International Inc. (2020a);
[a] Canadian Food Inspection Agency (2017a),
[b] Canadian Food Inspection Agency (2017c), BDL = below detection limit, N/D = not detected, N/A = not available.

(Canadian Food Inspection Agency 2017a). For livestock feed applications, "not detected" has been set as a limit for *Salmonella* (Canadian Food Inspection Agency 2017c). This shows that this material is safe and that it can be used in agricultural soil and livestock feed applications.

In addition to pathogens, Table 2.13 shows that Alberta low-rank coal contains small amounts of soil microorganisms, i.e. 1.3×10^3 cfu/g heterotrophic plate count (HPC) bacteria, 600 cfu/g fungal enumeration, and zero nematodes. No limits on these parameters have been set for the soil application (Canadian Food Inspection Agency 2017a). These analyses have been completed according to APHA 9215D (1992), Douey and Wilson (2004), and European and Mediterranean Plant Protection Organization (2013). For livestock feed applications, limits have been set for mycotoxins, i.e. toxins produced by fungi, not on the fungi themselves (Canadian Food Inspection Agency 2017c). The current maximum limit for mycotoxins is 20 ppb (see Table 2.14). This indicates that Alberta low-rank coal is safe for agricultural soil and livestock feed applications.

It is also very unusual to find toxic organics in natural mined materials, including low-rank coal. However, it is still important to analyze the samples accordingly. As shown in Table 2.14, Alberta low-rank coal contains a maximum polychlorinated dibenzo dioxins/furans toxicity equivalent (PTE) of 0.391 ppt. Dioxins/furans have been analyzed according to EPA Method 1613 Revision B (1994). Limits have been set at 4 ppt, 5.355 mg/ha for soil (based on product addition over 45 years), and 0.75 ppt/total feed for feed (Canadian Council of Ministries of the Environment 2002b and Canadian Food Inspection Agency 2017a, 2017b, and 2017c). Also included under this category are toxins produced by certain fungi grown in soil, decaying vegetation, or grains (i.e. aflatoxins B1, B2, G1, and G2). This table shows that this material contains less than 1 ppb of each type of aflatoxin (analyzed according to US Pharmacopoeia 2014). Alberta low-rank coal is therefore safe for agricultural soil and livestock feed applications.

Table 2.13 Microorganisms and nematodes.

Parameter	# of data	Value	Limit for soil[a]	Limit for feed[b]	Unit
HPC bacteria	1	1.3×10^3	N/A	N/A	cfu/g
Fungal enumeration	1	600	N/A	N/A[c]	cfu/g
Nematodes	1	0	N/A	N/A	#/g

Source: Canadian Humalite International Inc. (2020a);
[a] Canadian Food Inspection Agency (2017a),
[b] Canadian Food Inspection Agency (2017c),
[c] current maximum limit for aflatoxins at 20 ppb total feed, N/A = not available.

Table 2.14 Toxic organics.

Parameter	# of data	Value ppt	Limit for soil ppt[a]	Limit for soil mg/ha[b]	Limit for livestock[c] ppt total feed
PTE	3	0.391 (0.0496)	4	5.355	0.75

Parameter	# of data	Value ppb	Limit for livestock[d] ppb total feed
Aflatoxin B1	1	BDL	20
Aflatoxin B2	1	BDL	20
Aflatoxin G1	1	BDL	20
Aflatoxin G2	1	BDL	20

Source: Canadian Humalite International Inc. (2020a);
note:
[a] soil quality guidelines for environmental health (Canadian Council of Ministries of the Environment 2002b),
[b] maximum acceptable cumulative additions to soil over 45 years (Canadian Food Inspection Agency 2017a),
[c] current action level, smallest value (Canadian Food Inspection Agency 2017b),
[d] Canadian Food Inspection Agency (2017c), in bracket is standard deviation, BDL = below detection limit at 1.0 ppb.

Microtox bioassay is typically requested for materials for which possible spills to the environment may affect the life of biota. In this case, it is relevant due to the use of this material as a soil amendment or feed supplement. This analysis is related to the toxicity characteristics of the material itself, not to the impurities contained within. It uses bioluminescent bacteria of the *V. fischeri* species as an indicator (Microbics Corporation 1992, ISO No. 11348-1-2-3 2007, and Gaudet 1994). Sufficient reduction in light emission indicates a severity of toxicity that may affect the survival of organisms. It has been found that 19.0 mg/l of Alberta low-rank coal (at 1:1 volume/volume dilution) reduces light emission by 55% and 62% after 5 and 15 minutes of exposure, respectively. This is defined as the 50% inhibitory concentration (IC_{50}).

Further analyses of the sample have found the median effective concentration after 5 and 15 minutes to be over 82% original; this is interpreted as a "pass." EC_{50} (or effective concentration median) is the concentration of a stressor that is estimated to be effective in producing a biological response, other than mortality, in 50% of the test organisms over a specific time interval (Canadian Council of Ministries of the Environment 1999). These analyses show that the material is nontoxic (see Table 2.15). One significant challenge here is that a dark color in the material being tested (including low-rank coal) may affect the reading for the worse. When necessary, corrections have to be completed for dark color samples (Ashworth et al. 2010). This finding shows that Alberta low-rank coal is safe with respect to its toxicity to living organisms.

The European Agency for the Evaluation of Medicinal Products (1999) presents data specifically for lignite (brown coal). Table 2.16 indicates that this material has a low oral toxicity with respect to livestock feed, i.e. a median lethal dose (LD_{50}) of 11 500 mg/kg body weight for an oral administration in rats. LD_{50} is the dose of material that is estimated to be lethal to 50% of the test organisms (Canadian Council of Ministries of the Environment 1999). It is also defined as the amount of material, given at once, which causes the death of 50% (one half) of a group of test animals. It is one way to measure the short-term poisoning potential (acute toxicity) of the material (Canadian Centre for Occupational Health and Safety 2019a).

In a 30-day study, no clinical disturbances have been reported for rats orally fed at rates of up to 1000 mg/kg body weight. The same result has also been found for dogs at a rate of 300 mg/kg for 90 days. Na humate salt at rates of up to 15 000 μg/ml

Table 2.15 Microtox bioassay.

Parameter	# of data	Value	Comment
IC_{50} after 5 minutes	1	19.0 mg/l	Dark material color
IC_{50} after 15 minutes	1	19.0 mg/l	Dark material color
EC_{50} after 5 minutes	1	> 82 %	Pass
EC_{50} after 15 minutes	1	> 82 %	Pass

Source: Canadian Humalite International Inc. (2020a).

Table 2.16 General toxicity.

Parameter	Subject	Rate	Unit
LD_{50}	Rats	11 500	mg/kg body weight
30-day study	Rats	≤ 1000	mg/kg body weight
90-day study	Dogs	300	mg/kg body weight
Cell aberration study	Rabbits, hamsters	≤ 15 000	μg/ml
Mutagenicity study	*Salmonella*	≤ 0.5	%

Source: The European Agency for the Evaluation of Medicinal Products (1999).

has not shown any aberrations in kidney cells of rabbits and hamsters, nor any mutagenicity in *S. typhirium* TA98 and TA100 at rates of up to 0.5%. This material has commonly been used in horses, ruminants, swine, and poultry (at rates of 500 to 2000 mg/kg body weight) for the treatment of diarrhea, dyspepsia, and acute intoxication. This suggests that low-rank coal (including Alberta low-rank coal), at its typical rates, is safe when applied as a livestock feed supplement (The European Agency for the Evaluation of Medicinal Products 1999).

The parameters of total extractable hydrocarbons are more related to the practices of soil remediation than to those of agricultural soil and livestock feed. It is, however, still worth discussing, since both practices are sometimes interconnected. Total extractable hydrocarbons indicate the presence of toxic petroleum hydrocarbons in the material (Canadian Council of Ministries of the Environment 2008). The challenge here is that material of plant origin, including low-rank coal, naturally contains biogenic organic compounds, which are often mistaken for petroleum hydrocarbons (Kelly-Hooper et al. 2013, Nelson and Garrard 2015, and Rae et al. 2015). Table 2.17 (Column 3) shows that according to the basic analysis, Alberta low-rank coal contains a very small amount of fraction 2 (F_2) of hydrocarbons (i.e. C10 to C16), but significant amounts of fraction 3 (F_3) of hydrocarbons (i.e. C16 to C34) and fraction 4 (F_4) of hydrocarbons (i.e. C34 and above). Column 4 shows the results of analysis after a cleanup step where Si gel is applied to the material used in the basic analysis. All analyses have been completed according to Canadian Council of Ministries of the Environment (2001). The material still shows some F_3, which is in the range of diesel. Since this material is excavated from a mine site located far away from any industrial activities, it is more likely to contain biogenic organic compounds, not petroleum hydrocarbons.

A further analysis using the Gas Chromatography (GC) method reveals a predominantly even over odd C predominance at C16 to C28 (Canadian Humalite International Inc. 2020a). According to Wang et al. (2012), most biogenic organic compounds, containing fatty acids and fatty alcohols (as well as sterols), show the

Table 2.17 Total extractable hydrocarbons.

Parameter	# of data	First analysis	Second analysis	Unit
F_2 (C10–C16)	1	15	12	mg/kg
F_3 (C16–C34)	1	2200	430	mg/kg
F_4 (≥ C34)	1	1400	38	mg/kg

Source: Canadian Humalite International Inc. (2020a).

kind of C patterns and atomic ranges in the GC analysis. This shows that low-rank coal contains mainly biogenic organic compounds, with negligible or no petroleum hydrocarbons at all. It can therefore be considered a non-hazardous material. However, the regulation regarding soil remediation practice does not differentiate between biogenic organic compounds and petroleum hydrocarbons (Canadian Council of Ministries of the Environment 2008). This may result in the restriction of low-rank coal applications in soil remediation. Applications for agricultural soil and livestock feed are not directly affected.

As shown in Table 2.18, Alberta low-rank coal is a weak acid (mean acidity, pH = 3.7) and is slightly lighter than water (mean density = 880 kg/m^3, individual

Table 2.18 Alberta low-rank coal.

Parameter	# of data	Value	Unit
pH	133	3.5–4.1[a] mean = 3.7 (0.10)	–
Density	133	790–920[a,b] mean = 880 (32)	kg/m^3
Moisture	133	28–45[a,c] mean = 37 (4.2)	%
Ash	922	16.9–27.9[d] mean = 22.8 (2.33)	%
Water solubility	133	1.0–7.4[a] mean = 4.2 (1.87)	%

Source: Canadian Humalite International Inc. (2020a);
note:
[a] including moisture analysis,
[b] rounded to closest 10 kg/m^3,
[c] rounded to closest 1%,
[d] dry matter analysis, in brackets are standard deviations.

values depend on their moisture). Raw material moisture contents vary greatly depending on the weather or season and other factors in the mine site during material collection (mean value = 37%). Alberta low-rank coal has low to medium ash (mean value = 22.8%) with a low water solubility at 20 °C (mean value = 4.2%). All analyses have been completed according to ASTM D2234/D2234M-17 (2017), ASTM D7430-18a (2018), ASTM D2013/D2013M-18 (2013), ASTM D4972-18 (2018), ASTM D7263-09(2018)e2 (2018), ASTM D3302/D3302M-17 (2017), ASTM D3174-12 (2012), and EEC A.6 (1992).

2.4 Safety Documentation

Overall, Alberta low-rank coal, or low-rank coal in general, is safe with respect to its handling and field application. However, important properties of the material with respect to its safety must be recorded in the form of a safety data sheet (SDS). This refers to the Globally Harmonized System of Classification and Labeling of Chemicals (GHS), which is aligned in Canada with the Workplace Hazardous Materials Information System (WHMIS). It recognizes three major groups (physical, health, and environmental), along with their class and category sub-sections. Any possible hazards are then to be communicated in label elements, which include hazard and precaution statements, signal words, and pictograms. In Canada, all of them have to be implemented by the middle of 2015 (Canadian Centre for Occupational Health and Safety 2019b).

Another important act, and correlative set of regulations, is the Transportation of Dangerous Goods (TDG) by Transport Canada. This is adopted by all provinces and promotes public safety when dangerous goods are being handled, ordered for transport, or transported by road, rail, or water, by establishing safety requirements. Dangerous goods are the ones listed in Schedules 1 and 3 (having UN numbers corresponding to the UN Dangerous Goods List) or the ones that meet the classification criteria outlined in Part 2 of the regulations. Low-rank coal, lignite, leonardite, and humalite are not listed in these schedules and do not meet the criteria; therefore they are not considered to be dangerous goods (Canadian Centre for Occupational Health and Safety 2019c).

SDS generated for this material shows that Alberta low-rank coal is non-hazardous Under Section 2 (Hazard Identification). This material has a "warning" label element (see Figure 2.2) with a low hazard classification (Canadian Centre for Occupational Health and Safety 2019b and Canadian Humalite International Inc. 2020b). The label element is then followed by the following statements related to the product's hazard's classification:

1) acute toxicity: oral, dermal, inhalation (Category 4) – may be harmful if swallowed;

Figure 2.2 Warning label element. (*See color plate section for color representation of this figure*).

2) skin irritation (Category 3) – causes mild skin irritation;
3) eye irritation (Category 2A) – causes eye irritation; and
4) specific target organ toxicity: respiratory tract irritation (Category 3) – (dust) may cause respiratory irritation.

This material is nonreactive and stable under ambient conditions. No ecological concerns or specific methods for disposal have been mentioned. With respect to the Department of Transportation's regulations, it is not considered to be a hazardous material. The "may be harmful if swallowed" statement under "acute toxicity" is based on the worst-case scenario where an excessive amount of the material is consumed. The SDS then provides precautions and controls to prevent such hazards from happening, as well as providing measures to take if they do happen. The following precautions have been listed in this SDS:

1) if medical advice is needed, have product container or label at hand;
2) keep out of reach of children;
3) read label before use;
4) do not breathe dust;
5) do not get in eyes, on skin, or on clothing;
6) do not eat, drink, or smoke when using this product;
7) use only outdoors or in a well-ventilated area; and
8) wear personal protective equipment (PPE), i.e. protective gloves, eye protection, and respiratory protection (with a target for respirable dust at ≤ 2 mg/m^3): an approved NIOSH/Mine Safety and Health Administration (MSHA) respirator with a dust filter, approved for Si bearing dust.

The following engineering controls and accidental release measures have been listed in this SDS:

1) local exhaust ventilation and confinement of handling system may be required to control dust;
2) sweep and contain in containers, then notify appropriate authorities when release to environment occurs; and
3) use media appropriate to surround fire and use protective clothing and breathing equipment if the material is on fire.

The following first-aid measures (under the worst-case scenario) have been listed in this SDS:

1) rinse mouth if swallowed;
2) rinse skin with water/shower if on skin or hair;
3) remove person to fresh air and keep comfortable for breathing if inhaled;
4) rinse cautiously with water for several minutes if in eyes; and
5) take off contaminated clothing and wash it before reuse if on clothing.

2.5 Closure

Low-rank coal is a type of material containing lower calorific values (19 300 kJ/kg or less) than those of anthracite, bituminous, and subbituminous. It can be considered to be a non-hazardous material. This is not a scientific definition, but rather a commercial and practical term, indicating that this material can safely be utilized in agriculture.

Synonyms have been used as substitutes for low-rank coal, including brown coal, soft coal, young coal, lignite, leonardite, humalite, low-heat coal, low-calorific coal, low-energy coal, weathered subbituminous, weathered lignite, oxidized subbituminous, and oxidized lignite. Although they may not be identical to one another, these materials contain the same or similar properties to those of low-rank coal. Humalite by definition includes both weathered subbituminous coal (low-rank coal) and weathered carbonaceous shale materials originating from Alberta (Hoffman et al. 1993). However, when limiting low-rank coal to that material being collected from the mine site, humalite is in fact identical to Alberta low-rank coal.

Brown coal refers to its color and soft coal to its texture (i.e. HGI). Both of these descriptions are true in most practical cases. Lignite, leonardite, and humalite are grouped according to their calorific values, as well as low-heat coal, low-calorific coal, and low-energy coal. Weathered subbituminous, weathered lignite, oxidized subbituminous, and oxidized lignite all point to low-calorific values, and therefore are synonyms for low-rank coal. The terms "weathered" and "oxidized" have been used interchangeably to indicate the presence of higher O contents within the material, resulting in reduced calorific values. Young coal refers to its coalification process, geological formation, and coal deposition. In most cases, it is an acceptable synonym for low-rank coal.

The following names are commonly used, although they are incorrect: coal, bituminous, subbituminous, anthracite, graphite, low-ash coal, low-quality coal, weathered coal, weathered bituminous, oxidized coal, oxidized bituminous, peat, peat moss, charcoal, and biochar. Although some of them may look similar to low-rank coal, they either contain high calorific values or are not coal materials in the first place. Categorization as low-ash coal indicates a good quality of coal, which applies to all coal materials. Low-quality coal, if used in relation to its low-calorific value, is an acceptable synonym for this material. More typically, however, this name implies that the material has a significant amount of impurities (ash, metals, pathogens, and toxic organics), which are not beneficial for most applications. As will be shown in Chapter 3, soil organic matter (or "humus"), humic substances (HS, or "humic"), humin, humic acids (HA), and fulvic acids (FA) are components within low-rank coal, while humates include low-rank coal and another non-coal material (i.e. carbonaceous shale). Table 2.19 lists both correct and incorrect synonyms for low-rank coal.

Table 2.19 Synonyms for low-rank coal.

Correct	Incorrect
Brown coal, soft coal, young coal	Coal, bituminous, subbituminous
Lignite, leonardite, humalite	Anthracite, graphite
Low-heat coal, low-calorific coal, low-energy coal	Low-ash coal, low-quality coal
Weathered subbituminous, weathered lignite	Weathered coal, weathered bituminous
Oxidized subbituminous, oxidized lignite	Oxidized coal, oxidized bituminous
	Peat, peat moss, charcoal, biochar, soil organic matter (humus)[a], humic substances[a], humin[a], humic acids[a], fulvic acids[a], humates[b]

Sources: Canadian Humalite International Inc. (2020a and 2020b);
note:
[a] components within low-rank coal,
[b] includes low-rank coal and another material.

Alberta low-rank coal is typically found relatively close to the ground surface compared to anthracite, bituminous, and subbituminous. Both low-rank coal and other coal materials can be located at the same site, or adjacently, but have distinct physical properties. Its texture is soft and brittle, and contains relatively high moisture contents. In contrast, other coal materials are typically shiny black in color, have a hard and solid texture, and have much lower moisture contents. The aggregate sizing varies, which does not determine its type or quality. It is more related to the weather or season (i.e. moisture) and other factors at the mine site, as well as the way the material is collected.

In short, low-rank coal, including Alberta low-rank coal (humalite), is unique, different from other coal materials, and therefore is not to be confused with them. Most importantly, this material can be practically and safely handled and processed into commercial products intended for agricultural usages.

Reference

Agus, F. and Waters, P.L. (1971). Determination of the grindability of coals, shales and other minerals by a modified Hardgrove-machine method. *Fuel* 50(4): 405–431.

Alberta Energy (2017). Coal types. http://www.alberta.ca/coal-types.asp (accessed 13 August 2020).

American Chemical Society (2020). Chemical abstract services. http://www.cas.org (accessed 20 January 2020).

APHA 9215D (1992). Heterotrophic plate count: membrane filter method. In: *Standard Methods for the Examination of Water and Wastewater*, 18th ed. (eds. A.E. Greenberg, L.S. Clesceri, and A.D., Eaton), 9-38 to 9-39. Washington, DC: American Public Health Association.

Arnold, B.J. (2013). Coal formation. In: *The Coal Handbook Towards Cleaner Production Vol. 1: Coal Production* (ed. D. Osborne), 31–52. Philadelphia, PA: Wood Haven Publishing.

Ashworth, J., Nijenhuis, E., Glowacka, B., et al. (2010). Turbidity and color correction in the Microtox^{TM} bioassay. *The Open Environmental Pollution and Toxicology Journal* 2: 1–7.

ASTM D121-15 (2015). *Standard terminology of coal and coke*. West Conshohocken, PA: ASTM International.

ASTM D2013/D2013M-18 (2013). *Standard practice for preparing coal samples for analysis*. West Conshohocken, PA: ASTM International.

ASTM D2234/D2234M-17 (2017). *Standard practice for collection of a gross sample of coal*. West Conshohocken, PA: ASTM International.

ASTM D2492-02(2012) (2012). *Standard test method for forms of sulfur in coal*. West Conshohocken, PA: ASTM International.

ASTM D2797/D2797M-11a (2011). *Standard practice for preparing coal samples for microscopical analysis by reflected light.* West Conshohocken, PA: ASTM International.

ASTM D2798-11a (2011). *Standard test method for microscopical determination of the vitrinite reflectance of coal.* West Conshohocken, PA: ASTM International.

ASTM D3172-13 (2013). *Standard practice for proximate analysis of coal and coke.* West Conshohocken, PA: ASTM International.

ASTM D3174-12 (2012). *Standard test method for ash in the analysis sample of coal and coke from coal.* West Conshohocken, PA: ASTM International.

ASTM D3175-18 (2017). *Standard test method for volatile matter in the analysis sample of coal and coke.* West Conshohocken, PA: ASTM International.

ASTM D3176-15 (2015). *Standard practice for ultimate analysis of coal and coke.* West Conshohocken, PA: ASTM International.

ASTM D3302/D3302M-17 (2017). *Standard test method for total moisture in coal.* West Conshohocken, PA: ASTM International.

ASTM D388-18a (2018). *Standard classification of coals by rank.* West Conshohocken, PA: ASTM International.

ASTM D4972-18 (2018). *Standard test methods for pH of soils.* West Conshohocken, PA: ASTM International.

ASTM D5003-19 (2019). *Standard test method for Hardgrove Grindability Index (HGI) of petroleum coke.* West Conshohocken, PA: ASTM International.

ASTM D5192-09(2015) (2015). *Standard practice for collection of coal samples from core.* West Conshohocken, PA: ASTM International.

ASTM D5373-16 (2016). *Standard test methods for determination of carbon, hydrogen and nitrogen in analysis samples of coal and carbon in analysis samples of coal and coke.* West Conshohocken, PA: ASTM International.

ASTM D5709-09(2015) (2015). *Standard test method for sieve analysis of petroleum coke.* West Conshohocken, PA: ASTM International.

ASTM D5865-13 (2013). *Standard test method for gross calorific value of coal and coke.* West Conshohocken, PA: ASTM International.

ASTM D6357-11 (2011). *Test methods for determination of trace elements in coal, coke, & combustion residues from coal utilization processes by Inductively Coupled Plasma atomic emission, Inductively Coupled Plasma mass, & graphite furnace Atomic Absorption spectrometry.* West Conshohocken, PA: ASTM International.

ASTM D6722 (2011). *Standard test method for total mercury in coal and coal combustion residues by direct combustion analysis.* West Conshohocken, PA: ASTM International.

ASTM D7263-09(2018)e2 (2018). *Standard test methods for laboratory determination of density (unit weight) of soil specimens.* West Conshohocken, PA: ASTM International.

ASTM D7348-13 (2013). *Standard test methods for loss on ignition (LOI) of solid combustion residues*. West Conshohocken, PA: ASTM International.
ASTM D7430-18a (2018). *Standard practice for mechanical sampling of coal*. West Conshohocken, PA: ASTM International.
ASTM E1226-12a (2012). *Standard test method for explosibility of dust clouds*. West Conshohocken, PA: ASTM International.
ASTM E1491-06(2012) (2012). *Standard test method for minimum autoignition temperature of dust clouds*. West Conshohocken, PA: ASTM International.
ASTM E1515-14 (2014). *Standard test method for minimum explosible concentration of combustible dusts*. West Conshohocken, PA: ASTM International.
ASTM E2019-03(2013) (2013). *Standard test method for minimum ignition energy of a dust cloud in air*. West Conshohocken, PA: ASTM International.
Bend, S.L. (1992). The origin, formation and petrographic composition of coal. *Fuel* 71(8): 851–870.
Bowen, B.H. and Irwin, M.W. (2008). Coal characteristics. *CCTR Basic Facts File #8* 1–27. West Lafayette, IN: Indiana Center for Coal Technology Research, Purdue University.
Canadian Centre for Occupational Health and Safety (2019a). What is a LD_{50} and LC_{50}? http://www.ccohs.ca/oshanswers/chemicals/ld50.html (accessed 31 July 2019).
Canadian Centre for Occupational Health and Safety (2019b). Global harmonized system (GHS). http://www.ccohs.ca/oshanswers/ chemicals/ghs.html (accessed 31 July 2019).
Canadian Centre for Occupational Health and Safety (2019c). Transportation of dangerous goods (TDG)-overview. http://www. ccohs.ca/oshanswers/legisl/tdg/tdg_overview.html (accessed 31 July 2019).
Canadian Council of Ministries of the Environment (1999). Canadian environmental quality guidelines. *Glossary* 1–10. Winnipeg, MB.
Canadian Council of Ministries of the Environment (2001). Reference method for the Canada-wide standard for petroleum hydrocarbons in soil-tier 1 method. *Publication No. 1310* 1–33. Winnipeg, MB.
Canadian Council of Ministries of the Environment (2002a). Agricultural. *Soil Quality Guidelines for the Protection of Environmental and Human* 1–9. Hull, QC.
Canadian Council of Ministries of the Environment (2002b). Polychlorinated Dibenzo-p-Dioxins and Polychlorinated Dibenzofurans (PCDD/Fs). *Soil Quality Guidelines for the Protection of Environmental and Human Health* 1–13. Hull, QC.
Canadian Council of Ministries of the Environment (2008). *Canada-Wide Standards for Petroleum Hydrocarbons (PHC) in Soil* 1–8. Winnipeg, MB.
Canadian Food Inspection Agency (2017a). Safety guidelines for fertilizers and supplements. *T-4-93* 1–11. Ottawa, ON.
Canadian Food Inspection Agency (2017b). *Proposal-Maximum Chemical Contaminant Levels in Livestock Feeds* 1–30. Ottawa, ON.

Canadian Food Inspection Agency (2017c). *Proposal-Contaminant Standards for Aflatoxins, Deoxynivalenol, Fumonisins, Ergot Alkaloids and Salmonella in Livestock Feeds* 1–18. Ottawa, ON.

Canadian Humalite International Inc. (2020a). Raw material and products. *Laboratory Analyses and Technical Data* 1–656. Edmonton, AB.

Canadian Humalite International Inc. (2020b). Raw material and products. *Safety Data Sheets* 1–55. Edmonton, AB.

Center for Climate and Energy Solutions (2017). Coal. http://www.c2es.org/energy/source/coal (accessed 08 May 2019).

Christensen, D., Crawford, C., and Szabo, R. (2002). Enumeration of coliforms, faecal coliforms and of E. Coli in foods using the MPN method. *MFHPB-19* 1–14. Ottawa, ON: Health Products and Food Branch, Government of Canada.

Csoke, B., Bokanyi, L., Bohm, J., and Petho, Sz. (2003). Selective grindability of lignites and their application for producing an advanced fuel. *Applied Energy* 74: 359–368.

Dalhousie University (2002). *Leonardite Explosibility Testing* 1–4. Halifax, NS.

Douey, D. and Wilson, P. (2004). Enumeration of yeasts and moulds in foods. *MFHPB-22* 1–9. Ottawa, ON: Health Products and Food Branch, Government of Canada.

EEC A.6 (1992). Water solubility. Dir 92/69/EEC (O.J. L383 A). In: *Official Journal of the European Communities No. L 383/113* (member K. Van Miert), 1–9. Aberdeen, UK: Commission Directive 92/69/EEC.

EPA Method 1613 Revision B (1994). *Tetra-through octa-chlorinated dioxins and furans by isotope dilution HRGC/HRMS*. Washington, DC: Office of Water Engineering and Analysis Division, US Environmental Protection Agency.

European and Mediterranean Plant Protection Organization (2013). Diagnostic PM 7/119 (1) nematode extraction. *European and Mediterranean Plant Protection Organization Bulletin* 43(3): 471–495.

Fike Corporation (2015). Humalite dust. *Fike Report No. SUM11510491-Rev0* 1–6. Blue Springs, MO.

Fowkes, W.W. and Frost, C.M. (1960). *Leonardite: a lignite byproduct. Report on Investigations 5611* 1–3. Washington, DC: US Bureau of Mines.

Gaudet, I.D. (1994). *Standard Procedure for Microtox Analysis* 1–28. Edmonton, AB: Western Canada Microtox Users Committee, World Collaborative Microtox Users Council.

Gillmore, D.W., Wright, C.C., and Walker Jr., P.L. (1956). Grindability data on Pennsylvania anthracites. *Proceedings of the Anthracite Conference* 1–3. University Park, PA: The Pennsylvania State University.

Government of Alberta (2020). Occupational health and safety code. *Occupational Health and Safety Act* 2.1–2.20, 4.1–4.32, and S1–S34. Edmonton, AB.

Hoffman G.L., Nikols, D.J., Stuhec, S., and Wilson, R.A. (1993). Evaluation of leonardite (humalite) resources of Alberta. *Open File Report 1993-18* 1–18. Edmonton, AB: Alberta Research Council.

ISO No. 11348-1-2-3 (2007). *Water quality: determination of the inhibitory effect of water samples on the light emission of Vibrio fischeri (Luminescent bacteria test), Pt. 1: method using freshly prepared bacteria, Pt. 2: method using liquid-dried bacteria, Pt. 3: method using freeze-dried bacteria*. Geneva, Switzerland: International Organization for Standardization.

Kelly-Hooper, F., Farwell, A.J., Pike, G., et al. (2013). Is it clean or contaminated soil? Using petrogenic versus biogenic GC-FID chromatogram patterns to mathematically resolve false petroleum hydrocarbon detections in clean organic soils: a crude oil-spiked peat microcosm experiment. *Environmental Toxicology and Chemistry* 32(10): 2197–2206.

Kohanowski, N.N. (1970). Leonardite in North Dakota. *North Dakota Quarterly* 38: 36–42.

Langenberg, W., MacDonald, D., Kalkreuth, W., and Strobl, R. (1989). Coal quality variation in the Cadomin-Luscar coalfield, Alberta. *Earth Sciences Report 89-1* 6–16. Edmonton, AB: Alberta Research Council.

MacDonald, D.E., Langenberg, C.W., and Gentzis, T. (1989). A regional evaluation of coal quality in the foothills/mountains region in Alberta. *Earth Science Report 89-2* 2–16. Edmonton, AB: Alberta Research Council.

McCabe, P.J., Strobl, R.S., MacDonald, D.E., et al. (1988). Coal resources of the Horseshoe Canyon formation and laterally equivalent strata, to a depth of 400 m, in the Alberta plains area. *Open File Report 1988-03* 23–28. Edmonton, AB: Alberta Research Council.

Microbics Corporation (1992). Detailed solid-phase test protocol. In: *Microtox Manual-A Toxicity Handbook* 153–178. Carlsbad, CA.

Mukhopadhyay, P.K. (1994). Petrographic and molecular characterization and its applications to basin modelling. In: *Vitrinite Reflectance as a Maturity Parameter. Applications and Limitations. ACS Symposium Series No. 570* (eds. P.K. Mukhopadhyay and W.G. Dow), 1–24. Washington, DC: American Chemical Society.

National Research Council (2005). *Mineral Tolerance of Animals*, 2nd rev., 10–14. Washington, DC.

National Safety Council (2016). OSHA publishes final rule on silica. http://www.safetyandhealthmagazine.com/articles/13942-osha-publishes-final-rule-on-silica (accessed 10 July 2019).

Nelson, G. and Garrard, A. (2015). Pyrogenic vs petrogenic source determination: diagnostic PAH ratios. *ESAA Presentation* 1–36. Edmonton, AB: Environmental Services Association of Alberta.

NIOSH Method No. 7500 (2003). Silica, crystalline, by XRD (filter redisposition). In: *NIOSH Manual of Analytical Methods (NMAM)-Method Numbers 6001-8000*, 4th ed., 1–9. Atlanta, GA: Centers for Disease Control and Prevention.

Occupational Safety and Health Administration (2009). Hazard communication guidance for combustible dust. *Report No. OSHA 3371-08* 1–7. Boston, MA.

Occupational Safety and Health Administration (2019). Safety and health topics: silica, crystalline. http://www.osha.gov/dsg/topics/ silicacrystalline/ (accessed 30 January 2019).

Palmer, L. (2012). Where is all the coal? http://slate.com/technology/2012/ 11/ coal-reserves-in-the-united-states-geology-explains-why-we-have-so-much-coal.html (accessed 9 May 2019).

Puskar, M.A. (1991). Recognition and evaluation of chemical hazards. In: *The Work Environment: Occupational Health Fundamentals Vol. 1* (ed. D.J. Hansen), 3–20. Boca Raton, FL: CRC Press.

Rae, D., Smith, E., and Braun, J. (2015). An Introduction to biogenic hydrocarbon issues at federal contaminated sites. *Presentation* 1–25. Ottawa, ON: Real Property Institute of Canada.

Reid, A.R. (2009). Isolation and identification of Salmonella from food and environmental samples. *MFHPB-20* 1–15. Ottawa, ON: Health Products and Food Branch, Government of Canada.

Siemers, C.T. and Wadell, J.S. 1977. Humate deposits of the Menefee formation (upper Cretaceous), northwestern New Mexico. *28th Field Conference* 1–22. Socorro, NM: New Mexico Geological Society.

Speight, J.G. (2015). *Handbook of Coal Analysis,* 2nd ed., 1–26. Hoboken, NJ: John Wiley and Sons.

State of California (2018). Respirable crystalline silica standards. *Cal/OSHA* 1–3. Sacramento, CA: Department of Industrial Relations.

The European Agency for the Evaluation of Medicinal Products (1999). Humic acids and their sodium salts. *Summary Report* 1–2. London, UK.

US Bureau of Mines (1992). Crystalline silica primer. *Special Publication* 1–25. Washington, DC.

US Energy Information Administration (2018). Coal-energy explained, your guide to understanding energy. http://www.eia.gov/ energyexplained/index.php?page=coal_home (accessed 11 December 2018).

US International Trade Commission (2020). Harmonized tariff schedule: harmonized tariff schedule of the United States (2019) Revision 17. https://hts.usitc.gov/ (accessed 18 March 2020).

US Pharmacopoeia (2014). Chemical tests. *{561} Articles of Botanical Origin* 1–10. Rockville, MD.

Van Krevelen, D.W. (1950). Graphical-statistical method for the study of structure and reaction processes of coal. *Fuel* 29: 269–284.

Wang, Z., Yang, C., Yang, Z., et al. (2012). Fingerprinting of petroleum hydrocarbons (PHC) and other biogenic organic compounds (BOC) in oil-contaminated and background soil samples. *Journal of Environmental Monitoring* 14: 2367–2381.

3

Humic Analyses

3.1 Overview

As shown in Chapter 2, low-rank coal is a good-quality material containing low quantities of ash, very low quantities of selected metals, and negligible amounts of toxic organics and pathogens. Most importantly, it can be considered a non-hazardous material with respect to its handling and field application. The key factor for whether low-rank coal can feasibly be manufactured into agricultural products is its content of humic substances (HS, or "humic"). HS are components of soil organic matter (OM, or "humus"), which can be found abundantly in nature (i.e. soil or water), albeit in much less concentrated amounts. Analyses for HS for low-rank coal are therefore identical to those of soil and water. Before going into more detail, the relationship between humus and HS, as well as their components, must firstly be understood.

Humus is defined as all OM found in soil, except undecayed plant and animal tissue, their partial decomposition products, and soil biomass (Stevenson 1994). Canadian Food Inspection Agency (2020) defines humus as a homogeneous and friable mixture of biochemically stable, partially decomposed OM. Humus is formed from soil organic debris (plant or animal remains) through a unique process called humification. The end products of this process (the humified OM) are newly synthesized components with distinct properties from the original materials (the non-humified OM) (Wershaw 2000).

Humification has been defined by Zech et al. (1997) as an accumulation over time of lipid-derived alkyl compounds, aromatic compounds with carboxyl (COOH), and phenol substituents, along with a depletion of carbohydrates and amino acids. To be more specific, humification follows four main pathways: lignin modification (Waksman 1938 as cited in Stevenson 1994), reactions of quinone with nitrogen (N) compounds (Martin et al. 1975), microbial decomposition of

Low-Rank Coal Applications in Agriculture: Humic Analyses, Products, and Performance, First Edition. L. Edwin Liem.

aromatic compounds (Martin and Haider 1971, and Haider and Martin 1967), and sugar-amine condensation (Maillard 1913 as cited in Kawamura 1983 and Ikan et al. 1986).

According to Hayes and Swift (2018), humus can be divided into compounds belonging to recognizable classes (such as carbohydrates, peptides, and others) and amorphous polymeric brown-colored components (which are defined as HS). HS can therefore be defined as the non-identifiable portions of humus containing a series of high-molecular weight substances, which are rich in oxygen (O)-containing functional groups (Stevenson 1994). According to Roybal and Barker (1986), humic material has been characterized as a non-pure substance formed during the complete or partial decay of OM. HS are also defined as naturally occurring, biogenic, heterogeneous organic substances, characterized as being yellow to black in color, refractory, and of high molecular weight (Aiken et al. 1985). A more recent definition states that HS are the major organic constituents of soil OM and the aquatic environment, consisting of complex heterogeneous mixtures of carbon (C)-based substances formed by biochemical reactions during the decay and transformation of plant and microbial remains (Association of American Plant Food Control Officials 2018 and Canadian Food Inspection Agency 2020). ISO 19822:2018(E) (2018) defines HS in a singular form "humic substance" as a major organic constituent of natural OM, consisting of complex heterogeneous mixtures of C-based substances formed by biochemical reactions during the decay and transformation of plant and microbial remains.

HS are primarily composed of three main fractions: humic acids (HA), fulvic acids (FA), and humin. They are operationally defined by their solubility in dilute alkali and acid solutions. Humin is the component that is not soluble at any acidity (pH) levels. HA are defined as portions of the alkali extracted from HS that are not soluble in strongly acidic solution. They precipitate from the alkali extract in acid solutions of pH 2 or less (Association of American Plant Food Control Officials 2018 and Canadian Food Inspection Agency 2020). Organic Materials Review Institute (2017) identifies HA as humic acid derivatives, i.e. acids extracted from humates (the definition of humates is provided later in this section).

FA are HS components that are soluble at any pH (Aiken et al. 1985), or fractions of humates that are soluble at neutral to acid pH (Organic Materials Review Institute 2017). Canadian Food Inspection Agency (2020) defines FA as the extracted portion of HS that are soluble in both alkali and acid aqueous solutions. Association of American Plant Food Control Officials (2015) defines FA slightly differently, as fulvic fractions (FF) and hydrophobic fulvic acids (HFA). FF are portions of HS that are soluble in both alkali and acid aqueous solutions. HFA are portions of HS that are soluble in both alkali and acidic aqueous solutions that are separated from non-humic aqueous substances in FF by selective adsorption onto a hydrophobic resin of non-ionic macroporous acrylic ester of moderate polarity (DAX-8) at low pH. These definitions are also adopted by ISO 19822:2018(E) 2018. The new HFA definition therefore eliminates compounds that may have previously been classed as FA; that is, it eliminates hydrophilic fractions of FF, including fatty acids, polysaccharides, amino sugars or acids, proteins, and carbohydrates. Further breakdowns have also been suggested, including crenic and aprocrenic acids as components of FA, and brown (ulmic) humic acids, grey humic acids, and hymatomelanic acids as components of HA (Stevenson and Butler 1969).

For practical purposes, however, only HA and FA are taken into consideration. Due to its water insolubility and associated difficulties during analyses, humin has been ignored in most studies (Rice and MacCarthy 1988). To date, all humic-related analyses have been completed only on the water-soluble components (De Nobili et al. 1990, Stevenson 1994, Swift 1996, ISO 5073 2013, Tremblay and Gagne 2002, Van Zomeren and Comans 2007, ISO/TC 134 N 260 2014, Lamar et al. 2014, and ISO 19822:2018(E) 2018). The sum of HA and FA can be assumed to be the total HS (thus omitting humin from all analyses); therefore: HS = HA + FA. The science of humin is, however, still important in terms of research work, with possible practical usages in the future. A thorough evaluation of the characteristics and role of humin can be found in Rice and MacCarthy (1988) and Hayes et al. (2017).

There have been debates on the use of alkali to extract natural OM, resulting in the water-soluble components of HS. Details of such can be found in Janzen (2019), Lehmann and Kleber (2015), and Kleber and Lehmann (2019). One key argument here is that alkali extraction results in chemical alteration of natural OM. In this instance, humic fractions (most importantly, HA and FA) are merely laboratory-generated parameters, and are therefore not suitable for the study of natural OM. This argument has been challenged by Olk et al. (2019), who claim that alkali extraction does not change the structure of natural OM if completed in an O-free environment (i.e. using N_2 gas), and therefore humic fractions are a suitable approach to natural OM. Myneni (2019) discusses the use of Nuclear Magnetic Resonance (NMR) spectroscopy to provide more pristine conditions (i.e. without alkali extraction). However, this method is weak when dealing with samples with high iron (Fe) and low OM contents. Regardless of the outcome of these debates, they confirm the fact that natural OM, HS, HA, and FA do not have exact physical and chemical structures and compositions. Both their definitions and the analytical methods employed are approximate, and a basic understanding is necessary to grasp their appropriate applications and limitations. Later in this book (see Section 3.2), the alkali extraction approach, which results in humic fractions, is adopted. This is because, in the context of low-rank coal applications in agriculture, this method is practical to employ and maintains accuracy and consistency.

HS are ubiquitous in soil and nature in general. Association of American Plant Food Control Officials (2018) identifies sources of HS as commercially harvested terrestrial deposits, including leonardite, oxidized lignite, oxidized subbituminous coals, humalite, carbonaceous shale (including humic shale), peat, and sapropel. The most important humic-containing materials are humates because they are abundant in nature and contain high strengths of HA (Hoffmann et al. 1993). Humates are defined as stable decomposed OM (Organic Materials Review Institute 2017) or as the salts of HA (Siemers and Wadell 1977 and Association of American Plant Food Control Officials 2013). These include leonardite, lignite, subbituminous coal, and carbonaceous sediments (Roybal and Barker 1986). Carbonaceous sediments, which include carbonaceous shale, mudstone, and clay stone, are not coal materials. However, they can be in direct contact with coal seams and are therefore considered to be humates. The term "humalite" has been used to identify weathered subbituminous coal (low-rank coal) and weathered carbonaceous shale materials mined within the province of Alberta, Canada (Hoffmann et al. 1993). Considering only low-rank coal material being collected from the mine site, humalite is identical to Alberta low-rank coal.

Table 3.1 compares selected humates from North America: Alberta low-rank coal (mean values of seven data); lignite from North Dakota, Wyoming, and New Mexico (mean values of five data); and carbonaceous shale from Idaho (based on

Table 3.1 Selected humate materials.

Material	C %	H %	O %	N %
Alberta low-rank coal	49.9	2.9	24.9	1.3
North Dakota, Wyoming, and New Mexico lignite	47.5	3.1	21.0	0.9
Idaho carbonaceous shale	4.4	0.3	7.3	0.2

Material	S %	Ash %	OM %	Calorific value kJ/kg
Alberta low-rank coal	0.6	20.4	79.6	14 600
North Dakota, Wyoming, and New Mexico lignite	1.2	26.3	73.7	14 900
Idaho carbonaceous shale	0.3	87.5	12.5	1000

Source: Canadian Humalite International Inc. (2020a); *note:* all are mean values (based on dry matter analysis).

one datum). C, hydrogen (H), O, N, sulfur (S), and ash results have been obtained without moisture (or the dry matter analysis) using the ultimate analysis according to ASTM D2013/D2013M-18 (2013), ASTM D5373-16 (2016), ASTM D2492-02(2012) (2012), ASTM D3174-12 (2012), ASTM D3176-15 (2015), and O by difference. OM can simply be calculated as the sum of C, H, O, N, and S, with the remaining material being inorganics or ash. The calorific value analysis (including moisture, or as is analysis) is according to ASTM D2013/D2013M-18 (2013) and ASTM D5865-13 (2013).

This table illustrates that good-quality humates are the ones with high OM concentrations but with low calorific values (19 300 kJ/kg or less; see Section 2.1). Alberta low-rank coal (also known as humalite) has a calorific value of 14 600 kJ/kg, while lignite from North Dakota, Wyoming, and New Mexico has a calorific value of 14 900 kJ/kg (both are mean values). As a comparison, subbituminous has calorific values of 19 300 to 30 200 kJ/kg, bituminous of 24 400 to 34 900 kJ/kg, and anthracite of 30 200 to 34 900 kJ/kg (Bowen and Irwin 2008 and Center for Climate and Energy Solutions 2017). Luscar coal from Alberta has calorific values of 21 200 to 25 100 kJ/kg (Canadian Humalite International Inc. 2020a). It is interesting to note that while carbonaceous shale has a very low calorific value at 1000 kJ/kg, its compositions are totally different from those of other materials. It contains very high ash, meaning very low OM (12.5%) compared to well over 70% of those of Alberta low-rank coal and lignite. As shown later in Table. 3.8, it has very low HS.

3.2 Extraction and Analyses

From the above definitions, it is clear that the exact structures of HS are not known. Efforts have been made to extract HS from soil or other materials to identify their characteristics, then to separate them into different components. The extraction involves the use of alkali (i.e. strong bases) at high pH to solubilize the material in water and, when necessary, also includes reactions with acids, neutral salts, metals, and organic solvents. Elementary chemical compositions containing C, H, and O atoms have been identified by Achard, Sprengel, and Brazelius as cited in Stevenson (1994). It has commonly been accepted that C, H, O atoms form in a unique way with COOH, acidic-hydroxyl (OH), and carbonyl (C=O) groups attached to benzene rings, while N atoms may or may not be included in the structure (Hermann, Oder, and Shmook, as cited in Stevenson 1994). According to Leenheer and Croue (2003), HS have similar chemistries as those of dissolved OM, in which FA fall under the hydrophobic components.

Since their exact structures and formulations are not known, humic analyses can only be completed on specific parameters (i.e. surrogates), in which assumptions and restrictions on what to expect from the results have firstly to be made. Then, conversions from these parameters are made that yield the best estimated values for HA and/or FA. None of the known analyses is therefore exact and there is often great variation between analyses. Analyses on humates are even more challenging since, on top of HS, the raw material contains impurities that, at large amounts, may interfere with the results. Analyses on commercially processed (liquid and soluble) humic products are the most complicated due to the possible presence of nutrients, extracting chemicals, and others within, which may interfere with the analyses (Canadian Humalite International Inc. 2020a).

Regardless of the methods used for the analysis of HS, each can be completed with or without moisture within the material. For the dry matter analysis, the sample is dried prior to analysis at approximately 65 °C to remove all moisture without burning off its OM components. ASTM D3302/D3302M-17 (2017) can be used as the basis of the analysis. The result is presented with respect to its dry weight. In the case of liquid products, the analysis is usually completed as is (including moisture). The result is presented with respect to its original weight.

3.2.1 Extraction

In most cases, a humic material has to be extracted and made into a solution before a specific analysis can be completed. As shown later in this chapter, this applies for gravimetric, colorimetric, and organic C analyses (Sections 3.2.3 to 3.2.6). For physical properties and reactive functional groups (Sections 3.2.2 and 3.2.7), the extraction may or may not be needed during the analysis. Among other methods, HS (or OM in general) can be made water-soluble when the pH of the solution is raised.

A laboratory study was completed to evaluate the performance of different extracting chemicals. Low-rank coal (water solubility = 4.4%, analyzed according to EEC A.6 1992 and pH = 4.0, analyzed according to ASTM D4972-18 2018) was selected as the raw material. It was mixed with tap water at room temperature to reach 15% total OM within the mixture. Sodium hydroxide (caustic soda or NaOH), potassium hydroxide (caustic potash or KOH), ammonium hydroxide (NH_4OH), or calcium hydroxide (hydrated lime or $Ca(OH)_2$) was individually added to the mixture to achieve 7, 10, and 13 $\pm$ 0.1 pH. Chemical ratio was calculated in the following way: (dry weight of chemical $\div$ dry weight of humic material) $\times$ 100%. Each mixture was settled for one hour, then the supernatant was analyzed for OM. The performance of each chemical was evaluated based on the ratio of OM over the amount of chemical being added, i.e. OM/chemical ratio. The higher the value, the better the chemical performed. Throughout this study, OM was analyzed according to ASTM D7348-13 (2013).

The best extracting chemical was found to be NaOH, followed by KOH, NH_4OH, and $Ca(OH)_2$. To achieve a mixture's pH of 7 ± 0.1, 4.4% NaOH produced 9.4% OM, followed by 5.5% KOH producing 5.7% OM, 5.4% NH_4OH producing 5.8% OM, and 28.4% $Ca(OH)_2$ producing 8.9% OM. To achieve a mixture's pH of 10 ± 0.1, 6.6%, NaOH produced 10.2% OM, 13.3% KOH producing 9.2% OM, 23.4% NH_4OH producing 9.4% OM, and 133.3% $Ca(OH)_2$ producing 9.3% OM. For reasonable usage rates, only NaOH and KOH were able to achieve a mixture's pH of 13 ± 0.1 at 19.6 and 28.9% strength, producing 11.1 and 9.8% OM, respectively. pH values between 7 and 10 were found to be optimum for the extraction of this material in water, with OM/chemical ratios between 1.55 and 2.13 for NaOH, and between 0.69 and 1.04 for KOH (see Table 3.2).

For laboratory analyses, NaOH would be the better extracting chemical. However, for the agricultural application, Na would not be beneficial for soil or crops. A subsequent laboratory study was completed to evaluate the performance of KOH to produce the liquified OM from the same raw material at

Table 3.2 Extracting chemicals.

	pH = 7 ± 0.1			pH = 10 ± 0.1		
	Chemical ratio %	OM %	OM/chemical ratio	Chemical ratio %	OM %	OM/chemical ratio
NaOH	4.4	9.4	2.13	6.6	10.2	1.55
KOH	5.5	5.7	1.04	13.3	9.2	0.69
NH_4OH	5.4	5.8	1.07	23.4	9.4	0.40
$Ca(OH)_2$	28.4	8.9	0.31	133.3	9.3	0.07

	pH = 13 ± 0.1		
	Chemical ratio %	OM %	OM/chemical ratio
NaOH	19.6	11.1	0.57
KOH	28.9	9.8	0.34
NH_4OH	N/A	N/A	N/A
$Ca(OH)_2$	N/A	N/A	N/A

Source: Canadian Humalite International Inc. (2020b); *note:* N/A = not available.

a mixture's pH of 7 to 10 ± 0.1. In this study, the extraction temperature was raised to 65 ± 1 °C. Beyond this, a possible loss of OM might be experienced and was therefore avoided. OM at 14.1% was the highest strength of liquid solution that could be produced at 10.0 ± 0.1 pH. At 7.0 ± 0.1 pH, the highest OM strength was 6.3%. OM/chemical ratio values ranged from 0.88 to 1.02. These findings indicated that increasing the mixture temperature helped with the extraction of OM from the raw material (see Table 3.3).

It was also found that pH changed with temperature significantly at high values (see Figure 3.1 for pH vs. temperature profiles), but that changes were extremely small (almost unchanged) at neutral pH. A target pH at room temperature should be estimated when the solution was at a higher temperature during the mixing of all ingredients. For example, 10.05 pH for the finished solution at 20 °C would require having 9.76 pH at 65 °C during mixing.

Table 3.3 KOH performance.

Solution	Chemical ratio %	pH	OM %	OM/chemical ratio
Liquified OM	6.2–16.1	7–10 ± 0.1	6.3–14.1	0.88–1.02

Source: Canadian Humalite International Inc. (2020b).

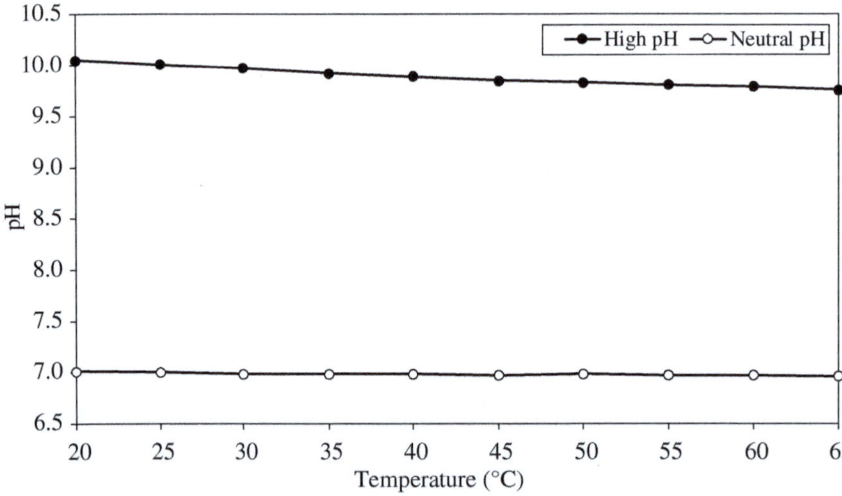

Figure 3.1 pH vs. temperature.

3.2.2 Physical Properties

This method uses physical properties, i.e. molecular weight, mass, or sizing as parameters to qualify humic materials. HS have molecular weights of 500 to 200 000 g/mole, in which leonardite has an average molecular weight of 18 700 g/mole (Beckett et al. 1987). According to Stevenson (1994), molecular weights for HS range from 2000 (FA) to 300 000 g/mole (humin). HA are somewhere in between. Using the adsorption-based approach, Tarasevich et al. (2016) have found a dry humic sample supplied by Aldrich, having a molecular mass of 1000 to 15 000 g/mole, with molecular sizes between 2.3 and 9.0 nm. The same material has previously been analyzed by Ochs et al. (1994) and Vermeer and Koopal (1998) using the gel permeation chromatography, finding a molecular mass between 1000 and over 100 000 g/mole, averaged at 21 000 g/mole. Using the ultrafiltration method, Baigorri et al. (2007) has observed > 99% of Aldrich material containing between 5000 and 100 000 g/mole in molecular masses, and 68% between 5000 and 10 000 g/mole.

Since smaller molecular weight fractions of HS are better uptaken by plants (Vaughan and Ord 1981) and are more biologically active (Vaughan and Linehan 1976), any humic materials containing molecular weights within the FA range (around 2000 g/mole) or at least close to the standard humic sample (around 21 000 g/mole, containing mainly HA) can be assumed to be of good quality. This is mainly true with respect to the scientific evaluation of HS. In real agricultural applications, commercial humic products with larger molecular weight components are sometimes preferred. They are better retained in soil, while products of smaller molecular weights tend to leach easily through the medium (see Chapters 4 and 6).

This method, however, is not practical and is also costly. In addition, it cannot quantify the amounts of HA and FA within humic materials. The information is useful when selecting a humic material to be the standard material that can be utilized for other analytical methods. Looking at its molecular weight, it is fair to assume that the material provided by Aldrich contains small to medium molecular weight molecules, and therefore is ideal for HA and FA analyses.

3.2.3 Gravimetric

This method uses gravimetric weights of specific soil OM components contained within humic materials as parameters to quantify HA and/or FA. This method is based on the classical technique described by Stevenson (1994) and Swift (1996). A strong base is firstly used to extract the alkaline soluble components (i.e. HA and FA) while leaving out the insoluble components (humin and inorganics). A strong acid is then used to precipitate HA while keeping FA within the analytical solution.

The California Department of Food and Agriculture (CDFA) method discards FA and analyzes only HA after stabilizing its weight in a drying oven. This method does not include ash corrections in the combustion oven (CDFA Method 1999), therefore the results tend to be overestimated (Lamar and Talbot 2009 and Canadian Humalite International Inc. 2020a). It can, however, be improved by washing the analytical solution with hydrochloric and hydrofluoric acids, thereby unlocking the bond between clay and OM, and thus reducing the ash content without altering the composition of HA (Sanchez-Monedero et al. 2002).

The Humic Products Trade Association (HPTA) method, later known as the International Organization for Standardization (ISO) 19822 keeps both precipitate (HA) and soluble (FF) components. FF is then run through a non-ionic resin (DAX-8) to capture the more hydrophobic component (as per Thurman and Malcolm 1981 and Leenheer and Croue 2003), which is identified as HFA. The solution is de-ashed by running it through a cation exchange resin of H form (IR-120). HA and HFA are then determined based on their weights gravimetrically including ash corrections (Lamar et al. 2014 and ISO 19822:2018(E) 2018). This method assumes components that are only soluble at high pH as HA and components that are soluble at any pH as well as more hydrophobic as HFA. The challenge is that there are non-humic materials that have similar properties as those of humic materials. Analyzing humic materials is straightforward, but commercial humic products may contain foreign materials having similar properties (also known as adulterants) that may interfere with the results. Lignosulfonates have been found to affect the quantification of HFA (Canadian Humalite International Inc. 2020a). As suggested by Lamar et al. (2014) and ISO 19822:2018(E) (2018), a pre-screening on the amount of S be required for suspect products. Lignosulfonates typically contain around 5.0% S, while humic materials have approximately 1.0% S or less (see Table 3.1). It has been proposed that any materials containing over 0.75% S be analyzed using the reactive functional groups method (see Sections 3.2.7 and 3.2.8). Still, it is very challenging to quantify the actual amount of HFA when lignosulfonates and HS are combined together.

Table 3.4 shows HA + FA (= HS) results for Aldrich and HPTA dry samples, analyzed using ISO 19822. The Aldrich sample is a mixture of plant parts, peat, and soft coal (Lamar and Talbot 2009). It has HA, FA, and HS contents of 73.5, 2.8, and 76.3%, respectively (all are mean values based on three data). The HPTA sample is lignite from North Dakota. It contains 74.0, 2.2, and 76.2% of HA, FA, and HS, respectively (all are mean values based on four data); in brackets are standard deviations. While containing some impurities (ash), they do not contain components that may influence HA and FA analyses. Both samples are very similar and consistent from one analysis to another.

Table 3.4 Standard humic samples using ISO 19822.

Sample	# of data	HA %	FA %	HS %
Aldrich	3	73.5 (0.86)	2.8 (0.62)	76.3 (1.21)
HPTA	4	74.0 (1.32)	2.2 (0.15)	76.2 (1.39)

Source: Canadian Humalite International Inc. (2020a); *note:* all are mean values, in brackets are standard deviations.

Table 3.5 Alberta low-rank coal using ISO 19822.

Material	# of data	HA and FA using ISO 19822%
Alberta low-rank coal	8	HA = 80.6 (5.11)
	8	FA = 2.6 (0.47)
	8	HA + FA (= HS) = 83.2 (5.05)

Source: Canadian Humalite International Inc. (2020a); *note:* all are mean values, in brackets are standard deviations, N/A = not available.

Alberta low-rank coal analyzed using ISO 19822 shows 80.6 and 2.6% HA and FA, respectively. Both are mean values of 8 samples (see Table 3.5). HS is calculated at 83.2%. This indicates that this material is similar to Aldrich and HPTA samples, implying that all of them originate from a similar origin of low-rank coal or lignite material.

Five humate samples from North America of low-rank coal and carbonaceous shale origins have been quantified using the CDFA method. Table 3.6 shows HA results for each. Alberta low-rank coal and lignite from North Dakota contain similar HA at 65.8 and 61.2%, respectively (both are mean values from ten and two samples, respectively). Lignite from New Mexico and Wyoming show lower contents at 26.4 and 16.4%, respectively. Carbonaceous shale from Idaho has the lowest HA at 1.3%. For Alberta low-rank coal, in comparison to ISO 19822 (Table 3.5), the CDFA method produces a lower result.

Overall, the CDFA method is practical and economical. ISO 19822 is less practical as it consumes a significant amount of analytical time and has a higher capital cost. However, due to its accuracy, there has been a movement to adopt this new method within the commercial community over the CDFA method (Humic Products Trade Association 2017). Recently, the CDFA has approved the use of the HPTA (or ISO 19822) test method seal for all HA products sold in California (Humic Products Trade Association 2020).

Table 3.6 Humates using CDFA method.

Material	# of data	HA using CDFA method %
Alberta low-rank coal	10	65.8 (10.07)
New Mexico lignite	1	26.4
Wyoming lignite	1	16.4
North Dakota lignite	2	61.2
Idaho carbonaceous shale	1	1.3

Source: Canadian Humalite International Inc. (2020a); *note:* all are mean values (for two or more data), in bracket is standard deviation.

3.2.4 Colorimetric

This method uses visible light absorption of organic molecules responsible for producing color (i.e. the chromophoric group) as a parameter to qualify and quantify humic materials. While spectra for humic materials may vary depending upon their degree of humification as well as pH and the structures of the chromophoric group (Tsutsuki and Kuwatsuka 1979, and Baes and Bloom 1990), specific wavelengths between 400 and 800 nm have commonly been selected for humic materials (Stevenson 1994). The material has to firstly be dissolved in water, using NaOH that can be combined with other chemicals, to completely dissolve all HA and FA prior to analysis using an ultraviolet (UV)-visible spectrophotometer (Mehlich 1984).

The quality of humic materials can be obtained from the E_4/E_6 ratio, in which E_4 is the absorbance at 465 nm and E_6 is the absorbance at 665 nm. E_4/E_6 is also called the index of molecular condensation. This index decreases with increasing molecular weight and condensation, indicating more aromatic components, while the opposite indicates more aliphatic structures. In short, this index is governed primarily by particle sizes and weights (Chen et al. 1977). Soil HA has been reported to have E_4/E_6 ratios of 3.0 to 5.0, while soil FAs are between 6.0 and 9.6 (Schnitzer 1971 and Schnitzer 1972). Canadian Humalite International Inc. (2020a) has found that based on 12 data, Alberta low-rank coal has E_4/E_6 ratios between 3.1 and 4.5, with a mean value of 3.5 (see Table 3.7). This indicates that this material contains mainly HA, which is congruent with results obtained using ISO 19822 (see Table 3.5).

Furthermore, the presence of humified or humic materials may be detected when analyzing it under a wide range of absorbance readings. The correlation between aromatic compounds and molecular weights, the relationship between humified and non-humified materials, as well as the degree and indicators of

Table 3.7 E_4/E_6 of HA, FA, and Alberta low-rank coal.

Material	E_4/E_6 using colorimetric method
HA	3.0–5.0
FA	6.0–9.6
Alberta low-rank coal	3.1–4.5, mean = 3.5 (0.45), based on 12 data

Sources: Schnitzer (1971, 1972) and Canadian Humalite International Inc. (2020a); *note:* in bracket is standard deviation.

humification process, have been well identified by Chin et al. (1994), Chen et al. (1977), Mahieu et al. (2002), Jiang et al. (2011), Weishaar et al. (2017), Chen et al. (2017), Zaccone et al. (2018), and Yakimenko et al. (2018). This approach is able to detect whether the material is of humic or non-humic origin. For example, when Alberta low-rank coal is analyzed using this approach and plotted in a log scale that is normalized to its dissolved organic C content, it shows a smooth spectral of frequencies of an almost straight line with a slope of approximately 1:290. As discussed later in this chapter (Section 3.3), a non-humified material is expected to produce a distinct pattern, and therefore can be excluded from the selection of a good-quality humic material (Canadian Humalite International Inc. 2020a).

The quantity of HS within humic materials can be obtained from the absorbance at 450 nm. A standard humic sample has firstly to be chosen, i.e. one with very high HS and very low contaminants. Dry humic samples from Aldrich and HPTA are well known for their purity (HS contents are listed in Table 3.4). Therefore, one of them can be selected as the standard material. The sample is firstly alkali-extracted to capture HS, then diluted in alkali solution. The solution is analyzed using a spectrophotometer at 450 nm, which is based upon the Beer-Lambert law:–log (light intensity after passing through solution ÷ initial light intensity) = absorptivity coefficient × analyte concentration × path length (HACH 2018). A relationship between HS (which are directly proportional to their weights) and spectrophotometer readings can be established as shown in Figure 3.2.

This figure shows readings at 450 nm for both Aldrich and HPTA standard HS samples at different dry sample weights. Their relationship is clearly linear, as shown in each the following equations: reading = 0.63 × weight (R^2 = 0.996) for the Aldrich sample and reading = 0.64 × weight (R^2 = 0.995) for the HPTA sample. Note: weight = sample dry weight. Based on ISO 19822, 2.2 g dry weight of the Aldrich sample contains 76.3% × 2.2 g = 1.68 g HS, corresponding to a reading of 1.4. Conversely, a reading of another humic sample directs one to a specific HS value. Knowing its original dry weight, the percentage of HS can then be calculated.

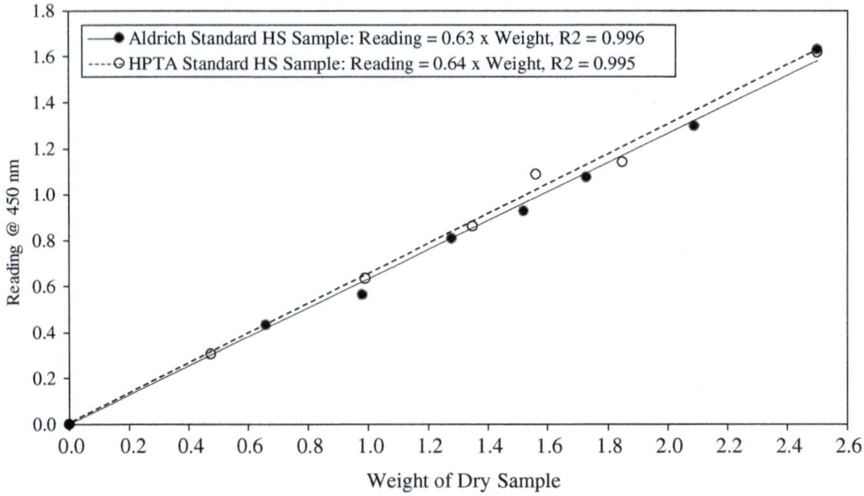

Figure 3.2 Aldrich and HPTA dry humic samples.

However, it is important to note that there is no test for absolute OM (including HS). There is only OM in terms of a given analytical procedure (Mehlich 1984).

Since this method assumes components that reflect light at 450 nm to be HS, it is only acceptable if the material contains mostly HS with few impurities. The challenge lies in the fact that there are other non-humic materials showing similar properties (i.e. the adulterants) producing similar readings. Analyzing raw humic materials may not be too challenging (assuming there are few impurities within the material). Processed products are more difficult to quantify due to the possible presence of foreign materials that may interfere with the readings (Canadian Humalite International Inc. 2020a). While the colorimetric method is not especially accurate, it is very practical and economical, and so has become the most widely used method in the industry.

The same humate samples from North America of low-rank coal and carbonaceous shale origins have been quantified using the colorimetric method. Alberta low-rank coal and lignite from North Dakota contain high HS at 91.8% and 79.3%, respectively (both are mean values of 461 and 2 samples, respectively). Lignite from New Mexico and Wyoming show lower HS at 65.3% and 39.1%, respectively. Carbonaceous shale from Idaho has the lowest HS at 5.6% (see Table 3.8). For Alberta low-rank coal, in comparison to ISO 19822 (Table 3.5) and the CDFA method (Table 3.6), the colorimetric method produces the highest result.

The colorimetric method cannot quantify HA on its own (separate from FA). However, it can still quantify FA on its own (separate from HA) by incorporating a different standard sample. A known material containing only FA with little

Table 3.8 Humates using colorimetric method.

Material	# of data	HA + FA (= HS) using colorimetric method %
Alberta low-rank coal	461	91.8 (6.20)
New Mexico lignite	1	65.3
Wyoming lignite	1	39.1
North Dakota lignite	2	79.3
Idaho carbonaceous shale	1	5.6

Source: Canadian Humalite International Inc. (2020a); *note:* all are mean values (for two or more data), in bracket is standard deviation.

impurities has firstly to be quantified using ISO 19822. Then, readings at different dry weights are to be completed using a spectrophotometer at a designated wavelength. Due to smaller molecular weights of OM being analyzed, a higher wavelength than 450 nm is necessary, i.e. 665 nm. The relationship between the reading and weight of dry sample is expected to be linear, similar to that of Figure 3.2, in which a general equation can be developed. Any samples can then be analyzed at the same wavelength, and FA values can be calculated using the equation (Canadian Humalite International Inc. 2020a).

A modification, as suggested by Lamar and Talbolt (2009), allows this method to quantify HA only. First of all, a known sample of the same or similar material as the tested material is selected. The analytical steps are identical to those of the standard colorimetric method, with additional steps to remove acid soluble components (FA) from the analytical solution using acidification and centrifuging. The solution is then analyzed using the spectrophotometer, in which HA can be quantified.

3.2.5 Combined Colorimetric and Gravimetric

The colorimetric method always gives higher results than those of the CDFA method because it quantifies HS, while the CDFA method only quantifies HA. It then becomes a common practice within the commercial community to assume FA as the difference between both methods, i.e. FA = (HA + FA) from the colorimetric method – HA from the CDFA method. This method is of course not accurate, since combining two analytical methods of different principles produces inherently false results (Canadian Humalite International Inc. 2020a).

Table 3.9 Humates using colorimetric – CDFA methods.

Material	FA using colorimetric – CDFA methods %
Alberta low-rank coal	26.0
New Mexico lignite	38.9
Wyoming lignite	22.7
North Dakota lignite	18.1
Idaho carbonaceous shale	4.3

Source: Canadian Humalite International Inc. (2020a); *note:* values are by difference between colorimetric and CDFA methods.

Taking the difference between both methods from Tables 3.8 and 3.6, FA can be calculated as follows: 26.0% for Alberta low-rank coal, 38.9% for lignite from New Mexico, 22.7% for lignite from Wyoming, 18.1% for lignite from North Dakota, and 4.3% for carbonaceous shale from Idaho (see Table 3.9). For Alberta low-rank coal, in comparison to ISO 19822 (Table 3.5), this approach produces a result that is inflated to 10 times that of the actual value (FA = 26.0 instead of 2.6%), indicating that this approach is not valid and cannot be accepted.

A better approach can be taken by combining the modified colorimetric method, as suggested by Lamar and Talbolt (2009), and the original colorimetric method (see Section 3.2.4). In this case, FA = (HA + FA) from the colorimetric method – HA from the modified colorimetric method. While this approach cannot directly quantify FA, it at least combines two methods that are based on the same analytical principle of the colorimetric concept.

With the adoption of the new gravimetric method (i.e. ISO 19822), any of these practices should not be encouraged. This new method uses a valid principle that is able to directly quantify FA with a high accuracy and consistency.

3.2.6 Organic Carbon

This method uses organic C as a parameter to quantify humic materials. The one proposed by Van Zomeren and Comans (2007) has some similarities to those of ISO 19822. The material is dissolved in water, adjusted for pH, and undergoes physical processes to distinguish HA, FA, hydrophilic acids, and hydrophobic neutral OM as per Leenheer (1981). HA are not soluble in water at low pH. FA and hydrophobic neutral OM are soluble in water at low pH and are adsorbed by DAX-8, while hydrophilic acids are the ones passing through. FA are desorbed from the resin at high pH. Different from ISO 19822, HA, FA, and hydrophilic

Table 3.10 Alberta low-rank coal using ISO 5073.

Analysis	Value %
Total HA (extracted using an alkaline salt)	64.8
Free HA (extracted using a strong base)	63.5

Source: Canadian Humalite International Inc. (2020a); *note:* based on 1 datum.

acids are quantified based on their organic C contents obtained as total organic C. Hydrophobic neutral OM = (adsorbed FA + hydrophobic neutral OM) – desorbed FA. This method is quicker than that of ISO 19822 since no de-ashing and gravimetric quantification of the analytical solution are required. However, it is still time consuming and prone to errors mainly during DAX-8 adsorption and desorption steps.

ISO 5073 (2013) quantifies total HA and free HA as components extracted using an alkaline salt and a strong base, respectively. Free HA are therefore comparable to HA that have previously been defined. Organic C in the extractant is chemically oxidized and the excess oxidizer is reacted with an acidic salt and compared to a standard reference (blank). Total HA and free HA are each calculated as the amount of salt consumed multiplied by a certain factor. This factor has been determined specifically for brown coal and lignite materials, assuming that they all have similar properties, most importantly on the O:C ratios and the contents of impurity. This method is quite practical and economical given the type of material to be analyzed, but may not be applicable for other materials. Based on 1 datum, Alberta low-rank coal contains 64.8% total HA and 63.5% free HA (see Table 3.10, Canadian Humalite International 2020a). In comparison to others (see Tables 3.5, 3.6, and 3.8), this method produces the lowest result, which is very close to that of the CDFA method.

3.2.7 Reactive Functional Groups

This method uses reactive functional groups to qualify humic materials (Stevenson 1994). Included in these are COOH and phenol-OH groups. Good-quality humic materials contain high percentages of these components. Analyses can be completed on each, or in combined groups, using the traditional wet chemistry, NMR spectroscopy, or Fourier Transform Infra-Red (FTIR) spectroscopy. Traditional wet chemistry determines total acidity as the sum of COOH and acidic OH groups. Total acidity is analyzed by firstly reacting the material with a base, then reacting the residual base using an acid (Schnitzer 1972). Cation exchange capacity measures the ability of material to capture cations, which is similar to total acidity. It is analyzed by firstly reacting the material with an ammonium salt, then determining the exchanged ammonium using the aeration or acid method according to EPA Method 9080 (1986). The COOH group can individually be analyzed by firstly reacting the material with an acid complexed metal, then reacting the residual acid with a base. The phenol OH group can then be calculated as the difference between total acidity and COOH group (Schnitzer 1972). Traditional wet chemistry is rigorous and rather impractical, but economical.

The NMR method is a form of absorption spectroscopy that uses interaction of electromagnetic radiation with another element, in the case of soil OM, with C-13. The results are expressed as chemical shifts with respect to a specific reference. Both solid and liquid states of spectrophotometry can be applied for dry (in situ) and liquid (water-soluble, mainly at high pH) samples. Certain functional groups show unique chemical shifts; therefore, their amounts can be determined. FTIR analysis is a form of absorption spectroscopy that uses stretching and bending movements of molecular analysis and chemical bonds of a molecule. A detector records an absorption peak of a wavelength when the energy of infrared light is absorbed and its vibration increases (Silverstein et al. 2005). Both methods are very accurate, but are time intensive and costly. Efforts have been made to improve accuracy, analytical time, and cost of the solid-state NMR spectroscopy with some success (Trembley and Gagne 2002).

The following are COOH, phenol-OH, and total acidity results of different humic-containing materials (note: total acidity = COOH + phenol-OH). As summarized by Schnitzer (1977), reference HA and FA extracted from different types of soil have 560 to 890 meq/100 g and 890 to 1420 meq/100 g total acidity, respectively. Different samples of humates and soil have been analyzed by Schnitzer et al. (2001) and Schnitzer and Dinel (2001), as presented in Table 3.11. Two samples of Alberta low-rank coal contain 490 and 700 meq/100 g total acidity, while two lignite samples from North Dakota contain 590 and 670 meq/100 g total acidity. Based on one sample, lignite from New Mexico and Wyoming, each contains 680 and 690 meq/100 g total acidity, respectively, while carbonaceous shale from Idaho contains 520 meq/100 g total acidity. As a comparison, reference samples of HA soil (Alberta) contain 660 and 700 meq/100 g total acidity, while a reference sample of FA soil (Prince Edward Island) contains 1090 meq/100 g total acidity.

This table shows that humates from five different regions are similar to each other with respect to their COOH, phenol-OH, and total acidity contents. Interestingly, carbonaceous shale from Idaho, regardless its high ash in comparison to other lignite materials, also contains similar functional groups. Note: each of these analyses has been completed on an ash-free basis. Comparing their total acidity with those of reference HA and FA obtained from soil, it can be concluded that all of these materials contain mainly HA. For Alberta low-rank coal, this is congruent with the previous results (see Tables 3.5 to 3.8, and 3.10).

Table 3.11 COOH, phenol-OH, and total acidity.

Material	# of data	COOH meq/100 g
Alberta low-rank coal	2	200 and 320
New Mexico lignite	1	310
North Dakota lignite	2	310 and 340
Wyoming lignite	1	360
Idaho carbonaceous shale	1	270
Reference HA soil (Alberta)	2	450 and 490
Reference HA soil	N/A	N/A
Reference FA soil (Prince Edward Island)	1	780
Reference FA soil	N/A	N/A
Material	**Phenol-OH meq/100 g**	**Total acidity meq/100 g**
Alberta low-rank coal	290 and 380	490 and 700
New Mexico lignite	370	680
North Dakota lignite	280 and 330	590 and 670
Wyoming lignite	330	690
Idaho carbonaceous shale	250	520
Reference HA soil (Alberta)	210 and 210	660 and 700
Reference HA soil	N/A	560–890
Reference FA soil (Prince Edward Island)	310	1090
Reference FA soil	N/A	890–1420

Sources: Schnitzer et al. (2001), Schnitzer and Dinel (2001), and Schnitzer (1977); *note:* N/A = not available.

3.2.8 Comparison Among Analytical Methods

Without going into too much detail, it is important to reiterate that all analytical methods on HA and/or FA analyze certain soil OM components within humic materials, in which certain parameters (i.e. surrogates) are to be measured.

Regardless what they are, in most cases, only the ones that are soluble in water are considered. That is, only HA and/or FA can be analyzed, leaving out humin and inorganics. From these parameters, assumptions and restrictions on what to expect from the end results can be determined. The physical properties method measures molecular weight, mass, or sizing. The gravimetric method measures weights. The colorimetric method measures visible light absorption. The organic C and reactive functional groups methods measure concentrations. Conversions are then made towards these parameters, in which HA and/or FA values can be generated.

Since HS are defined as the non-identifiable components of humus (Stevenson 1994), none of these methods is accurate in terms of having the ability to completely represent HS, as well as HA and/or FA. For example, it has commonly been believed that using the reactive functional groups method, COOH, phenol-OH, and C=O groups are the most accurate parameters to be converted to HA and/or FA. While HA and/or FA indeed consist of these functional groups, their actual compositions and proportions are not identified in a precise manner and vary from one material to another. An extensive HA and FA database has been generated from different origins of soil OM (International Humic Substances Society 2020). This is indeed helpful in providing general trends and characteristics for various humic materials. Still, the exact values of HA and/or FA cannot be quantified. The same challenges are experienced with others, including the physical properties, colorimetric, gravimetric, and organic C methods.

Most importantly, there are several definitions for HA and FA being suggested, including those by the Canadian Food Inspection Agency (CFIA), the Association of American Plant Food Control Officials (AAPFCO), ISO, and the Organic Materials Review Institute (OMRI). Definitions proposed by AAPFCO and ISO are specific and are directly related to the gravimetric method (i.e. ISO 19822). HA and/or FA results obtained from any other methods may not correspond well with the ones generated using this method. In addition, while typical humic materials contain small impurities that may interfere with the analyses, processed products may contain added components that may skew the results due to having similar functional groups to those of HA and/or FA.

Hereon, the gravimetric method (i.e. ISO 19822) can be adopted as the standard method to quantify HA and/or FA (as HFA under this method) of humic materials. The reactive functional groups have been included to support this method. For example, when any humic materials contain over 0.75% S, FTIR analysis is to be conducted to screen the presence of lignosulfonates (see Section 3.2.3). The colorimetric method is the second-best method because of its practicality; it is recommended that it be used with ISO 19822. As shown later in Section 3.3, under this method, the UV-visible spectral performance can be used to screen out non-humic from true humic products.

3.2.9 Further Studies on Analytical Methods

Two original studies were completed on the colorimetric and gravimetric methods. Analyses on HS, HA, HFA, OM, ash, and pH were done in-house. The ultimate, Inductively Coupled Plasma (ICP), and FTIR analyses were completed by third-party laboratories.

The first study was a straightforward laboratory analysis on the effects of pH on spectrophotometer readings of a liquid humic product. This was completed due to discrepancies in HS claims by various commercial humic products. Not only had these discrepancies been confusing, they also created significant regulatory issues during product registration. Work completed by Tsutsuki and Kuwatsuka (1979) was the basis of this study, that pH influenced chemical structures of HS affecting the light absorbance behavior. The main objective of this laboratory work was to hypothesize that higher pH could inflate HS results for samples with the same OM content.

The second study was a detailed and thorough laboratory analysis on ISO 19822. It was conducted in order to address the time intensive analysis of this new method, as it could hinder its acceptance within the commercial community. The main objective of this work was to modify this method so that it would require a shorter analytical time while maintaining its validity, accuracy, and practicality.

3.2.9.1 Effects of pH on Colorimetric Method

A laboratory study was completed to evaluate the effects of pH on the spectrophotometer readings of a liquid humic sample. It was important to remember that high pH resulted in the dissolution of humic materials into soluble (HA and FA, or HS) and insoluble (humin and inorganic) components. This method, which was based on the interaction between the light and particulates within the solution, was limited to analyzing water-soluble materials. Typically, between 5 and 6 pH was critical; here a drastic separation of humic components was observed. Beyond 9 pH, the solution was generally stable, at least at its typical HS strength of approximately 12% (based on the colorimetric method).

It was stated that the dissolution of humic materials (or OM in general) might not stop at a certain pH value, and higher pH might also affect properties of the particulates with respect to their interaction with light. The main objective of this study was to hypothesize that the same amount of OM might produce different HS results depending on the pH values. Claims on commercial products had to be carefully evaluated.

A commercial liquid humic product was used as the stock solution. This product contained 11.7% HS and had 10.1 pH. From this stock solution, identical samples were prepared. Each sample was treated with acid or base to alter its pH, then analyzed for HS and OM. Should the hypothesis be correct, higher HS would be recorded for the same OM content at higher pH values. The following were chemical and equipment requirements, as well as detailed analytical steps including sample preparation, HS analysis, and OM analysis.

Chemical requirements:

1) Aldrich standard humic sample.
2) Liquid humic product, HS (HA + FA) = 11.7% (based on the colorimetric method, including moisture), pH = 10.1.
3) KOH of 45% strength.
4) Phosphoric acid (H_3PO_4) of 75% strength.
5) NaOH of 99% strength.
6) Deionized (DI) water.

Equipment requirements:

1) An analytical balance of 220 g capacity (± 0.1 mg precision).
2) Glass Erlenmeyer flasks (1000 ml).
3) Glass beakers (80 ml ± 5% accuracy).
4) Glass volumetric flasks (100 ml ± 0.08 ml accuracy).
5) Spectrophotometer cuvettes (5 ml) of plastic material.
6) A magnetic stir plate.
7) A magnetic bar.
8) A magnetic stick.
9) Plastic paraffin films.
10) A mechanical pipette of 1000 μL capacity (± 1 μL accuracy).
11) A pH meter (± 0.01 unit accuracy).
12) A spectrophotometer at 450 nm wavelength (± 0.005 unit accuracy).
13) A drying oven of 120 °C capacity (± 0.1 °C accuracy).
14) A combustion oven of 1000 °C capacity (± 1 °C accuracy).
15) A desiccator of 4 l capacity.

The liquid humic product was stored in a 1000 ml Erlenmeyer flask. Twenty 80 ml beakers were individually weighed. Fifty ml of this product was poured into each beaker. Either KOH or H_3PO_4 was added to the beaker. pH was then recorded, and each beaker was adjusted to 60 ml using DI water. The beaker was then mixed using a magnetic stir plate and magnetic bar for 15 minutes. The magnetic bar was taken out using the magnetic stick and weight of the beaker was recorded. Calculation: original weight = weight of full beaker – weight of empty beaker.

Using a cuvette filled with DI water, the spectrophotometer was zeroed at 450 nm absorbance. After five minutes of settling, using a mechanical pipette, 200 μL solution from each beaker was transferred to a 100 ml volumetric flask. One g of NaOH was added and the flask was filled up with DI water, providing 500:1 dilution ratio. The solution was poured into a cuvette, and its absorbance at 450 nm was recorded. Calculation: the percentage of HS = reading × 7 (generated from the Aldrich sample and a dilution factor specific for this application).

Each beaker was placed in the drying oven at 65 °C and left overnight until its weight became stable. It was then placed in a desiccator to cool to room temperature and its weight was recorded. Each beaker was put in the combustion oven at 500 °C and left overnight until its weight became stable. It was then placed in a desiccator to cool to room temperature and its weight was recorded. Calculation: the percentage of OM = {(weight at 65 °C - weight at 500 °C) ÷ original weight]} × 100%.

As expected, OM was approximately constant from one beaker to another. Since each beaker was corrected to a volume of 60 ml, its spectrophotometer reading was reliable across all samples. HS/OM values were then calculated, such that the results could be directly compared. It was found that at pH < 7.4, HS/OM values significantly dropped. For simplicity, HS/OM at pH = 7.4 was set to the value of one, and others were corrected accordingly.

The results were best interpreted by plotting HS/OM vs. pH (see Figure 3.3). From pH of 7.4 to 13.1 (no settling was found in each beaker), it was clearly observed that HS/OM increased with pH, indicating that the same OM would produce different HS depending on the pH. At this range of pH, a linear relationship between pH and HS/OM was found as HS/OM = 0.020 × pH + 0.862 (R^2 = 0.709). At lower pH (a sharp drop around pH 5.2), most HA precipitated, leaving out FA, resulting in low HS and OM readings.

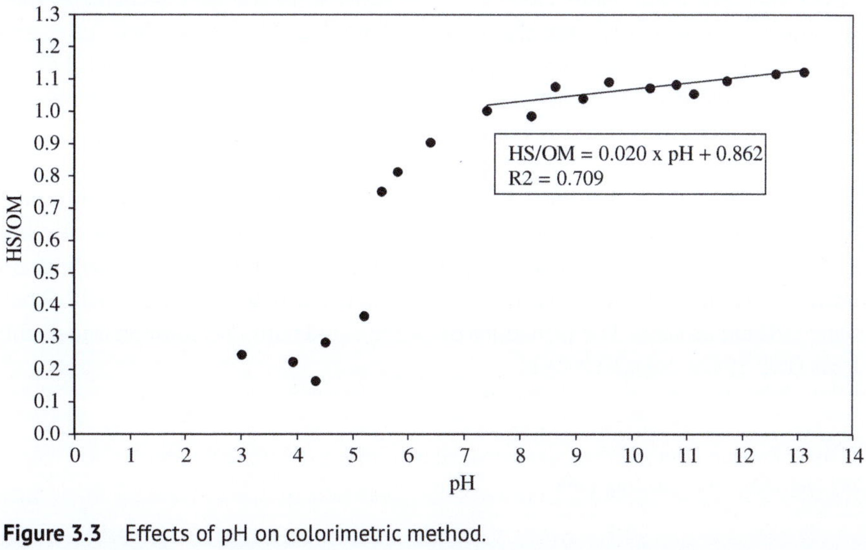

Figure 3.3 Effects of pH on colorimetric method.

In conclusion, high pH dissolved more HS components and/or changed their characteristics, resulting in higher spectrophotometer readings at 450 nm, as implied by Tsutsuki and Kuwatsuka (1979). It was interesting to note that the solution was treated with 1.0 g NaOH addition into the volumetric flask, thus raising its pH (to over 13), before a reading was made. This, however, was done after the solution was diluted using DI water at 500:1 within the volumetric flask (note: the liquid humic product had a pH of 10.1). While no further efforts were made in this study to investigate the actual mechanism, this laboratory work proved the above hypothesis to be correct: that pH indeed influenced HS results of liquid humic samples containing the same OM content. Humic claims on commercial products would have to carefully be evaluated with respect to their pH.

3.2.9.2 Modified ISO 19822[1]

A laboratory study was completed to modify ISO 19822 since this method consumed a significant amount of analytical time. The modification was not due to its fundamental science, but to its practicality with regard to several of its analytical steps. In this case, acetone was proposed as a desorbing agent to replace NaOH. Its use would eliminate several analytical steps as well as the use of two pieces of laboratory equipment. The main objective of this study was to evaluate the practicality of acetone as the desorbing agent so that the total analytical time could significantly be reduced without compromising its validity and accuracy.

3.2.9.2.1 Background The gravimetric analysis was one of the most accurate and practical methods to quantify fulvic components in humic-containing materials. Among several methods available in the industry, the HPTA method (Lamar et al. 2014) and ISO 19822 (ISO 19822:2018(E) 2018) were the best to date, as they included improvements over the previous methods (CDFA Method 1999 and Swift 1996). ISO 19822 was developed from the HPTA method with several changes to its analytical steps. In terms of principles, both were identical. Throughout this study, both were identified as "the standard method."

In the standard method, HS were defined as major organic constituents of natural OM consisting of complex heterogeneous mixtures of C-based substances formed by biochemical reactions during the decay and transformation of plant and microbial remains. Fulvic components were identified as those soluble in both alkali and acids solutions, and adsorbed to a hydrophobic resin at pH 1. They were defined as HFA. The principles of this method could be summarized as follows (ISO 19822:2018(E) 2018):

---

1 Its revised and shorter version was published in Liem et al. (2019), in which L.E. Liem was the main and corresponding author.

1) Alkali extraction was completed under anoxic conditions to reduce oxidation of the analytical sample during extraction.
2) Alkali extracted portions of HS that were insoluble in strongly acidic solutions and precipitated from alkali extract in acid solutions of pH 1 were defined as HA.
3) Alkali extracted portions of HS that were soluble in both alkali and acid aqueous solutions were defined as FF.
4) Materials composed of less than 0.75% elemental S soluble in aqueous alkaline and acid solutions and adsorbed at pH 1 onto a polymeric adsorbent resin of moderate polarity, designed for adsorption of amphiphilic compounds having molecular weights typical of FF, were defined as HFA. The resin of interest would be DAX-8 of methacrylic ester material, which was a hydrophobic resin of non-ionic macroporous acrylic ester of moderate polarity.
5) Ash-free gravimetric analyses were completed for both HA and HFA.

Since this study only discussed HFA, analytical steps related to HA were omitted.

In the standard method, HFA were adsorbed by DAX-8, desorbed using NaOH, cleaned from Na ions using IR-120, and then quantified gravimetrically including ash corrections (Lamar et al. 2004 and ISO 19822:2018(E) 2018). However, steps involving IR-120 produced a significant volume of analytical solution (containing FF), which had to be partially reduced using a Rotovap prior to the gravimetric analysis and required a significant time for the analysis, including its regeneration. This could become a disadvantage of this method with respect to its acceptance within the commercial agricultural community.

An effort was therefore made to modify this standard method so that the total analytical time could be significantly reduced. In this modified method, acetone replaced NaOH to desorp HFA from DAX-8. Acetone was selected because it was one of the most common and effective organic solvents available in the market with the following advantages: it was miscible in water, bound both polar and non-polar materials, and had a lower boiling point than that of water. Note: acetone was previously found to produce a low desorption rate of humic materials in DAX-2 resin of styrene divinyl benzene material (Mantoura and Riley 1975). This could be caused by the following factors: this resin was in fact less effective than DAX-8 with respect to related applications of FF (Aiken et al. 1979); a higher pumping rate was used during resin adsorption (Aiken et al. 1992); the humic material was not firstly dissolved in water at high pH; HA was not separated from HFA; and an analytical solution of slightly higher pH (approximately 2 instead of 1) was used (ISO 19822:2018(E) 2018).

In the modified method, NaOH was not added during DAX-8 desorption, and therefore IR-120 would not be required. Steps involving IR-120, including its regeneration, were all skipped. A smaller volume of the analytical solution with a lower boiling point was produced, so that a Rotovap would not be necessary. The solution could be directly quantified for HFA.

The substitution of NaOH with acetone might, however, have effects on the principles. The following items needed to be addressed and carefully evaluated in this study:

1) NaOH would exchange (read: desorp) all organic acid anions including hydrophobic acids (HyA) as in the case of HFA, while acetone would dissolve (read: desorp) all organic components resulting in the possible presence of hydrophobic neutrals (HyN) within the quantification of HFA (Mantoura and Riley 1975, Leenheer 1981, and Leenher and Croue 2003).
2) During DAX-8 regeneration, steps using DI water and hydrochloric acid (HCl) might not be sufficient due to the possible presence of HyN within the analytical solution.
3) Since IR-120 would not be required, Na ions added during alkali extraction would still be in the analytical solution, which might affect the accuracy of gravimetric analysis.

Table 3.12 Acetone properties.

Parameter	Value
Molecular weight	58.08 g/mole
Empirical formula	C_3H_6O
Boiling point	56 °C
Freezing point	−95 °C
Specific gravity	0.79
Water solubility	Total (miscible)
Solubility parameter	10.0
Polarity (water = 100)	35.5
pH	7.0
Acetone	99.8%
Water	0.2%
Methanol	< 0.02%
Alcohol	< 0.005%
Aldehyde	< 0.002%

Sources: Smallwood (1996) and Fisher Scientific Co. (2018).

3.2.9.2.2 Materials and Methods Acetone of laboratory grade was used in this study. It had a 99.8% strength and contained very small impurities, including water, methanol, alcohol, and aldehyde. It was miscible in water and had a neutral pH (see Table 3.12).

A lignite sample from North Dakota was used in this study. Its chemical compositions were analyzed by a third-party laboratory (Loring Laboratories Ltd. 2019) according to ASTM D2013/D2013M-18 (2013), ASTM D5373-16 (2016), ASTM D2492-02(2012) (2012), ASTM D3174-12 (2012), ASTM D3176-15

Table 3.13 Lignite sample properties.

Parameter	Value	Unit
C	56.5	%
H	4.6	%
O	22.4	%
N	1.4	%
S	1.2	%
Ash	13.9	%
Aluminum (Al)	5.72	%
Ca	0.52	%
Fe	3.17	%
K	0.29	%
Magnesium (Mg)	2.46	%
Na	1.41	%
Phosphorous (P)	< 0.01	%
Silica (Si)	0.70	%
Titanium (Ti)	0.35	%
Arsenic (As)	11	ppm
Barium (Ba)	403	ppm
Cadmium (Cd)	1	ppm
Chloride (Cl)	3	ppm
Chromium (Cr)	10	ppm
Copper (Cu)	75	ppm
Mercury (Hg)	0.09	ppm
Manganese (Mn)	445	ppm
Molybdenum (Mo)	9	ppm

(Continued)

Table 3.13 (Continued)

Parameter	Value	Unit
Nickel (Ni)	14	ppm
Lead (Pb)	28	ppm
Strontium (Sr)	1157	ppm
Thallium (Th)	165	ppm
Zinc (Zn)	50	ppm
Zirconium (Zr)	178	ppm

Source: Loring Laboratories Ltd. 2019; *note:* each result is based on dry matter analysis, sample's original moisture = 12.1%.

(2015), ASTM D6357-11 (2011), ASTM D6722 (2011), and O by difference. As shown in Table 3.13, the sample contained mainly OM with little impurities, including toxic metals.

Thirty-four runs were completed in this study, including 28 runs using the lignite sample and six blank runs using DI water. The standard method was used in 11 runs and the modified method was used in 23 runs. Four runs were analyzed for FTIR spectra, 28 runs were quantified gravimetrically for HFA, and 2 runs were evaluated using the ICP analysis for Na and Cl ions (see Table 3.14 for the experimental protocol). Run 1: Lignite-Standard-1 represented the first run completed in this study, i.e. the first replication of the lignite sample using the standard method; Run 16: Lignite-Modified75%-4 represented the sixteenth run completed in this study, i.e. the fourth replication of the lignite sample using the

Table 3.14 Experimental protocol.

Run	Identification	Sample	DAX-8 desorption	Analysis
1	Lignite-Standard-1	Lignite	0.1 M NaOH	HFA
2	Lignite-Modified25%-1	Lignite	25% acetone	HFA
3	Lignite-Modified50%-1	Lignite	50% acetone	HFA
4	Lignite-Modified75%-1	Lignite	75% acetone	HFA
5	Lignite-Standard-2	Lignite	0.1 M NaOH	HFA
6	Lignite-Standard-3	Lignite	0.1 M NaOH	HFA
7	Lignite-Modified25%-2	Lignite	25% acetone	HFA

Table 3.14 (Continued)

Run	Identification	Sample	DAX-8 desorption	Analysis
8	Lignite-Modified50%-2	Lignite	50% acetone	HFA
9	Lignite-Modified50%-3	Lignite	50% acetone	HFA
10	Lignite-Modified75%-2	Lignite	75% acetone	HFA
11	Lignite-Modified25%-3	Lignite	25% acetone	HFA
12	Lignite-Modified75%-3	Lignite	75% acetone	HFA
13	Lignite-Standard-4	Lignite	0.1 M NaOH	HFA
14	Lignite-Modified25%-4	Lignite	25% acetone	HFA
15	Lignite-Modified50%-4	Lignite	50% acetone	HFA
16	Lignite-Modified75%-4	Lignite	75% acetone	HFA
17	Lignite-Standard-5	Lignite	0.1 M NaOH	FTIR
18	Lignite-Standard-6	Lignite	0.1 M NaOH	HFA
19	Lignite-Standard-7	Lignite	0.1 M NaOH	HFA
20	Blank-Modified50%-1	DI water	50% acetone	HFA
21	Blank-Modified50%-2	DI water	50% acetone	FTIR
22	Lignite-Modified50%-5	Lignite	50% acetone	FTIR
23	Lignite-Modified50%-6	Lignite	50% acetone	HFA
24	Lignite-Modified50%-7	Lignite	50% acetone	HFA
25	Lignite-Standard-8	Lignite	0.1 M NaOH	HFA
26	Blank-Modified50%-3	DI water	50% acetone	FTIR
27	Lignite-Standard-9	Lignite	0.1 M NaOH	HFA
28	Blank-Modified50%-4	DI water	50% acetone	HFA
29	Lignite-Modified50%-8	Lignite	50% acetone	HFA
30	Lignite-Modified50%-9	Lignite	50% acetone	HFA
31	Lignite-Modified50%-10	Lignite	50% acetone	Na + Cl
32	Blank-Modified50%-5	DI water	50% acetone	HFA
33	Lignite-Standard-10	Lignite	0.1 M NaOH	Na + Cl
34	Blank-Standard-1	DI water	0.1 M NaOH	HFA

modified method at 75% acetone strength; Run 32: Blank-Modified50%-5 represented the thirty-second run completed in this study, i.e. the fifth replication of DI water using the modified method at 50% acetone strength; and so on.

Differences in analytical steps between the two methods were as follows: the standard method used NaOH at 0.1 M strength as a desorbing agent, IR-120 to remove Na ions from the analytical solution, and a Rotovap to partially evaporate the solution prior to its gravimetric analysis. For the regeneration of DAX-8, two columns of DI water + one column of 0.1 M HCl + one column of DI water were used in each of eleven runs. The modified method used acetone at 25%, 50%, and 75% strengths as the desorbing agent. IR-120 was not required and no partial evaporation using a Rotovap was needed. For the regeneration of DAX-8, two columns of DI water + one column of 0.1 M HCl + one column of DI water were used in each of 12 runs (Runs 2 to 4, 7 to 12, and 14 to 16), and two columns of DI water + one column of 0.1 M HCl + one column of 50% acetone + one column of DI water were used in each of 11 runs (Runs 20 to 24, 26, and 28 to 32).

The principles (or validity) of this modified method would be evaluated based on FTIR spectra and gravimetric results. These determined the nature of the organic components being analyzed (i.e. HyA and/or HyN). Many articles were published differentiating the FTIR spectra of both components generated from water and soil samples (Leenher 1981 and Polubesova et al. 2008). A more straightforward approach was taken in this study by considering only absorbance readings at 2922 and 1701 cm^{-1}, representing C–H bonds and C=O groups, respectively. The C–H group was assumed to be responsible for the negative charge behavior at pH levels of less than 7, as in the case of HyA. The C=O group was assumed to be responsible for the neutral charge behavior, as in the case of HyN (ChemAnalytical LLC. 2018a and 2019). Both analyses also determined the performance of DAX-8 regeneration steps.

The accuracy of this modified method would be evaluated using the gravimetric and ICP analyses. The gravimetric analysis determined the correct strength of acetone as the desorbing agent corresponding to HFA results produced by the standard method using 0.1 M NaOH. It also showed the amount of ash within the analytical solution, which was later confirmed using the ICP method on Na and Cl ions. The effect of ash on the HFA gravimetric analysis could therefore be evaluated.

The practicality of this modified method would be evaluated based on the time consumed for each analytical step. Changes made to DAX-8 regeneration steps might result in a slightly longer analytical time, while skipping steps involving the use of IR-120 and a Rotovap would result in a significant time reduction.

3.2.9.2.3 Laboratory Work Chemical requirements:

1) NaOH of laboratory grade, 95% minimum purity (Anachemia Canada Inc. 2018), made into 0.1 M solution.
2) HCl of laboratory grade, 31% strength (ClearTech Industries Inc. 2018), made into 0.1, 1.0, and 6.0 M solutions.

3) Silver nitrate ($AgNO_3$) of laboratory grade, 0.1 M strength (EMD Millipore Co. 2018).
4) N_2 gas of laboratory grade, 99.9% purity (Praxair Canada Inc. 2018).
5) Acetone of laboratory grade, 99.8% purity (Smallwood 1996 and Fisher Scientific Co. 2018), made into 25%, 50%, and 75% strengths.
6) Supelco Supelite DAX-8 resin 21567-U as the hydrophobic resin stored in methanol (Sigma Aldrich Co. 2018a).
7) Amberlite IR-120 resin H form 10322 as the cation exchange resin (Sigma Aldrich Co. 2018b).
8) DI water.

Equipment requirements:

1) A ceramic mortar set of 200 ml capacity.
2) A ceramic combustion crucible (100 ml).
3) An analytical balance of 220 g capacity, $\pm$ 0.1 mg precision.
4) A sieve analyzer with # 200 mesh screen.
5) Glass Erlenmeyer flasks (1000 ml).
6) Glass centrifuge tubes (50 ml).
7) A glass chromatography column of 4×25 cm^2 (for DAX-8 resin).
8) A glass chromatography column of 5×60 cm^2 (for IR-120).
9) A glass beaker (50 ml).
10) Plastic spectrophotometer cuvettes (5 ml).
11) A magnetic stir plate and a bar.
12) Plastic paraffin films.
13) A centrifuge, $\geq$ 1500 relative centrifugal force.
14) A pH meter, $\pm$ 0.01 unit accuracy.
15) A peristaltic pump and tubing, 5 l/minute pumping rate.
16) A spectrophotometer at 350 nm wavelength, $\pm$ 0.005 unit accuracy.
17) A Rotovap, 400 ml capacity.
18) A drying oven of 120 °C capacity, $\pm$ 0.1 °C accuracy.
19) A combustion oven of 1000 °C capacity, $\pm$ 1 °C accuracy.
20) A desiccator of 4 l capacity.
21) A plastic beaker (4 l).
22) A manual stirrer.
23) A timer.

Alkali extraction: approximately 5 g of the lignite sample was pulverized to pass through mesh # 200 of the US standard sieve size, placed in a crucible, then in a drying oven to achieve a constant weight at 62 ± 3 °C. Approximately 2.5 g of dry sample was weighed and put in a 1000 ml Erlenmeyer flask, followed by 0.1 M NaOH and DI water to make a 1000 mL final solution. Note that for each blank run, DI water of the same weight replaced the lignite sample. The flask was placed on a magnetic plate with a magnetic bar inside. Air was replaced with N_2 gas, the flask was covered with a plastic paraffin film, and the solution was mixed for sixteen hours. The content was transferred to centrifuge tubes and centrifuged for thirty minutes. Precipitates were discarded and supernatant (containing HA and FF) was collected in a 1000 ml Erlenmeyer flask. A pH probe was inserted into the solution and 6.0 M HCl was added until pH = 1.0 ± 0.1. The flask was covered with a plastic paraffin film and mixed for one hour. The pH was occasionally checked until it was stable for five minutes and readjusted when necessary. The solution was left unstirred for four hours. The contents were transferred to centrifuge tubes and centrifuged for thirty minutes. Precipitates (containing HA and not needed in this study) were discarded. The analytical solution (containing FF) was transferred to a 1000 ml Erlenmeyer flask, and analyzed for HFA.

HFA analysis: 50 ml DAX-8 was packed in 4 × 25 cm^2 column. It was firstly rinsed with DI water, then with 0.1M NaOH and 0.1 M HCl three times alternatingly, then soaked with DI water. HFA was adsorbed by pumping the analytical solution from the top of the column. Effluent was discarded from the bottom. The resin was washed using DI water after two column volumes. HFA was then desorbed back to the analytical solution as follows.

Standard method: NaOH of 0.1 M strength was pumped from the top until three column volumes. Effluent was collected from the bottom and poured from the top to 5 × 60 cm^2 column filled with 500 ml IR-120 to flow by gravity until two column volumes. The column was washed with 500 ml DI water, and added to the solution. The analytical solution was concentrated using a Rotovap to approximately 50 ml at 62 ± 3 °C. The solution was transferred to a ceramic crucible, placed inside a drying oven to achieve a constant weight at 62 ± 3 °C. It was then placed in a desiccator to cool to room temperature and its weight was recorded. The crucible was put in a combustion oven for four hours at 500 °C. It was then placed in a desiccator to cool to room temperature and its weight was recorded. Calculations: the percentage of ash = (material weight after combustion ÷ dry material weight) × 100%; the percentage of HFA = [(material weight after drying − ash) ÷ dry material weight] × 100%.

Modified method: acetone of the desired strength (25%, 50%, or 75%) was pumped from the top until three column volumes. Effluent was collected from the bottom in a 1000 ml Erlenmeyer flask, placed inside a drying oven to approximately 100 ml at 62 ± 3 °C, transferred to a ceramic crucible, then placed inside the oven to a constant weight at the same temperature. It was then placed in a desiccator to cool to room temperature and its weight was recorded. The crucible was put in a combustion oven for four hours at 500 °C, then in a desiccator to cool to room temperature, and its weight was recorded. Calculations: same as above.

At this point, NaOH was assumed to desorp only HyA from DAX-8 (leaving out HyN), which would be quantified later as HFA (= HyA). Acetone was assumed to desorp HyA + HyN from DAX-8, which would be quantified as HFA (= HyA + HyN). An analytical solution of the standard method would contain Na ions added during both the alkali extraction and DAX-8 desorption, while the analytical solution of the modified method would contain Na ions added during the alkali extraction only. It was assumed that IR-120 would remove all Na ions from the analytical solution.

DAX-8 regeneration for the standard method: two columns of DI water + one column of 0.1 M HCl + one column of DI water were pumped from the top. For the modified method of 12 runs (Runs 2 to 4, 7 to 12, and 14 to 16): two columns of DI water + one column of 0.1 M HCl + one column of DI water were pumped from the top. For the modified method of 11 runs (Runs 20 to 24, 26, and 28 to 32): two columns of DI water + one column of 0.1 M HCl + one column of 50% acetone + one column of DI water were pumped from the top. IR-120 regeneration for the standard method: the resin was poured to a 4 l beaker, covered with 1.0 M HCl, and left for 30 minutes with occasional stirring once every 5 minutes. It was rinsed with DI water, stirred for 15 seconds, and left for 5 minutes. These procedures were repeated until the rinse water was free from Cl ions as per the following steps. Ten ml of rinse water was put in a 50 ml beaker, then two drops of 0.1 M $AgNO_3$ were added until no precipitates or changes in color were observed. For the modified method, this resin was not required. and all of these steps could be skipped.

During the quantification of HFA, the analytical time to complete each step was recorded. Changes in the modified method over the standard method included the use of acetone as an additional step during DAX-8 regeneration, and the omission of IR-120 and use of a Rotovap.

3.2.9.2.4 Results Using FTIR analysis, spectra for Runs 17, 21, 22, and 26 were generated within absorbance ranges of 600 to 4000 cm^{-1} (see Figures 3.4 to 3.7). Run 21 had no identifiable (blank) spectra. Others showed the presence of polysaccharides, C=C, COO-stretch, C=O of COOH, aliphatic C-H stretch, C-H stretch, and O-H stretch, typically observed for samples containing fulvic components. pH values of these three runs were 2.6, 2.8, and 2.4, respectively (ChemAnalytical LLC. 2018a and 2019). For the identification of HyA and HyN, only readings at 2922 and 1701 cm^{-1} were considered, respectively. Run 17 showed 0.193 and 0.131 readings (ratio = 1.47), Run 22 showed 0.194 and 0.176 readings (ratio = 1.10), and Run 26 showed 0.100 and 0.070 readings (ratio = 1.43), respectively (see Table 3.15). For this specific analysis, it was assumed that for pH levels of less than 7, higher ratio values indicated that the organic components were more in line with the negative charge behavior, while lower values were more to the neutral side (ChemAnalytical 2018a and 2019).

Using the gravimetric analysis for the lignite sample, acetone at 50% strength (Runs 3, 8, 9, 15, 23, 24, 29, and 30) produced HFA results (mean = 2.31%, SD =

0.227%, and CV = 9.8%) corresponding to those of the standard method of Runs 1, 5, 6, 13, 18, 19, 25, and 27 (mean = 2.38%, SD = 0.203%, and CV = 8.5%). Acetone at 25% strength (Runs 2, 7, 11, and 14) produced lower results (mean = 1.88%, SD = 0.144%, and CV = 7.7%). Acetone at 75% strength (Runs 4, 10, 12, and 16) produced higher results (mean = 2.47%, SD = 0.343%, and CV = 13.9%). Note: SD = standard deviation, CV = coefficient of variation = (SD ÷ mean) × 100%. Run 20 produced 0.35% HFA. Run 28 produced 0.16% HFA. Run 32 produced 0.03% HFA. Run 34 produced 0.31% HFA (see Table 3.16).

For the gravimetric analysis of the lignite sample, acetone at 50% strength of Runs 3, 8, 9, 15, 23, 24, 29, and 30 produced higher ash contents (mean = 34.81%, SD = 9.058%, and CV = 26.0%) than those of the standard method of Runs 1, 5, 6, 13, 18, 19, 25, and 27 (mean = 16.03%, SD = 5.286%, and CV = 32.9%). Run 32: Blank-Modified50%-5 and Run 34: Blank-Standard-1 produced 92.34 and 64.68% ash, respectively (see Table 3.17). The ICP analysis found 11.2 mg/l Na and 43.7 mg/l Cl for Run 31: Lignite-Modified50%-10, while Run 33: Lignite-Standard-10 contained 1.9 and 14.6 mg/l, respectively (see Table 3.18, ALS Environmental Laboratories Ltd. 2019, analyzed according to EPA Method 6020A (SW-846) (1998).

From 28 runs of gravimetric analysis on HFA, the total analytical time to complete one analysis using the standard method was 57 hours. The addition of acetone during DAX-8 desorption in the modified method increased it by 15 minutes,

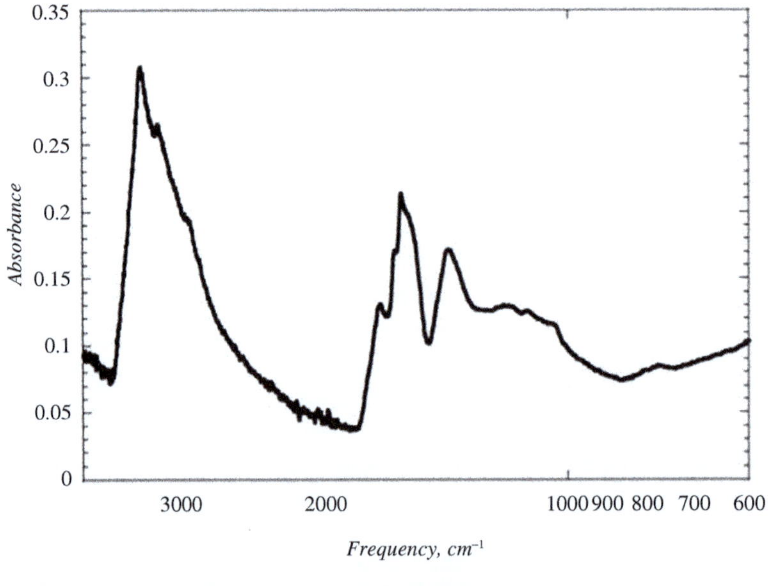

Figure 3.4 Run 17 (ChemAnalytical LLC. 2018a).

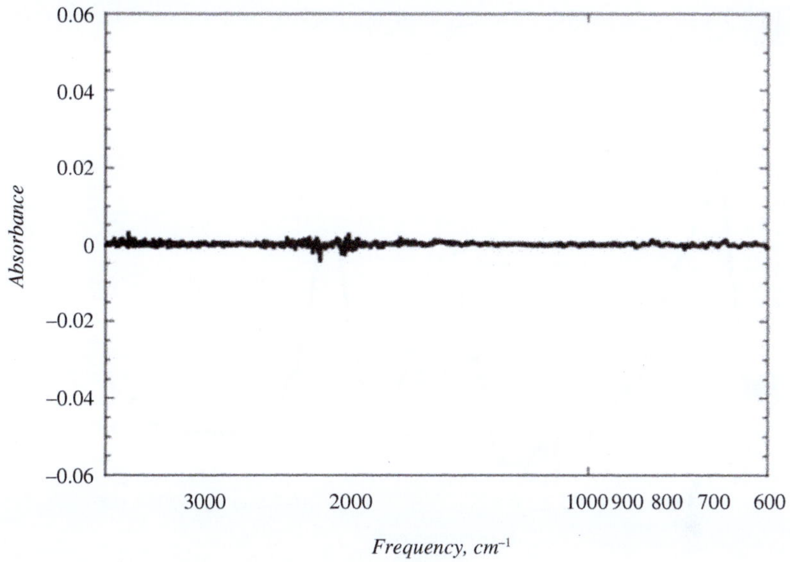

Figure 3.5 Run 21 (ChemAnalytical LLC. 2019).

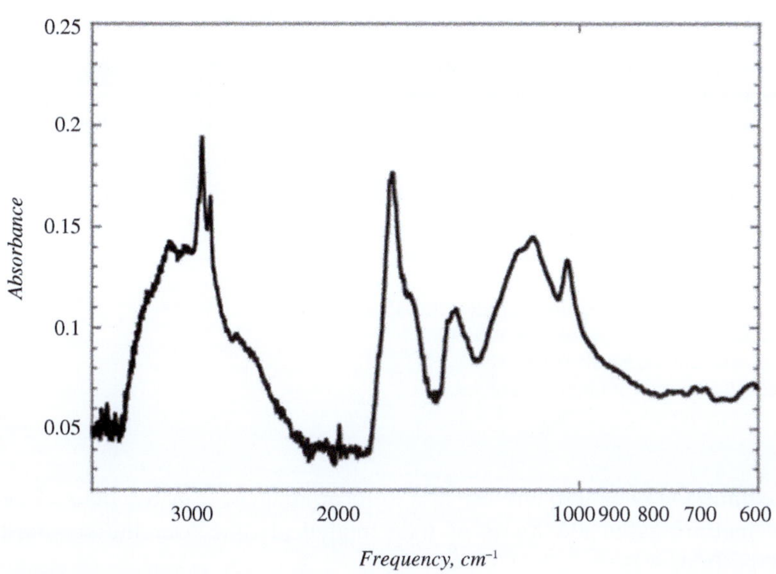

Figure 3.6 Run 22 (ChemAnalytical LLC. 2019).

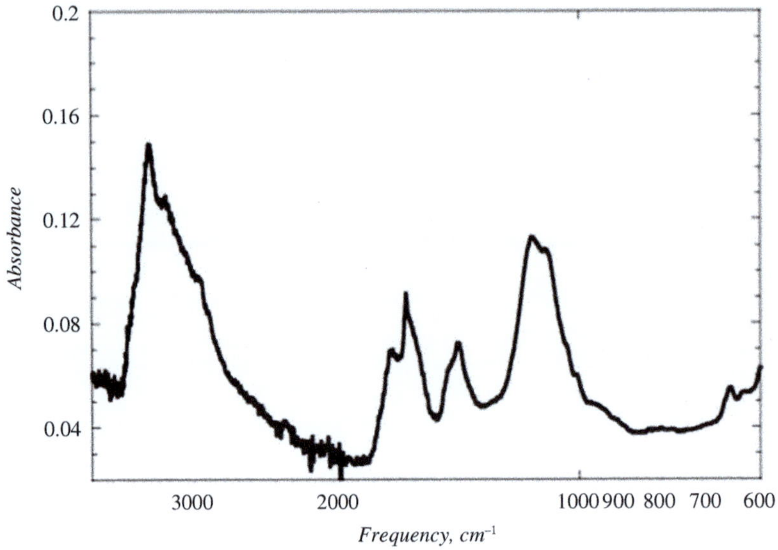

Figure 3.7 Run 26 (ChemAnalytical LLC. 2019).

Table 3.15 FTIR results.

			Group		
Run	**Identification**	**pH**	**C–H**	**C=O**	**Ratio**
17[a]	Lignite-Standard-5	2.6	0.193	0.131	1.47
21[b]	Blank-Modified50%-2			Blank	
22[c]	Lignite-Modified50%-5	2.8	0.194	0.176	1.10
26[d]	Blank-Modified50%-3	2.4	0.100	0.070	1.43

Note:
[a] completed after one run of Lignite-Modified75%,
[b] completed after one run of Blank-Modified50%,
[c] completed after one run Blank-Modified50%,
[d] completed after one run of Lignite-Standard.

while the omission of IR-120 and use of a Rotovap reduced it by 8½ hours. The modified method saved 8¼ hours of total analytical time over the standard method (see Table 3.19).

Table 3.16 HFA results.

Run	Identification
1, 5, 6, 13, 18, 19, 25, 27	Lignite-Standard-1, 2, 3, 4, 6, 7, 8, 9
2, 7, 11, 14	Lignite-Modified25%-1, 2, 3, 4
3, 8, 9, 15, 23, 24, 29, 30	Lignite-Modified50%-1, 2, 3, 4, 6, 7, 8, 9
4, 10, 12, 16	Lignite-Modified75%-1, 2, 3, 4
20[a]	Blank-Modified50%-1
28[b]	Blank-Modified50%-4
32[c]	Blank-Modified50%-5
34[d]	Blank-Standard-1

Run	Mean %	SD %	CV %
1, 5, 6, 13, 18, 19, 25, 27	2.38	0.203	8.5
2, 7, 11, 14	1.88	0.144	7.7
3, 8, 9, 15, 23, 24, 29, 30	2.31	0.227	9.8
4, 10, 12, 16	2.47	0.343	13.9
20[a]	0.35	N/A	N/A
28[b]	0.16	N/A	N/A
32[c]	0.03	N/A	N/A
34[d]	0.31	N/A	N/A

Note:
[a] completed after three consecutive runs of Lignite-Standard,
[b] completed after one run of Lignite-Standard,
[c] completed after three consecutive runs of Lignite-Modified50%,
[d] completed after one run of Lignite-Standard, SD = standard deviation, CV = coefficient of variation, N/A = not available.

3.2.9.2.5 Discussion Run 21: Blank-Modified 50%-2 was completed after Run 20: Blank-Modified50%-1. This meant that DAX-8 should be clean from any organic components prior to Run 21, other than the acetone that was used as the desorbing agent, as shown in FTIR spectra. The non-identifiable (blank) spectra for Run 21 indicated that acetone would not interfere with any spectral interpretations. Run 17: Lignite-Standard-5 and Run 22: Lignite-Modified50%-5 produced almost identical amounts of HyA as expected, since both contained the same sample. Interestingly, both runs also showed some amounts of HyN. This suggested that when using the standard method, both acidic and neutral

Table 3.17 Ash results.

Run	Identification		
1, 5, 6, 13, 18, 19, 25, 27	Lignite-Standard-1, 2, 3, 4, 6, 7, 8, 9		
3, 8, 9, 15, 23, 24, 29, 30	Lignite-Modified50%-1, 2, 3, 4, 6, 7, 8, 9		
32[a]	Blank-Modified-5		
34[b]	Blank-Standard-1		
Run	**Mean %**	**SD %**	**CV %**
1, 5, 6, 13, 18, 19, 25, 27	16.03	5.286	32.9
3, 8, 9, 15, 23, 24, 29, 30	34.81	9.058	26.0
32[a]	92.34	N/A	N/A
34[b]	64.68	N/A	N/A

Note:
[a] completed after three consecutive runs of Lignite-Modified50%,
[b] completed after one run of Lignite-Standard, SD = standard deviation, CV = coefficient of variation, N/A = not available.

Table 3.18 Na and Cl results.

Run	Identification	Na mg/l	Cl mg/l
31[a]	Lignite-Modified50%-10	11.2	43.7
33[b]	Lignite-Standard-10	1.9	14.6

Note:
[a] completed after two consecutive runs of Lignite-Modified50%,
[b] completed after one run of Blank-Modified 50%.

components were adsorbed by DAX-8 and desorbed using NaOH, which was later quantified as HFA. This finding was critical to showing the previous belief that NaOH only desorbed HyA from DAX-8 was not true. It desorbed both HyA and some HyN. Due to the use of acetone, the modified method resulted in more HyN being detected. This was congruent with the fact that acetone would desorp all organic components including HyA and some HyN. It also indicated that organic components being quantified by both standard and modified methods (defined as HFA) were actually a combination of mainly HyA and some HyN, which still met

Table 3.19 Analytical times of dry samples.

Analytical step	Analytical time hour[a] Standard	Modified	Modified method over standard method
Sample drying, alkaline extraction, acidification, centrifuging, settling, analytical solution through DAX-8, adsorption, desorption	28		No difference
DAX-8 regeneration	½	¾	
IR-120 usage, regeneration	1½	0	IR-120 not required
Rotovap usage, drying, ash corrections	27	20	Rotovap not required
Total analytical time	57	48¾	8¼ hours shorter

Note:
[a] average of 28 runs to the closest quarter hour.

the definition of HFA as "components soluble in both alkali and acids solutions, and adsorbed to a hydrophobic resin (DAX-8) at pH 1" (ISO 19822:2018(E) 2018).

Most interestingly, Run 26: Blank-Modified50%-3 (completed after Run 25: Lignite-Standard-8) showed some amounts of HyA and HyN. This suggested that DAX-8 adsorbed both of them, and the fact that it also contained HyA showed that regeneration steps using the standard method (two columns of DI water + one column of 0.1 HCl + one column of DI water) completed on the previous run (Run 25) were inadequate.

Using the gravimetric method, HFA results obtained from both standard and modified methods showed that acetone at 50% strength was the right strength to produce corresponding results as those of the standard method using 0.1 M NaOH. Acetone at 25% strength produced lower results, suggesting that the desorption of HyA (and some HyN) from DAX-8 was not complete. Acetone at 75% strength produced higher results, suggesting that more HyN was being desorbed from the resin. As shown earlier by FTIR analysis, HFA results obtained from both standard and modified methods included some amounts of HyN. This indicated that in terms of principles, the modified method did not deviate from the standard method and was therefore valid in this regard. Its accuracy would depend on the right strength of acetone being used.

For the modified method, implementing acetone at 50% strength within regeneration steps of DAX-8 (Runs 20 to 24, 26, and 28 to 32) or not implementing it (Runs 2 to 4, 7 to 12, and 14 to 16), did not have any effect on HFA results. This suggested that the 50% acetone used during desorption earlier, together with two columns of DI water + one column of 0.1 M HCl + one column DI water during the regeneration afterwards, provided an adequate regeneration for the resin.

Run 20: Blank-Modified50%-1 (completed after three consecutive runs of the lignite sample using the standard method) quantified some organic components at 0.35%. These were leftover HyN (and most likely some HyA, see the above FTIR analysis) originating from the three previous runs. For each run, it could be approximated at 0.35% ÷ 3 = 0.12%. The same explanation applied for Run 28: Blank-Modified50%-4 (completed after one run of the lignite sample using the standard method), that 0.16% organic components were leftover HyN (and some HyA) originating from the previous run. The fact that it also contained some HyA indicated that regeneration steps using the standard method completed on the previous runs (Runs 17, 18, 19, and 27) were inadequate.

Run 32: Blank-Modified50%-5 (completed after three consecutive runs of the lignite sample using the modified method at 50% strength) quantified very few organic components at 0.03%. These were leftover HyN originating from the three previous runs. For each run, it could be approximated at 0.03% ÷ 3 = 0.01%. This suggested that the modified method, due to the presence of acetone, provided an adequate (almost complete) regeneration on each of these three previous runs.

Run 34: Blank-Standard-1 (completed after 1 run of the lignite sample using the standard method) quantified some organic components at 0.31%. These were leftover HyA (and most likely some HyN, see the above FTIR analysis) originating from the previous run. This suggested that DAX-8 adsorbed both HyA and HyN. The fact that it also contained HyA indicated that regeneration steps using the standard method on the previous run were inadequate.

The gravimetric analysis found that the omission of IR-120 resulted in higher ash contents, as shown in Runs 3, 8, 9, 15, 23, 24, 29, and 30 (the modified method at 50% acetone strength, mean = 34.81%) in comparison to Runs 1, 5, 6, 13, 18, 19, 25, and 27 (the standard method, mean = 16.03%). However, these higher ash contents did not affect the accuracy of HFA quantification, shown by the fact that results from both methods were very well corresponded.

The gravimetric analysis for blank samples showed that Run 32: Blank-Modified50%-5 (completed after three consecutive runs of the lignite sample using the modified method at 50% acetone strength), produced 92.34% ash. For each run, it could be approximated at 92.34 ÷ 3 = 30.78%, which was smaller than 64.68% obtained from Run 34: Blank-Standard-1 (completed after one run of the lignite sample using the standard method). These results corresponded well with the previous results. The modified method provided a much better DAX-8 regeneration, i.e. Run 32 produced smaller organic components (HyA and HyN) and ash contents leftover from the previous runs, compared to those of Run 34.

Ionic analyses using the ICP method for Run 31: Lignite-Modified50%-10 (completed after two consecutive runs of the lignite sample using the modified method at 50% strength acetone) produced 11.2 and 43.7 mg/l Na and Cl, respectively; or $11.2 \div 3 = 3.7$ and $43.7 \div 3 = 14.6$ mg/l per run, respectively (note that since it contained lignite material, Run 31 was included in this calculation). Run 33: Lignite Standard-10 (completed after one run of DI water using the modified method at 50% acetone strength) produced lower results at 1.9 and 14.6 mg/l, respectively. This agreed with the previous findings that the omission of IR-120 resulted in higher ash contents. It was also found that IR-120 was not effective in removing Cl.

It could be speculated that the use of the standard method would result in some HyA and HyN being retained within DAX-8 after its regeneration, which would reach their maximum amounts and did not increase after one run or so. When using fresh DAX-8, the first result could be expected to be lower, which would be followed by higher and more stable results afterwards. A similar case, but at a smaller magnitude due to a better regeneration performance on DAX-8, could also be expected for the modified method.

The use of acetone as a desorbing agent reduced the total analytical time by eight and a quarter hours. This represented a reduction of approximately 14% of analytical time over the standard method. In addition, a lower capital cost on the laboratory equipment as well as chemicals could also be expected.

3.2.9.2.6 Conclusion and Recommendation Both standard and modified methods resulted in the presence of HyN within the gravimetric quantification of HFA; more HyN were detected using the modified method. This indicated that organic components being quantified were actually a combination of mainly HyA and some HyN (i.e. HFA = HyA + HyN), which still met the definition of HFA as being "components soluble in both alkali and acids solutions, and adsorbed to a DAX-8 at pH 1" (ISO 19822:2018(E) 2018).

The modified method using acetone at 50% strength produced HFA results that corresponded well with those of the standard method using NaOH at 0.1 M. This method resulted in higher ash contents within the analytical solution, but they did not affect its accuracy. The modified method provided an adequate (almost complete) DAX-8 regeneration, but the standard method did not. This method consumed a significantly shorter analytical time (14% less) due to the omission of IR-120 and the use of a Rotovap. A lower capital cost on the laboratory equipment and chemicals could also be expected.

This study demonstrated the potential of acetone as a desorbing agent for DAX-8 on the gravimetric quantification of HFA in lignite material (including low-rank coal in general). Humic materials other than lignite might have different proportions of HyA and HyN as indicated in Leenher (1981), Polubesova et al.

Table 3.20 Analytical times of liquid samples.

Analytical step	Analytical time hour[a] Standard	Modified	Modified method over standard method
Sample drying, alkaline extraction, acidification, centrifuging, settling, analytical solution through DAX-8, adsorption, desorption	13		No difference
DAX-8 regeneration	½	¾	
IR-120 usage, regeneration	1½	0	IR-120 not required
Rotovap usage, drying, ash corrections	27	20	Rotovap not required
Total analytical time	42	33¾	8¼ hours shorter

Note:
[a] estimated values to the closest quarter hour.

(2008), Nkambule et al. (2008), Kanokkantapong et al. (2006), Ilani et al. (2005), Ibarra et al. (1996), and Krumins (2012). The validity and accuracy of this modified method would have to be evaluated more thoroughly.

A detailed evaluation on the nature of organic components being quantified (i.e. HyA and/or HyN) using NMR analysis was recommended. A thorough evaluation on the capacity of DAX-8 using larger volumes of resin, higher strengths of HFA in the sample, and various methods of regeneration (including the use of acetone at a higher strength for both standard and modified methods), was also recommended.

End note: when quantifying liquid samples, the solution would be mixed for 1 hour, not 16 hours as in the case of solid samples. The standard method would consume approximately 42 hours of the total analytical time, while the modified method would take 33¼ hours, representing a total time saving of approximately 20% (see Table 3.20).

3.3 Organic Matter Product Comparison

There have been studies completed to analyze various commercial humic products using different methods. Among others, Shahbazi et al. (2019) quantified HA contents of twenty-two humic samples using the CDFA and colorimetric methods, ISO 19822 (identified as the method of Lamar et al. 2014), as well as ISO 5073.

Overall, the colorimetric method has been found to produce the highest results, followed by ISO 19822 and ISO 5073 (producing about the same results), then the CDFA method with the lowest results.

In this section, an original laboratory study has been completed on different OM products using the colorimetric method and ISO 19822; as well as the ultimate, FTIR, and UV-visible spectral analyses as the supporting methods. The colorimetric method and ISO 19822, as well as the UV-visible spectral analysis have been done in-house. The ultimate and FTIR analyses have been completed by third-party laboratories.

Table 3.21 lists the selected products. Dry HS is identical to Alberta low-rank coal with a different particulate sizing. Liquid HA is manufactured from dry HS, containing mainly HA and some FA. Liquid FA is manufactured from dry HS, containing mainly FA and some HA (see Sections 5.3 to 5.5). Other products listed are non-humic products, including biochar, calcium lignosulfonate, soybean extract, peat moss, and molasses. They show some similarities to those of HS and are sometimes compared to and/or mistaken for real humic products. Foreign materials having similar properties to those of HS, which are sometimes present within humic products, are also known as adulterants.

Dry HS and peat moss are raw products (receiving only a physical process during their manufacturing), while the rest are processed products (receiving physical, chemical, and in some cases biological processes during their manufacturing). Dry HS, biochar, calcium lignosulfonate, soybean extract, and peat moss are solid products, with moisture contents ranging from 5.2% (calcium lignosulfonate) to 28.1% (dry HS). Liquid HA, liquid FA, and molasses are liquid products, with moisture contents ranging from 14.4% (molasses) to 98.2% (Liquid FA). Moisture has been analyzed according to ASTM D3302/D3302M-17 (2017). Dry HS, biochar, and peat moss are not water-soluble. Liquid HA and liquid FA have the highest solubility in water, followed by calcium lignosulfonate, soybean extract, and molasses. Biochar is powdery, different from the earthy nature of dry HS. Calcium lignosulfonate is light brown and emits a distinct smell of S in comparison to the musty odor of dry HS. Soybean extract is yellow and smooth. Peat moss is light brown and fibrous. Molasses produces a unique (sweet) odor.

While non-humic products show distinct physical appearances, detailed and more comprehensive analyses are still required to differentiate them from the true humic products. The main objective of this study is to evaluate the performance of each analytical method on different products, as well as to find a practical way to differentiate non-humic from the true humic products.

Based on the dry matter analysis, dry HS can be expected to contain high HA and low FA, as in the case of low-rank coal in general. Liquid HA should be high in HA and low in FA, while liquid FA should be medium in HA but high in FA, as a result of their manufacturing processes. Peat moss has the same origin as that of

Table 3.21 Commercial products (physical appearances).

Product	Type	Color	Phase
Dry HS	Raw	Dark brown	Solid
Liquid HA	Processed	Black	Liquid
Liquid FA	Processed	Dark red	Liquid
Biochar	Processed	Black	Solid
Calcium lignosulfonate	Processed	Light brown	Solid
Soybean extract	Processed	Yellow	Solid
Peat moss	Raw	Light brown	Solid
Molasses	Processed	Black	Liquid

Product	Moisture %	Water solubility	Expected HS
Dry HS	28.1	Low	High
Liquid HA	90.7	Very high	High
Liquid FA	98.2	Very high	High
Biochar	10.2	Very low	None
Calcium lignosulfonate	5.2	High	None
Soybean extract	7.6	High	None
Peat moss	27.5	Low	Medium
Molasses	14.4	High	None

Source: Canadian Humalite International Inc. (2020a).

low-rank coal in its younger and un-compacted form (see Section 2.1 on the coalification process), and therefore is expected to have medium HA and low FA. Other products should not contain any humic components.

Table 3.22 shows elemental results for each product, based on the dry matter analysis. Dry HS, liquid HA, and liquid FA have O:C of 0.5, 0.6, and 1.2, respectively. Low O:C indicates fewer COOH group within the product. This is to be expected as dry HS, in addition to HA and FA, contains humin (humic component with the least COOH group) as well as inorganics. Liquid HA contains mainly HA and some FA, with very little humin and some inorganics. Liquid FA contains mainly FA (humic component with the most COOH group) with some HA, and practically no humin and very little inorganics. Both liquid products contain K that is added during manufacturing, which contributes as the main component of

Table 3.22 Commercial products (elemental analysis).

	C %	H %	O %	N %	S %	Ash %	O:C
Dry HS	50.4	2.9	23.9	1.5	0.7	20.6	0.5
Liquid HA	48.0	2.8	27.6	0.9	0.4	20.3	0.6
Liquid FA	24.6	1.9	28.5	0.7	1.2	43.1	1.2
Biochar	75.2	1.1	8.1	1.2	0.3	14.1	0.1
Calcium lignosulfonate	43.0	4.4	26.7	0.3	7.2	18.4	0.6
Soybean extract	46.1	6.0	27.9	14.2	0.7	5.1	0.6
Peat moss	52.2	4.2	33.5	1.6	0.9	7.6	0.6
Molasses	40.0	6.0	51.4	0.5	0.1	2.0	1.3

Source: Canadian Humalite International Inc. (2020a); *note:* each result is based on dry matter analysis.

ash. It is unclear why liquid FA has the highest ash and second highest S levels (after calcium lignosulfonate). Note: the analyses have been completed according to ASTM D2013/D2013M-18 (2013), ASTM D5373-16 (2016), ASTM D2492-02(2012) (2012), ASTM D3174-12 (2012), ASTM D3176-15 (2015), and O by difference.

Biochar has the lowest O:C at 0.1, indicating that this product contains a very small amount of COOH group due to the nature of its manufacturing (Humic Products Trade Association 2015). Calcium lignosulfonate, soybean extract, and peat moss have a similar O:C as that of liquid HA at 0.6, but slightly higher than that of dry HS at 0.5 and are therefore difficult to differentiate in this regard. In the case of peat moss, this makes sense since this material is in its earlier phase of the coalification process from decayed vegetation and thus contains more COOH group, as well as less ash, compared to that of dry HS. However, calcium lignosulfonate contains a significant amount of S (as a result of its manufacturing) and soybean extract contains a very high amount of N, thus clearly differentiating themselves from true humic-containing products. Molasses has the highest O:C at 1.3, indicating that this product fits better in the aliphatic group. Liquid FA contains a similar ratio (at 1.2), indicating that it contains a significant amount of the COOH group.

Figures 3.8 to 3.15 show FTIR spectra, i.e. frequency (cm^{-1}) vs. absorbance (no unit), for each product. It is not the intention of this book to provide detailed interpretation for each spectra. Instead, they are presented to show differences so that these commercial products can be identified in a practical way. Therefore, only the following functional groups, which are relevant to this discussion, are

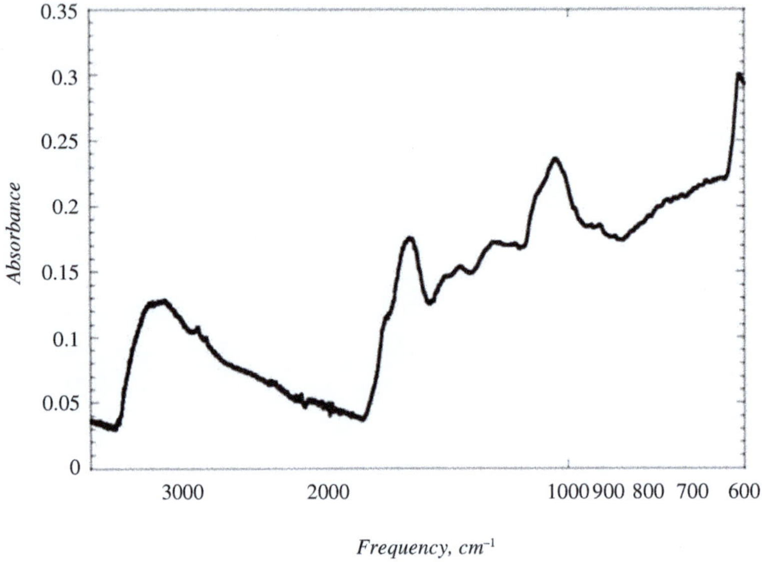

Figure 3.8 Spectra for dry HS (ChemAnalytical LLC. 2018b).

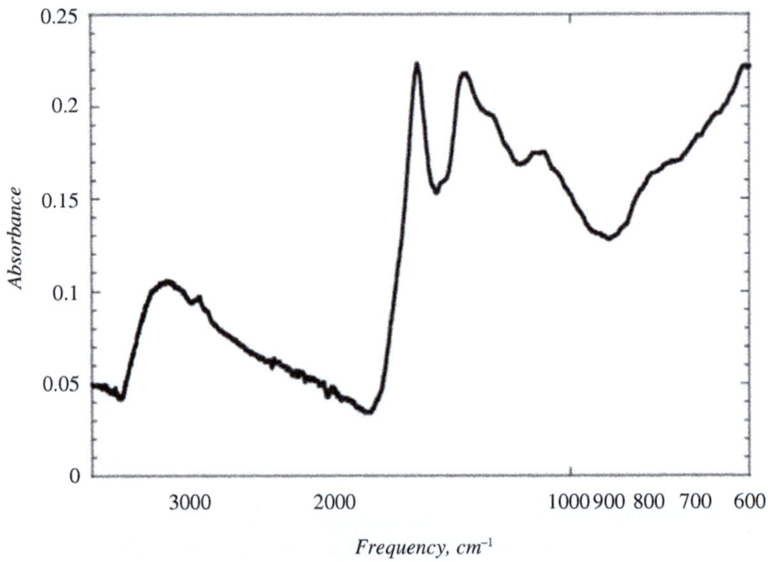

Figure 3.9 Spectra for liquid HA (ChemAnalytical LLC. 2018b).

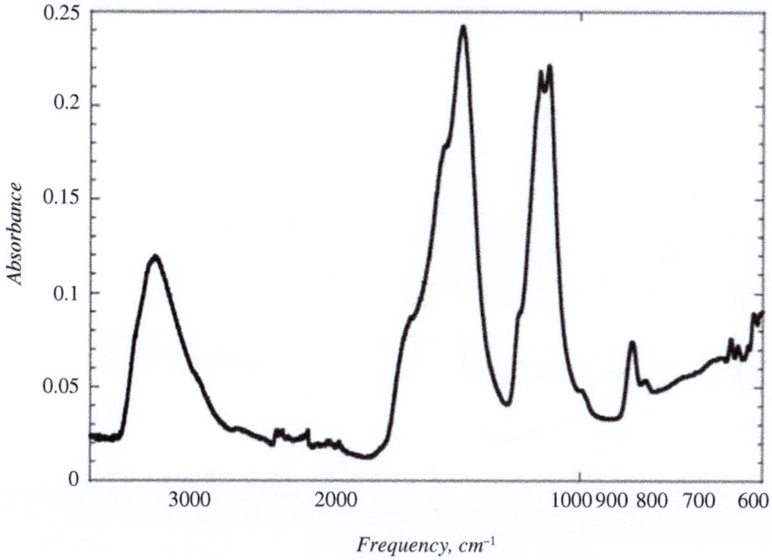

Figure 3.10 Spectra for liquid FA (ChemAnalytical LLC. 2018b).

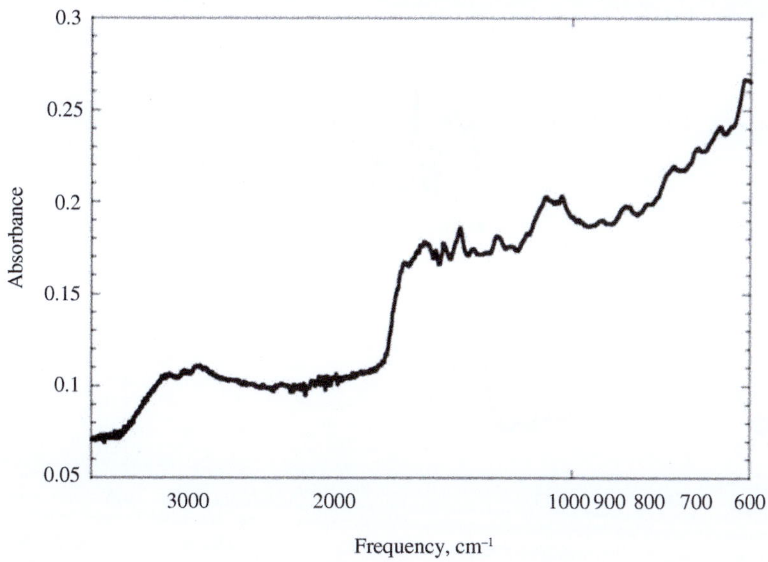

Figure 3.11 Spectra for biochar (ChemAnalytical LLC. 2018b).

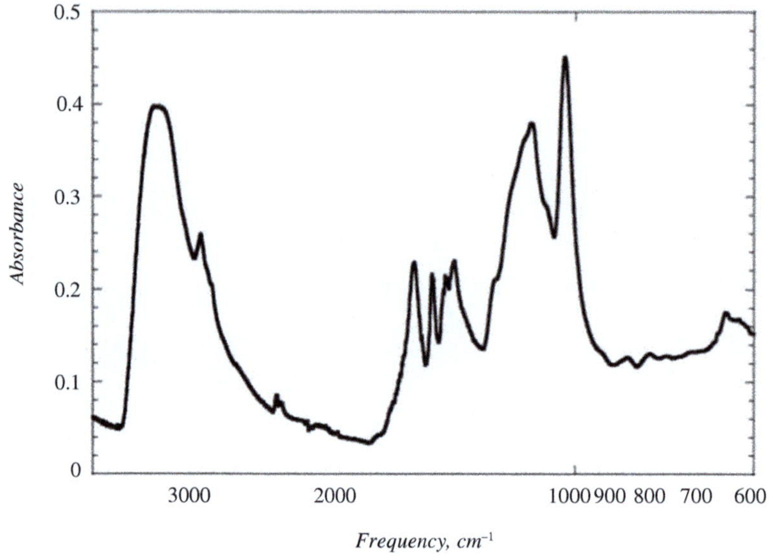

Figure 3.12 Spectra for calcium lignosulfonate (ChemAnalytical LLC. 2018b).

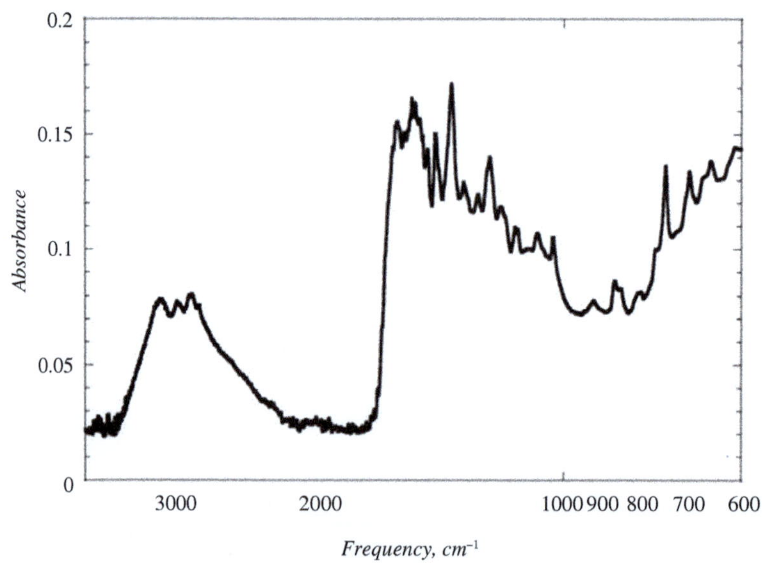

Figure 3.13 Spectra for soybean extract (ChemAnalytical LLC. 2018b).

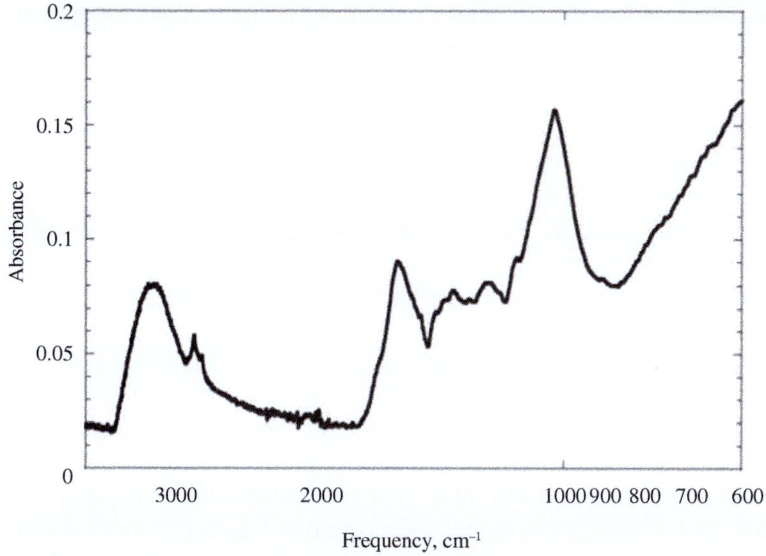

Figure 3.14 Spectra for peat moss (ChemAnalytical LLC. 2018b).

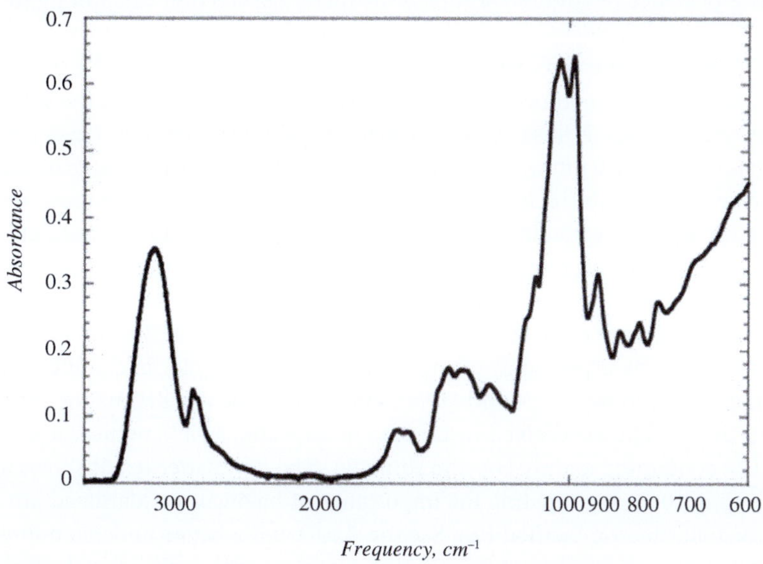

Figure 3.15 Spectra for molasses (ChemAnalytical LLC. 2018b).

addressed (ChemAnalytical LLC. 2018b, Rodriguez-Lucena et al. 2009, Artz et al. 2008, Coates 2000, and Ibarra et al. 1996):

1) ratio of aromatic and aliphatic H contents (peaks at 700 to 900 cm^{-1} ÷ peaks at 2700 to 3000 cm^{-1});
2) polysaccharides (peaks at 1030 to 1080 cm^{-1});
3) sulfonic groups (peaks at 620 to 660 cm^{-1});
4) amines (multiple peaks at 1130 to 1190, 1150 to 1210, 1250 to 1340, 1280 to 1350, and 1310 to 1360 cm^{-1}); and
5) COOH salts (multiple peaks at 1300 to 1420 and 1550 to 1610 cm^{-1}).

Biochar (Figure 3.11) shows a very high ratio of aromatic and aliphatic H contents, indicating the presence of fused benzene ring structures. Other products show lower ratios, especially molasses (Figure 3.15), which is dominant in the aliphatic group. Figure 3.8 (dry HS), Figure 3.9 (liquid HA), Figure 3.14 (peat moss), and Figure 3.10 (liquid FA) show some resemblance due to the nature of the material. Peat moss contains more polysaccharides. Liquid FA, which is manufactured from dry HS, shows more COOH salts and polysaccharides. COOH acid salts are observed in all products except soybean extract (Figure 3.13) and molasses (Figure 3.15), indicating that these two are from different origins. Polysaccharides are also observed in all other products (mainly molasses) except biochar. The presence of sulfonic groups is distinctly observed in calcium lignosulfonate (Figure 3.12). Soybean extract shows a strong presence of amines.

Table 3.23 shows results obtained using the colorimetric method. To give a simplified explanation, this method quantifies HS based on the darkness of the solution when the product is dissolved in water at high pH, measured using a spectrophotometer at 450 nm wavelength. Values not in brackets are results based on the as-received analysis (including moisture); the ones in brackets are based on the dry matter analysis (excluding moisture). The dry matter result has been calculated from the as-received result (or vice versa) by incorporating its moisture content in the product.

Dry HS has the highest HS at 69.5% (as-received analysis) and 97.3% (dry matter analysis). This is to be expected as dry HS is dark brown in color and soluble at high pH. Liquid HA shows 11.1% and 119.4% HS (as-received and dry matter analyses, respectively). The as-received result is to be expected as it is manufactured from dry HS, containing mainly HA and some FA. The dry matter result shows a value exceeding 100%, highlighting the importance of having a standardized protocol for the colorimetric method (see Section 3.2.4) and a better understanding on the effect of pH on HS readings (see Section 3.2.9.1), particularly when dealing with liquid humic products. Liquid FA shows 0.9% and 50.3% HS (as-received and dry matter analyses, respectively), which is reasonable as it is manufactured from

Table 3.23 Commercial products (HA, FA, and HS).

Product	HA + FA (= HS) Colorimetric %		
Dry HS	69.5 (97.3)		
Liquid HA	11.1 (119.4)[a]		
Liquid FA	0.9 (50.3)		
Biochar	4.5 (5.0)		
Calcium lignosulfonate	7.1 (7.5)		
Soybean extract	1.0 (1.1)		
Peat moss	16.0 (22.1)		
Molasses	0.8 (0.9)		
	HA ISO 19822 %	**HFA ISO 19822 %**	**HA + HFA ISO 19822 %**
Dry HS	49.7 (69.6)	1.6 (2.3)	51.3 (71.9)
Liquid HA	7.0 (74.8)	0.3 (2.9)	7.3 (77.7)
Liquid FA	0.3 (14.5)	1.0 (56.9)	1.3 (71.4)
Biochar	0.0 (0.0)	1.9 (2.1)	1.9 (2.1)
Calcium lignosulfonate	0.0 (0.0)	35.6 (37.6)	35.6 (37.6)
Soybean extract	0.0 (0.0)	14.2 (15.2)	14.2 (15.2)
Peat moss	15.2 (21.0)	4.1 (5.7)	19.3 (26.7)
Molasses	0.0 (0.0)	1.8 (2.1)	1.8 (2.1)

Source: Canadian Humalite International Inc. (2020a);
note:
[a] see comment, as-received results are not in brackets, dry matter results are in brackets.

dry HS, containing mainly FA and some HA. Results for dry HS and liquid HA are somehow inflated, while the one for liquid FA seems to be underestimated.

Biochar is black in color but is not soluble in water at high pH. It shows a reading of 5.0% (dry matter analysis). Calcium lignosulfonate and soybean extract are soluble in water at high pH, and are light brown and yellow in color, respectively. Readings have been recorded at 7.5% and 1.1% (dry matter analyses, respectively). Molasses is soluble in water at high pH, but it is dark yellow in color (after being diluted prior to analysis), thus producing a very low reading at 0.9% (dry matter analysis). All of these products should produce no HS readings, since they are not humic-based products. While results for soybean extract and molasses are somewhat overestimated, HS results for biochar and calcium lignosulfonate are clearly inflated.

Peat moss is light brown in color and slightly soluble in water at high pH, giving a significant reading of 22.1% (dry matter analysis). The humification of peat has been characterized by Klavins et al. (2008), which confirms that peat moss is related to low-rank coal and contains some amounts of HS. Among all of these eight products, true HS results are likely to be reported only for dry HS, liquid HA, liquid FA, and peat moss.

Table 3.23 also shows individual and HS results obtained using ISO 19822. Values not in brackets are results based on the as-received analysis (including moisture), the ones in brackets are based on the dry matter analysis (excluding moisture). The dry matter result has been calculated from the as-received result (or vice versa) by incorporating its moisture content within the product. This method quantifies HA as the amount (weight) of OM components within the product that is soluble in water at high pH. This method quantifies FA as the amount (weight) of OM components within the product that is soluble in water at low pH and is more hydrophobic in nature (i.e. being adsorbed by DAX-8 resin), defined as HFA. Ash corrections are applied for both HA and HFA analyses.

Liquid HA contains higher HA and HFA at 74.8% and 2.9% than those of dry HS at 69.6% and 2.3%, respectively (dry matter analysis). Liquid FA contains lower HA at 14.5% but higher HFA at 56.9% than those of liquid HA, respectively (dry matter analysis). These are in agreement that liquid HA is manufactured from dry HS, containing mainly HA and some HFA, with most inorganics and humin have been taken out. Liquid FA is manufactured from dry HS, containing mainly HFA and some HA.

Biochar shows zero HA and 2.1% HFA (both dry matter analyses). This product does not contain any humic components and is not soluble in water at any pH. The HA result is accurate, while the HFA result is inflated as it captures OM components other than HFA of similar properties. Calcium lignosulfonate and soybean extract each show zero HA, as expected, but significant HFA at 37.6% and 15.2%, respectively (dry matter analyses). Both products are highly soluble in water at any pH and contain significant amounts of hydrophobic OM components, which are not HFA but have similar properties. As shown in Sections 3.2.3 and 3.2.8, ISO 19822 includes a preliminary screening for any commercial humic products containing over 0.75% S. S levels higher than this may indicate the presence of lignosulfonates, in which a further analysis is required (Lamar et al. 2014 and ISO 19822:2018(E) 2018). This approach, however, cannot differentiate the presence of lignosulfonates when blended with true HFA. With respect to soybean extract and other non-humic products, ISO 19822 does not have any preliminary screening steps established yet.

Peat moss shows significant HA and HFA at 21.0% and 5.7%, respectively (dry matter analysis). This product is slightly soluble in water at any pH and contains some hydrophobic OM components, which is consistent with the nature of the product as a precursor of coal materials. Both results are therefore reasonable. Molasses shows zero HA and 2.1% HFA (dry matter analysis) because while this product is soluble in water at any pH, it contains some hydrophobic OM components having similar properties to those of FA. Since it is not a source of HS, the result is clearly inflated.

In terms of their quality, these products can also be compared based on their UV-visible spectral performance. After dissolving the product in water using NaOH under N_2 gas (an O-free) environment, adjusting it to pH of 6 ± 0.1, followed by centrifuging, the wavelength vs. % absorbance patterns can be generated using a spectrophotometer (see Figure 3.16). In this figure, the percentage absorbance for each product is presented in a log scale normalized to its dissolved organic C content (note: dissolved organic C has been measured as total organic C contained within the solution prior to the spectral analysis). Each of dry HS, liquid HA, and liquid FA produces almost a linear line. Liquid FA shows the steepest slope (at approximately 1:190), followed by liquid HA (at approximately 1:220), then dry HS (at approximately 1:290). This indicates that liquid FA has the lowest reading at higher wavelengths, followed by liquid HA, and then dry HS.

Recalling the E_4/E_6 ratio (absorbance at 465 nm divided by absorbance at 665 nm) as the representation of molecular sizes and weights (Chen et al. 1977) from Section 3.2.4, a straightforward comparison can be made for dry HS, liquid HA, and liquid FA, since they are from the same low-rank coal origin (see Table 3.24). Low E_4/E_6 ratios correspond to relatively large molecular sizes and weights, while high ratios do the opposite. Comparing to soil HA ($E_4/E_6 = 3.0$ to 5.0) and soil FA ($E_4/E_6 = 6.0$ to 9.6), as compiled by Schnitzer (1971) and Schnitzer (1972), the table shows that dry HS ($E_4/E_6 = 4.9$) contains all humic components (HA, FA, and humin), with significantly reduced humin contents for liquid HA ($E_4/E_6 = 6.5$), and mainly FA contents for liquid FA ($E_4/E_6 = 10.7$), as expected. Note: humin has the largest molecular weight, followed by HA, then FA. E_4/E_6 ratios for other products may also be calculated (i.e. 1.7, 23.0, 5.3, 1.5, and 4.3 for soybean extract, calcium lignosulfonate, peat moss, biochar, and molasses, respectively). However, the results are not relevant since they are not humic products. The result for peat moss may be an exception, since its origin is close to low-rank coal.

Additional indices may be considered to differentiate humic and non-humic products. The E_2/E_3 ratio (absorbance at 250 nm divided by absorbance at 365 nm) relates to the substance's aromaticity and molecular weight (Jiang et al. 2011).

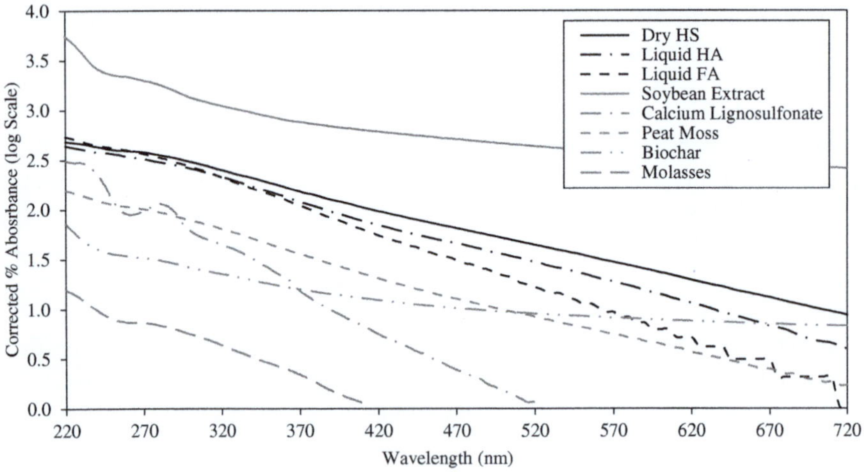

Figure 3.16 Normalized % absorbance (log scale) vs. wavelength.

Table 3.24 E_4/E_6 of humic products.

Product	E_4/E_6	Humic components	Molecular weight
Dry HS	4.9	HA, FA, and humin	↓
Liquid HA	6.5	Mainly HA, some FA	Decreasing
Liquid FA	10.7	Mainly FA, some HA	

E_2/E_4 (absorbance at 280 nm divided by absorbance at 472 nm) correlates to the proportion of lignins and other substances at its early stage of humification process and E_2/E_6 (absorbance at 280 nm divided by absorbance at 665 nm) concerns the relationship of humified and non-humified substances (Yakimenko et al. 2018). Molar extinction coefficients at 280, 465, and 540 nm represent aromatic substances with respect to molecular weight, humification indicators, and peat humification index, respectively (Chin et al. 1994, Mahieu et al. 2002, Chen et al. 1977, and Zaccone et al. 2018). SUV254 (specific UV absorbance at 254 nm, normalized to its dissolved organic C) correlates to the substance's aromaticity and compost maturity (Weishaar et al. 2003 and Chen et al. 2017).

Some of these indices can be calculated for each product, tabulated, and compared to one another (see Table 3.25). In the case of humic products (dry HS, liquid HA, and liquid FA), a general trend similar to that of Table 3.24 (on E_4/E_6) can be observed for E_2/E_3, E_2/E_4, and E_2/E_6. SUV254 (presented as relative to dry

Table 3.25 Indices from UV-visible spectrophotometer.

Product	E_4/E_6	E_2/E_3	E_2/E_4	E_2/E_6	SUV254[a]
Dry HS	4.9	2.4	5.5	24.9	1.0
Liquid HA	6.5	2.7	6.6	39.7	0.9
Liquid FA	10.7	3.5	11.5	107.7	1.0
Soybean extract	1.7	2.9	3.6	6.2	5.5
Calcium lignosulfonate	23.0	6.7	48.8	976.0	0.2
Peat moss	5.3	2.9	7.7	37.8	0.3
Biochar	1.5	2.2	3.1	4.3	0.1
Molasses	4.3	3.3	12.1	47.3	< 0.1

Note:
[a] presented as relative to dry HS.

HS) shows an almost constant value for each humic product as expected. Unfortunately, all of these indices cannot show clear differences and similarities between humic and non-humic products, at least from a practical point of view. These findings therefore indicate that the UV-visible spectral performance can only be utilized as a visual approach as a preliminary method to screen out non-humic products.

True humic products should have spectral patterns, as shown in Figure 3.16 by those of dry HS (for raw humic material containing HA, FA, and humin), liquid HA (containing mainly HA and some FA), and liquid FA (containing mainly FA and some HA). Other non-humic products have totally different patterns. Only peat moss produces an almost linear line, having a similar pattern to that of humic products (see Klavins et al. 2008 on the humification of peat). Calcium lignosulfonate shows the most aberration, having distinct peaks and valleys at lower wavelength readings. Some peaks and valleys are also visible for soybean extract, biochar, and molasses.

It is also possible to subtract the minimum absorbance of a set of data for a sample from all data in that set to obtain zeroed absorbances. These values are then divided by the absorbance of that sample at a specified wavelength, to produce the scaled absorbance (Bianchi et al. 2011). The reference zero absorbance is best selected at 290 nm since readings below 250 nm may be erroneous due to the presence of nitrates, nitrites, bromates, and other inorganic ions that absorb UV in this range; additionally, readings from both humic and non-humic products are more divergent beyond 300 nm. Then, a scaled absorbance difference plot can be generated by taking the difference between the scaled absorbance and

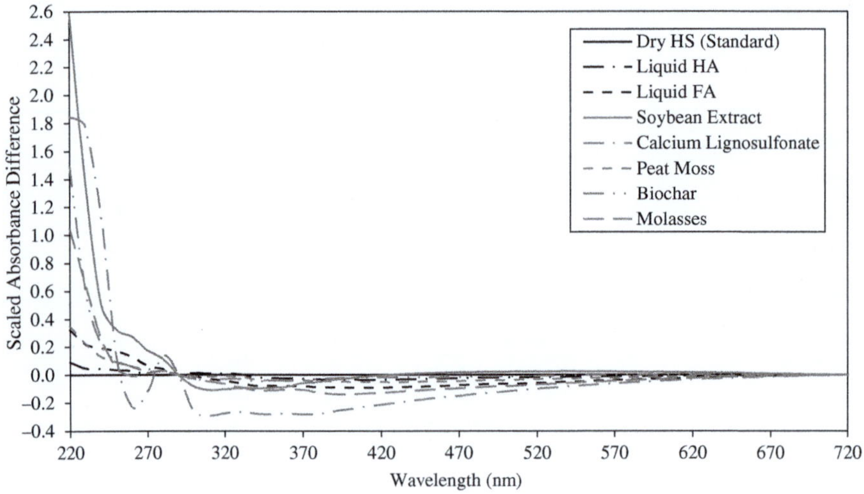

Figure 3.17 Scaled absorbance difference.

absorbance readings of a standard humic product (dry HS has been selected as the standard).

Figure 3.17 shows the scaled absorbance difference at 290 nm. This figure offers an alternate visualization of differences and similarities among humic and non-humic products than that of Figure 3.16. As expected, liquid HA shows the least

Table 3.26 Maximum and minimum readings.

Product	Maximum reading	Minimum reading	Maximum/minimum reading[a]
Dry HS	N/A (used as standard)		
Liquid HA	0.09	–0.04	2.3
Liquid FA	0.32	–0.09	3.6
Soybean extract	2.59	–0.11	23.5
Calcium lignosulfonate	1.85	–0.29	6.4
Peat moss	0.35	–0.05	7.0
Biochar	1.49	–0.04	37.3
Molasses	1.04	–0.14	7.4

Note:
N/A = not available;
[a] in absolute value.

difference compared to the standard, followed by liquid FA. Peat moss, as expected, shows a very similar pattern to that of liquid FA, which is consistent with the fact that peat moss comes from the same origin as low-rank coal. Calcium lignosulfonate again shows the most aberration from humic products, followed by soybean extract, biochar, and molasses.

Table 3.26 shows maximum and minimum readings (as well as their ratios, presented in absolute values) for each product. In comparison to dry HS which is used as standard, liquid HA and liquid FA have the smallest ratios at 2.3 and 3.6, respectively. Calcium lignosulfonate, peat moss, and molasses have higher values ranging from 6.4 to 7.4. Soybean extract and biochar have the highest ratios at 23.5 and 37.3, respectively. These values can be used as indicators in a preliminary process to screen adulterants out from true humic products and, when necessary, can be verified using other methods.

This laboratory study shows that both the colorimetric method and ISO 19822 inherit some inaccuracies, and therefore have to be used carefully, and are best when used side by side. With respect to the quantification of HS of true humic products, the colorimetric method performs relatively well (specifically for dry samples), considering the very short analytical time that it requires. ISO 19822 gives the most accurate results for HA and FA, but it is rather impractical and very time consuming. For non-humic products having similar properties to those of HS, the colorimetric method produces overestimated results, especially when used to analyze biochar and calcium lignosulfonate. ISO 19822 has a major weakness with calcium lignosulfonate and soybean extract. In this case, the ultimate, FTIR, and UV-visible spectral analyses can be used as the supporting methods.

3.4 Closure

Since no exact physical and chemical structures and compositions are available for HS, HA, or FA, analyses on both their quantity and quality are all approximates. With respect to the commercial agricultural community, the quantification of HA, and especially FA, is the main focus. To date, the colorimetric method and ISO 19822 are the best. Table 3.27 summarizes HA and FA analytical methods, including their critical factors.

The colorimetric method quantifies HS as the combined values of HA and FA. It is fast and relatively accurate specifically for dry materials containing HS only (without adulterants). For materials having a similar light absorbance to that of HS, this method may not be accurate. The colorimetric method is ideal to use during work at the mine site, in which a quick decision is required to determine whether the material is acceptable for collection or not. This method is also at its best when an instant result is requested during product manufacturing.

Table 3.27 Comparison among analytical methods.

Analytical method	Analyzed parameters	Method accuracy	Analytical time	Analytical principle
Colorimetric method	HS (= HA + FA)	Relatively[b]	Fast	Color
ISO 19822	Individual HA	Very	Slow	Gravimetric
Modified ISO 19822[a]	or FA		Medium	

Note:
[a] also known as the acetone method, [b] specifically for dry humic products; ultimate, FTIR, and UV-visible spectral analyses can be used as supporting methods to screen out non-humic products from true humic products.

ISO 19822 quantifies individual HA or FA gravimetrically. It is slow but very accurate for materials containing HA and FA only (without adulterants). It is ideal to quantify HA or FA when a high accuracy is necessary, as in the case of generating product technical data sheets. Utilizing ISO 19822 is the only way to quantify FA. For materials with similar adsorption properties on DAX-8 as those of FA, this method may not be accurate.

The modified ISO 19822 (also known as the acetone method), originally proposed in this book, is an improvement over ISO 19822. It has the same validity and accuracy as those of ISO 19822, but is more practical in that it significantly reduces the total analytical time. This method is, however, only valid for the analysis of lignite material (including low-rank coal in general).

Ideally, the colorimetric method and ISO 19822 are best used side by side. Both complement each other, especially when dealing with adulterants. In this case, the ultimate, FTIR, and UV-visible spectral analyses can be used as the supporting methods to screen-out non-humic from true humic products.

Reference

Aiken, G.R., McKnight, D.M., Thorn, K.A., and Thurman, E.M. (1992). Isolation of hydrophilic organic acids from water using nonionic macroporous resins. *Organic Geochemistry* 18(4): 567–573.

Aiken, G.R., McKnight, D.M., Wershaw, R.L., and MacCarthy, P. (1985). An introduction to humic substances in soil, sediment, and water. In: *Humic Substances in Soil, Sediments, and Water: Geo-Chemistry, Isolation and Characterization* (eds. G.R. Aiken, D.M. McKnight, R.L. Wershaw, and P. MacCarthy), 1–9. New York, NY: Wiley-Interscience.

Aiken, G.R., Thurman, E.M, and Malcolm, R.L. (1979). Comparison of XAD macroporous resins for the concentration of fulvic acid from aqueous solution. *Analytical Chemistry* 51(11): 1799–1803.

ALS Environmental Laboratories Ltd. (2019). Certificate of analysis. *Lab Work Order # L2201248 & L2223040-1/2* 1–6. Edmonton, AB.

Anachemia Canada Inc. (2018). *Sodium Hydroxide 95% Minimum* 1. Montreal, QE.

Artz, R.R.E., Chapman, S.J., Robertson, A.H.J, et al. (2008). FTIR spectroscopy can predict organic matter quality in regenerating cutover peatlands. *Soil Biology and Biochemistry* 40(2): 515–527.

Association of American Plant Food Control Officials (2013). 2013 midyear meeting. *Terms and Definitions Committee Agenda* 1–7. New Orleans, LA.

Association of American Plant Food Control Officials (2015). 2015 winter annual meeting report. *Terms and Definitions Committee Agenda* 17–23. Jacksonville, FL.

Association of American Plant Food Control Officials (2018). 2018 winter annual agenda. *Terms and Definitions Committee Agenda* 20–25. Savannah, GA.

ASTM D2013/D2013M-18 (2013). *Standard practice for preparing coal samples for analysis*. West Conshohocken, PA: ASTM International.

ASTM D2492-02(2012) (2012). *Standard test method for forms of sulfur in coal*. West Conshohocken, PA: ASTM International.

ASTM D3174-12 (2012). *Standard test method for ash in the analysis sample of coal and coke from coal*. West Conshohocken, PA: ASTM International.

ASTM D3176-15 (2015). *Standard practice for ultimate analysis of coal and coke*. West Conshohocken, PA: ASTM International.

ASTM D3302/D3302M-17 (2017). *Standard test method for total moisture in coal*. West Conshohocken, PA: ASTM International.

ASTM D4972-18 (2018). *Standard test methods for pH of soils*. West Conshohocken, PA: ASTM International.

ASTM D5373-16 (2016). *Standard test methods for determination of carbon, hydrogen and nitrogen in analysis samples of coal and carbon in analysis samples of coal and coke*. West Conshohocken, PA: ASTM International.

ASTM D5865-13 (2013). *Standard test method for gross calorific value of coal and coke*. West Conshohocken, PA: ASTM International.

ASTM D6357-11 (2011). *Test methods for determination of trace elements in coal, coke, & combustion residues from coal utilization processes by Inductively Coupled Plasma atomic emission, Inductively Coupled Plasma mass, & graphite furnace Atomic Absorption spectrometry*. West Conshohocken, PA: ASTM International.

ASTM D6722 (2011). *Standard test method for total mercury in coal and coal combustion residues by direct combustion analysis*. West Conshohocken, PA: ASTM International.

ASTM D7348-13 (2013). *Standard test methods for loss on ignition (LOI) of solid combustion residues*. West Conshohocken, PA: ASTM International.

Baes, A.U. and Bloom, P.R. (1990). Fulvic acid ultraviolet-visible spectra: influence of solvent and pH. *Soil Science Society of America Journal* 54(5): 1248–1254.

Baigorri, R., Fuentes, M., Gonzales-Gaitano, G., and Garcia-Mina, J.M. (2007). Simultaneous presence of diverse molecular patterns in humic substances in solution. *Journal of Physical Chemistry B* 111: 10577–10582.

Beckett, R., Jue, Z., and Giddings, J.C. (1987). Determination of molecular weight distributions of fulvic and humic acids using flow field-flow fractionation. *Environmental Science and Technology* 21(3): 289–295.

Bianchi, T.S., Cook, R.L., Perdue, E.M., et al. (2011). Impacts of diverted freshwater on dissolved organic matter and microbial communities in Barataria Bay, Louisiana, U.S.A. *Marine Environmental Research* 72(5): 248–257.

Bowen, B.H. and Irwin, M.W. (2008). *Coal Characteristics CCTR Basic Facts File #8* 1–27. West Lafayette, IN: Indiana Center for Coal Technology Research, Purdue University.

Canadian Food Inspection Agency (2020). Primary fertilizer and supplement materials. *Guidance Document Repository (GDR)* 11–14. Ottawa, ON.

Canadian Humalite International Inc. (2020a). Raw material and products. *Laboratory Analyses and Technical Data* 1–656. Edmonton, AB.

Canadian Humalite International Inc. (2020b). Processing chemicals and equipment. *Technical Data* 1–124. Edmonton, AB.

CDFA Method (1999). *Humic acid method # HUMIC HA4-JC*. Sacramento, CA: California Department of Food and Agriculture.

Center for Climate and Energy Solutions (2017). Coal. http://www.c2es.org/energy/source/coal (accessed 08 May 2019).

ChemAnalytical LLC. (2018a). *Analysis of a Liquid Sample by Fourier Transform Infra-Red (FT-IR) Spectra Measurements and Analysis* 1–5. Pittsford, NY.

ChemAnalytical LLC. (2018b). *Analysis of Solid Powder and Liquid Samples by Fourier Transform Infra-Red (FT-IR) Spectra Measurements and Analysis* 1–18. Pittsford, NY.

ChemAnalytical LLC. (2019). *Analysis of Liquid Samples by Fourier Transform Infra-Red (FT-IR) Spectra Measurements and Analysis* 1–6. Pittsford, NY.

Chen, Y., Jiang, Z., Zhang, X., et al. (2017). Variation in the humification degree of dissolved organic matter from cattle manure during composting as analyzed by ultraviolet-visible and fluorescence spectroscopy. *Journal of Environmental Quality* 46(6): 1489–1999.

Chen, Y., Senesi, N., and Schnitzer, M. (1977). Information provided on humic substances by E_4/E_6 ratios. *Soil Science Society of America Journal* 41(2): 352–358.

Chin, Y.-P., Aiken, G., and O'Loughlin, E. (1994). Molecular weight, polydispersity, and spectroscopic properties of aquatic humic substances. *Environmental Science and Technology* 28(11): 1853–1858.

ClearTech Industries Inc. (2018). *Hydrochloric Acid 31%* 1. Saskatoon, SK.

Coates, J. (2000). Interpretation of infrared spectra, a practical approach. In: *Encyclopedia of Analytical Chemistry* (ed. R.A. Meyers), 1–23. Chichester, UK: John Wiley and Sons.

De Nobili, M., Contin, M., and Leita, L. (1990). Alternative method for carboxyl group determination in humic substances. *Canadian Journal of Soil Science* 70: 531–536.

EEC A.6 (1992). Water solubility. Dir 92/69/EEC (O.J. L383 A). In: *Official Journal of the European Communities No. L 383/113* (member K. Van Miert), 1–9. Aberdeen, UK: Commission Directive 92/69/EEC.

EMD Millipore Co. (2018). *Silver Nitrate Solution 0.10 N* 1. Darmstadt, Germany.

EPA Method 6020A (SW-846) (1998). *Inductively Coupled Plasma-mass spectrometry*. Washington, DC: US Environmental Protection Agency.

EPA Method 9080 (1986). *Cation-exchange capacity of soils (ammonium acetate)*. Washington, DC: US Environmental Protection Agency.

Fisher Scientific Co. (2018). *Acetone Certified ACS 99.8%* 1. Fair Lawn, NJ.

HACH (2018). What is the scientific principle of colorimetry? *Document ID TE1598* 1–2. Loveland, CO.

Haider, K. and Martin, J.P. (1967). Synthesis and transformation of phenolic compounds by Epicoccum nigrum in relation to humic acid formation. *Soil Science Society of America Proceedings* 31(6): 766–772.

Hayes, M.H.B. and Swift, R.S. (2018). An appreciation of the contribution of Frank Stevenson to the advancement of studies of soil organic matter and humic substances. *Journal of Soils and Sediments* 18: 1212–1231.

Hayes, M.H.B., Mylotte, R., and Swift, R.S. (2017). Humin: its composition and importance in soil organic matter. *Advances in Agronomy* 143: 47–138.

Hoffman G.L., Nikols, D.J., Stuhec, S., and Wilson, R.A. (1993). Evaluation of leonardite (humalite) resources of Alberta. *Open File Report 1993-18* 1–18. Edmonton, AB: Alberta Research Council.

Humic Products Trade Association (2015). Biochar and humic substances: a comparison. *Report of the HPTA Science Committee* 1–7. Spring Green, WI.

Humic Products Trade Association (2017). California Department of Food and Agriculture (CDFA) considers switching to the HPTA approved test method. *News and Events, November 2017* 1. Spring Green, WI.

Humic Products Trade Association (2020). CDFA approves use of HPTA test method seal. *News and Events, March 2020* 1. Williston, ND.

Ibarra, J.V., Munoz, E., and Molner, R. (1996). FTIR study of the evolution of coal structure during the coalification process. *Organic Geochemistry* 24(6/7): 725–735.

Ikan, R., Rubinsztain, Y., Ioselis, P., et al. (1986). Carbon-13 cross polarized magic-angle samples spinning nuclear magnetic resonance of melanoidins. *Organic Geochemistry* 9(4): 199–212.

Ilani, T., Schulz, E., and Chefetz, B. (2005). Interactions of organic compounds with wastewater dissolved organic matter: role in hydrophobic fractions. *Journal of Environmental Quality* 34: 552–562.

International Humic Substances Society (2020). IHSS samples-a retrospective, source materials for IHSS samples, elemental compositions and stable isotopic ratios of IHSS samples, acidic functional groups of IHSS samples, amino acids compositions of IHSS samples, carbohydrate compositions of IHSS samples, ^{13}C NMR estimates of carbon distribution, and 12 absorbance spectra charts. http://www.humic-substances.org/#products (accessed 4 February 2020).

ISO 19822:2018(E) (2018). *Fertilizers and soil conditioners-determination of humic and hydrophobic fulvic acids concentrations in fertilizer materials*. Geneva, Switzerland: International Organization for Standardization.

ISO 5073 (2013). *Brown coals and lignites-determination of humic acids*. Geneva, Switzerland: International Organization for Standardization.

ISO/TC 134 N 260 (2014). *Fertilizers and soil conditioners-determination of humic and fulvic acid concentrations in fertilizer materials*. Geneva, Switzerland: International Organization for Standardization.

Janzen, H. (2019). The future of humic substances research: preface to a debate. *Journal of Environmental Quality* 48: 205–206.

Jiang, J., Yu, H., Xi, B., et al. (2011). UV-visible spectroscopic properties of dissolved fulvic acids extracted from salined fluvo-aquic soils in the Hetao irrigation district, China. *Soil Research* 49: 670–679.

Kanokkantapong, V., Marhaba, T.F., Panyapinyophol, B., and Pavasant, P. (2006). FTIR evaluation of functional groups involved in the formation of haloacetic acids during the chlorination of raw water. *Journal of Hazardous Materials* 2: 188–196.

Kawamura, S. (1983). Seventy years of the Maillard reaction. In: *American Chemical Society Symposium Series Vol. 215, Ch. 1: the Maillard Reaction in Food and Nutrition* (eds. G.R. Waller and M.S. Feather), 3–18. Washington, DC: American Chemical Society.

Klavins, M., Sire, J., Purmalis, O., and Melecis, V. (2008). Approaches to estimating humification indicators for peat. *Mires and Peat* 3(Article 7): 1–15.

Kleber, M. and Lehmann, J. (2019). Humic substances extracted by alkali are invalid proxies for the dynamics and functions of organic matter in terrestrial and aquatic ecosystems. *Journal of Environmental Quality* 48: 207–216.

Krumins, J., Klavins, M., Seglins, V., and Kaup, E. (2012). Comparative study of peat composition by using FT-IR spectroscopy. *Material Science and Applied Chemistry* 26: 106–114.

Lamar, R., Olk, D.C., Mayhew, L., and Bloom, P.R. (2014). A new standardized method for quantification of humic and fulvic acids in humic ores and commercial products. *Journal of AOAC (Association of Official Analytical Chemists) International* 97(3): 721–730.

Lamar, R.T. and Talbolt, K.H. (2009). Critical comparison of humic acid test methods. *Communications in Soil Science and Plant Analysis* 40: 2309–2322.

Leenheer, J.A. (1981). Comprehensive approach to preparative isolation and fractionation of dissolved organic carbon from natural waters and wastewaters. *Environmental Science and Technology* 15(5): 578–587.

Leenheer, J.A. and Croue, J.-P. (2003). Characterizing dissolved aquatic organic matter. *Environmental Science and Technology* 1: 19–26.

Lehmann, J. and Kleber, M. (2015). The contentious nature of soil organic matter. *Nature* 528: 60–68.

Liem, L.E., Serben, T., and Cano, M. (2019). Gravimetric quantification of hydrophobic fulvic acids in lignite material using acetone. *Journal of AOAC (Association of Official Analytical Chemists) International* 102(6): 1901–1907.

Loring Laboratories Ltd. (2019). Ultimate & ICP analyses. *Certificate of Assay File 62765* 1–2. Calgary, AB.

Mahieu, N., Olk, D.C., and Randall, E.W. (2002). Multinuclear magnetic resonance analysis of two humic acid fractions from lowland rice soils. *Journal of Environmental Quality* 31(March-April): 421–430.

Mantoura, R.F.C. and Riley, J.P. (1975). The analytical concentration of humic substances from natural waters. *Analytica Chimica Acta* 76: 97–106.

Martin, J.P. and Haider, K. (1971). Microbial activity in relation to soil humus formation. *Soil Science* 111(1): 54–63.

Martin, J.P., Haider, K., and Bondietti, E. (1975). Properties of model humic acids synthesized by phenoloxidase and autooxidation of phenols and other compounds formed by soil fungi. In: *Proceedings of International Meeting Humic Substances, Niieuwresluis (1972)* 171–186. Wageningen, Holland: PUDOC.

Mehlich, A. (1984). Photometric determination of humic matter in soils, a proposed method. *Communications in Soil Science and Plant Analysis* 15(12): 1417–1422.

Myneni, S.C.B. (2019). Chemistry of natural organic matter-the next step: commentary on a humic substances debate. *Journal of Environmental Quality* 48: 233–235.

Nkambule, T.I., Krause, R.W.M., Mamba, B.B., and Haarhoff, J. (2008). Characterisation of natural organic matter (NOM) and its removal using cyclodextrin polyurethanes. *Water Institute of Southern Africa Biennial Conference* 1–4. Sun City, South Africa: Water Institute of Southern Africa.

Ochs, M., Cosovic, B., and Stumm, W. (1994). Coordinative and hydrophobic interaction of humic substances with hydrophilic Al_2O_3 and hydrophobic mercury surfaces. *Geochimica et Cosmochimica Acta* 58(2): 639–650.

Olk, D.C., Bloom, P.R., Perdue, E.M., et al. (2019). Environmental and agricultural relevance of humic fractions extracted by alkali from soils and natural waters. *Journal of Environmental Quality* 48: 217–232.

Organic Materials Review Institute (2017). Crop, livestock, processing & handling. *OMRI Generic Materials List* 4–80. Eugene, OR.

Polubesova, T., Chen, Y., Navon, R., and Chefetz, B. (2008). Interactions of hydrophobic fractions of dissolved organic matter with Fe^{3+}- and Cu^{2+}-montmorillonite. *Environmental Science and Technology* 42(13): 4797–4803.

Praxair Canada Inc. (2018). *Nitrogen 99.999%* 1. Mississauga, ON.

Rice, J.A. and MacCarthy, P. (1988). Comments on the literature of the humin fraction of humus. *Geoderma* 43: 65–73.

Rodriguez-Lucena, P., Lucena, J.J., and Hernandez-Apaolaza, L. (2009). Relationship between the structure of Fe-lignosulfonate complexes determined by FTIR spectroscopy and their reduction by the leaf Fe reductase. *The Proceedings of the International Plant Nutrition Colloquium XVI* 1–6. Davis, CA: University of California.

Roybal, G.H. and Barker, J.M. (1986). Geology and production of humate and weathered coal in New Mexico. *Transactions of American Institute of Mining, Metallurgical and Petroleum Engineers* 280: 2105–2111.

Sanchez-Monedero, M.A., Roig, A., Cegarra, J., et al. (2002). Effects of HCl-HF purification treatment on chemical composition and structure of humic acids. *European Journal of Soil Science* 53: 375–381.

Schnitzer, M. (1971). Characterization of humic constituents by spectroscopy. In: *Soil Biochemistry Vol. 2* (eds. A.D. McLaren and J. Skujins), 60–95. New York, NY: Marcel Dekker.

Schnitzer, M. (1972). Chemical, spectroscopic, and thermal methods for the classification and characterization of humic substances. In: *Proceedings of International Meetings on Humic Substances* 293–310. Wageningen, Holland: PUDOC.

Schnitzer, M. (1977). Recent findings on the characterization of humic substances extracted from soils from widely differing climatic zones. In: *Proceedings of a Symposium on Soil Organic Matter Studies* 117–132. Vienna, Austria: International Atomic Energy Agency.

Schnitzer, M. and Dinel, H. (2001). Comparative study of humic acid-like material L31(H) and a humic acid from the Melfort series. *Technical Report* 1–14. Ottawa, ON: Eastern Cereals and Oilseed Research Centre, Agriculture and Agri-Food Canada.

Schnitzer, M., Dinel, H., Pare, T., et al. (2001). Some chemical and spectroscopic characteristics of six organic ores. *Technical Report* 1–22. Ottawa, ON: Eastern Cereals and Oilseed Research Centre, Agriculture and Agri-Food Canada.

Shahbazi, K., Marzi, M., and Tabakhian, S. (2019). The comparative evaluation of humic acid determining methods in humic-based commercial fertilizers. *Archives of Agronomy and Soil Science* 65(12): 1720–1732.

Siemers, C.T. and Wadell, J.S. 1977. Humate deposits of the Menefee formation (upper Cretaceous), northwestern New Mexico. *28th Field Conference* 1–22. Socorro, NM: New Mexico Geological Society.

Sigma Aldrich Co. (2018a). Supelite DAX-8 resin Supelco. *Certificate of Analysis* 1. Bellefonte, PA.

Sigma Aldrich Co. (2018b). Amberlite IR120 H resin Supelco. *Certificate of Analysis* 1. Bellefonte, PA.

Silverstein, R.M., Webster, F.X., and Kiemle, D.J. (2005). *Spectrometric Identification of Organic Compounds*, 7th ed., 72–126 and 204–244. New York, NY: John Wiley and Sons.

Smallwood, I.M. (1996). *Handbook of Organic Solvent Properties* 171–173. New York, NY: Halsted Press is an imprint of John Wiley and Sons.

Stevenson, F.J. (1994). *Humus Chemistry: Genesis, Composition, Reactions*, 2nd ed., 24–58, 212–235, 259–284, and 303–324. New York, NY: Wiley-Interscience.

Stevenson, F.J. and Butler, J.H.A. (1969). Chemistry of humic acids and related pigments. In: *Organic Geochemistry Methods and Results* (eds. G. Eglinton and M.T.J. Murphy), 534–557. Berlin, Germany: Springer-Verlag.

Swift, R.S. (1996). Organic matter characterization. In: *Methods of Soil Analysis Part 3 Chemical Method, Ch. 35* (eds. D.L. Sparks, A.L. Page, P.A. Helmke, R.H. Loeppert, P.N. Soltanpour, M.A. Tabatabai, C.T. Johnston, and M.E. Sumner), 1011–1069. Madison, WI: Soil Science Society of America/American Society of Agronomy.

Tarasevich, Y.I., Tryfonova, M.Y., Dolenko, S.A., and Aksenenko, E.V. (2016). Adsorption-based approach to determine the size and mass of humic acids molecules. *Adsorption Science and Technology* 34(2-3): 125–133.

Thurman, E.M. and Malcolm, R.L. (1981). Preparative isolation of aquatic humic substances. *Environmental Science and Technology* 15(4): 463–466.

Tremblay, L. and Gagne, J.-P. (2002). Fast quantification of humic substances and organic matter by direct analysis of sediments using DRIFT spectroscopy. *Analytical Chemistry* 74: 2985–2993.

Tsutsuki, K. and Kuwatsuka, S. (1979). Chemical studies on soil humic acids: VII. pH-dependent nature of the ultraviolet and visible absorption spectra of humic acids. *Soil Science and Plant Nutrition* 25(3): 373–384.

Van Zomeren, A. and Comans, R.N.J. (2007). Measurement of humic and fulvic acid concentrations and dissolution properties by a rapid batch procedure. *Environmental Science and Technology* 41: 6755–6761.

Vaughan, D. and Linehan, D.J. (1976). The growth of wheat plants in humic acid solutions under axenic conditions. *Plant and Soil* 44(2): 445–449.

Vaughan, D. and Ord, B.G. (1981). Uptake and incorporation of ^{14}C-labelled soil organic matter by roots of Pisum sativum L. *Journal of Experimental Botany* 32(129): 679–687.

Vermeer, A.W.P. and Koopal, L.K. (1998). Adsoprtion of humic acids to mineral particles. 2. Polydispersity effects with polyelectrolyte adsorption. *Langmuir* 14: 4210–4216.

Weishaar, J.L., Aiken, G.R., Bergamaschi, B.A., et al. (2003). Evaluation of specific ultraviolet absorbance as an indicator of the chemical composition and reactivity of dissolved organic carbon. *Environmental Science and Technology* 37(20): 4702–4708.

Wershaw, R.L. (2000). The study of humic substances-in search of a paradigm. In: *Humic Substances: Versatile Components of Plants, Soils and Water* (eds. E.A. Ghabbour and G. Davies), 1–7. Cambridge, UK: The Royal Society of Chemistry.

Yakimenko, O., Khundzhua, D., Izosimov, A., et al. (2018). Source indicator of commercial humic products: UV-Vis and fluorescence proxies. *Journal of Soils and Sediments* 18: 1279–1291.

Zaccone, C., Plaza, C., Ciavatta, C., et al. (2018). Advances in the determination of humification degree in peat since Achard (1786): applications in geochemical and paleoenvironmental studies. *Earth-Science Reviews* 185 (October 2018): 163–178.

Zech, W., Senesi, N., Guggenberger, G., et al. (1997). Factors controlling humification and mineralization of soil organic matter in the tropics. *Geoderma* 79(1-4): 117–161.

4

Applications in Agriculture

4.1 Overview

Numerous studies have been completed on the science of organic matter (OM) and humic substances (HS, or "humic"), including humic acids (HA) and fulvic acids (FA). With respect to their applications in agriculture, scholastic articles have been published explaining different mechanisms for how OM and HS enhance soil, crops, and livestock productivity. Physical, chemical, and biological changes in soil have been evaluated in great detail, most importantly on troubled soils. The effects of HA and FA on the availability of the applied nutrients have been investigated, as well as their metabolism and usages by crops. These additions result in improvements on the growing components and health of crops, and most importantly on their yields. The same observation has also been made for livestock, with positive changes in the availability, metabolism, and usages of applied nutrients. Enhanced growth, health, and yield of livestock have been reported.

As detailed by Follett et al. (1981), OM benefits soil and plants through the following actions:

1) storing (i.e. binding/retaining) anions essential for plant growth;
2) increasing the cation exchange capacity (CEC) of soil (especially in the case of humus);
3) buffering the soil against rapid changes in acidity (pH), alkalinity, salinity, pesticides, and toxic metals;
4) protecting the surface of soil against erosion, increasing filtration, reducing water runoff, increasing water holding capacity, and increasing surface wetness;
5) supplying food for beneficial soil organisms;

Low-Rank Coal Applications in Agriculture: Humic Analyses, Products, and Performance, First Edition. L. Edwin Liem.

6) reducing extremes of soil surface temperature;
7) decreasing surface crust formation;
8) reducing the crystallization and hardening of the laterite layer of soil; in the case of humus, it complexes with iron (Fe) and aluminum (Al);
9) supplying small quantities of all essential plant nutrients;
10) making phosphorus (P) and micronutrients more readily available over a wider range of pH (mainly for humus);
11) optimizing the application rates of herbicide; and
12) decreasing the bulk density of soil.

With respect to the livestock application, animals in nature actively consume soil in different forms from the environment. The ingested soil may be a source of minerals for the grazing ruminant (Healy 1968 and Mayland et al. 1975). Topsoil is also a rich source of OM and HS (Stevenson 1994). The use of peat as a feed supplement has been reviewed and summarized by Trckova et al. (2005). The humic components of peat have been characterized by Klavins et al. (2008). As documented by Thompson (1932) and cited in Taylor et al. (2014), soft coal and other coal materials have also been fed to swine to improve their growth.

As discussed in Sections 3.1 and 3.2, HS are components of soil OM with unique properties. The important sources of HS in nature include leonardite, oxidized lignite, oxidized subbituminous coals, and humalite. These materials are in the category of low-rank coal containing a high strength of HS and have minimum amounts of impurities.

Numerous short summaries of commercial HA and FA products used in soil, crop, and livestock applications are presented in this chapter. The focus is on the practical aspect of their use, i.e. improvement in soil, crop, and livestock productivity. All information is obtained from public domains, mainly generated by research and/or teaching institutions, and therefore is trustworthy and accurate. It is not the purpose of this book to reinvent the wheel on the science of low-rank coal and HS. Instead, its objective is to show that the knowledge behind HS, HA, and FA is indeed scientific and that the manufacturing of low-rank coal (including Alberta low-rank coal, i.e. humalite) into commercial products and their applications within the agricultural sector are justified. Not all information contained within a specific article is included in each summary; only the relevant information is presented. For example, results obtained from non-humic products (being evaluated side by side or even simultaneously with humic products during the same study) are excluded from the discussion. Results generated from humic products that are not practical or contain similar trends as those of others are also skipped.

Also discussed in this chapter are issues related to the regulation of commercial HA and FA products. While HA and FA have scientifically been proven to benefit soil, crops, and livestock, their commercial products must still be registered before they can be marketed and sold under a given jurisdiction or to meet a certain target market. The main objective here is to make sure that the product claims are legitimate and that recommended application rates are efficient, with minimum risks of injury to the target soil, crops, and livestock.

4.2 Benefits of Humic Substances

Up to 39 samples of Alberta low-rank coal (humalite) were analyzed for macro and micronutrients (Follett et al. 1981), i.e. nitrogen (N), P, potassium (K), sulfur (S), calcium (Ca), magnesium (Mg), boron (B), chloride (Cl), copper (Cu), Fe, manganese (Mn), molybdenum (Mo), and zinc (Zn), as well as ash. This material was found to be low in macronutrients (containing at most 1.4% for each component) and, with the exception of Fe (at 5701 ppm), also low in micronutrients (containing at most 300 ppm for each component). Ash (at 22.0%) represented inorganic components, most likely sand and clay, as well as trace elements. Humalite had significant amounts of OM (78.0%), indicating a high content of HS. All were mean values, analyzed according to ASTM D2013/D2013M-18 (2013), ASTM D5373-16 (2016), ASTM D2492-02(2012) (2012), ASTM D3174-12 (2012), ASTM D3176-15 (2015), and ASTM D6357-11 (2011). Based on this nutritional analysis, humalite and humic products would be classed as soil amendments, not fertilizers (see Table 4.1). When applied to crops, they would be best applied together with nutrients in order for their benefits to be observed.

Other than the presence of nutrients in soil, the concentration of soil solution (i.e. soluble salts) would also affect the germination or establishment of crops. High soluble salts were found to elevate osmotic pressure in soil causing the injury of seeds and grown crops, due to plasmolysis or actual toxicity. Plasmolysis happened when the salt concentration outside root cells exceeded the cellular salt concentration. Cellular water moved out of the cell, causing the crop to exhibit symptoms similar to drought stress (Rader Jr. et al. 1942 and Havlin et al. 2005). The severity of soluble salts in soil would commonly be represented as the salt index (SI), defined as: (increase in osmotic pressure of soil solution due to the application of a definite weight of a fertilizer mixture or fertilizer material ÷ increase due to application of the same weight of sodium nitrate ($NaNO_3$) under identical conditions) × 100. The effect of $NaNO_3$ was given the value of 100, while effects of others would be expressed relative to 100 (Rader Jr. et al. 1942).

SI values of common fertilizers, soil amendments, and three commercial humic products were then compared (see Table 4.2). Dry HS was a dry humic product (identical to Alberta low-rank coal or humalite, containing 64.7% HS), liquid FA contained 0.3% HA and 0.7% FA (total HS = 1.0%), and liquid HA contained 12.0% HS. Each analysis was completed based on ISO 19822, including moisture. All

Table 4.1 Soil and crops nutritional analysis for humalite.

Parameter	# of data	Content	Unit
Macronutrients			
N	33	1.4 (0.21)	%
P	14	0.0 (0.01)	%
K	34	0.2 (0.16)	%
S	33	0.6 (0.14)	%
Ca	34	1.3 (0.71)	%
Mg	35	0.4 (0.34)	%
Micronutrients			
B	29	300 (252.8)	ppm
Cl	19	71 (62.8)	ppm
Cu	38	9 (14.6)	ppm
Fe	35	5701 (3211.5)	ppm
Mn	34	186 (141.0)	ppm
Mo	39	2 (1.1)	ppm
Zn	37	17 (9.7)	ppm
Others			
Ash	33	22.0 (8.32)	%
OM	33	78.0	%

Source: Canadian Humalite International Inc. (2020a); in brackets are standard deviations.

humic products contained very small SI values (at 0.6, 1.5, and 2.8, respectively) compared to those of common fertilizers and soil amendments. Limestone was the closest one, having 4.7 SI. This would be an advantage of blending humic products with fertilizers to reduce the combined SI. And considering all benefits of HS as discussed in this section, the productivity of soil and crops could still be maintained (or even improved) at reduced nutrient inputs with the incorporation of humic products within the fertilization program.

Soil would typically consist of OM, soil particles (referring to soil texture, i.e. sand, silt, and clay), and pore spaces containing air and water. Soil pores were critical for water infiltration and percolation into the root zone and subsoil, and air exchange with the atmosphere. Biological and chemical processes also occurred within soil pores that would need both air and water. A common problem related to soil pores was soil compaction, either on the surface (soil degradation due to soil tillage, raindrops, and irrigation water impact) or subsurface

Table 4.2 SI of selected materials.

Material	SI
Fertilizers	
Mono potassium phosphate, MKP (KH_2PO_4)	8.1
Ammonium poly phosphate, APP ($(NH_4PO_3)_n(OH)_2$)	20.0
Mono ammonium phosphate, MAP ($NH_4H_2PO_4$)	26.7
Di ammonium phosphate, DAP ($(NH_4)_2HPO_4$)	29.2
Potassium sulfate (K_2SO_4)	42.6
Ammonia (NH_3)	47.1
Urea ammonium nitrate, UAN (28% N)	63.0
Calcium nitrate ($Ca(NO_3)_2$)	65.0
Potassium thiosulfate ($K_2S_2O_3$)	68.0
Potassium nitrate (KNO_3)	69.5
Urea ($CO(NH_2)_2$)	74.4
Ammonium sulfate ($(NH_4)_2SO_4$)	88.3
Ammonium thiosulfate $(NH_4)_2S_2O_3$)	90.4
Sodium nitrate ($NaNO_3$)	100.0
Ammonium nitrate (NH_4NO_3)	104.1
Potassium chloride, potash (KCl)	116.1
Soil amendments	
Dry HS, Alberta low-rank coal, or humalite (64.7% HS)	0.6
Liquid FA (0.3% HA + 0.7% FA = 1.0% HS)	1.5
Liquid HA (12.0% HS)	2.8
Calcium carbonate, limestone (CaCO3)	4.7
Calcium sulfate, gypsum ($CaSO_4.2H_2O$)	8.1
Magnesium sulfate, Epsom salt ($MgSO_4.7H_20$)	44.0
Manure salts	91.9–112.7

Sources: Killorn and Voss (1986), Havlin et al. (2005), and Canadian Humalite International Inc. (2020a); *note:* results for humic products include moisture.

(hardpan tillage-induced compaction). Subsurface compaction was worst for soil with high silt and clay contents, affecting water and root penetration (McKenzie 2010). Soil compaction resulted in reduced crop yields, even with tillage. Among other strategies to alleviate this problem, increasing soil OM by adding organic soil amendments had been recommended. Soil with high OM contents was more resistant to compaction and recuperated better from slight compaction damage (Duiker 2004). Humalite, one type of low-rank coal, had been a strong candidate among organic soil amendments due its high strength of soil OM and HS, as well as very low impurities (Canadian Humalite International Inc. 2020a). The use of this material was found to improve the structure of heavy clay soil, breaking up larger aggregates into smaller ones, resulting in better air and moisture circulation within the matrix and easier soil tilling (Liem et al. 2003a and 2003b).

The addition of organic soil amendments would improve soil moisture contents beneficial for the growth of crops and soil organisms. HS had a great affinity to water, i.e. 5 to 10 times greater than that of inorganic soil components. Its capacity was directly proportional to its humic content (Scheffer and Ulrich 1960 and Linser 1956, as cited in Chen and Schnitzer 1976). The sorption mechanism was through the hydrogen (H)-bonding between the H atoms of water and the oxygen (O) atoms of functional groups on the surface of HS, and between the H atoms of the functional groups and the O atoms of water (Chen and Schnitzer 1976). The mechanisms were the monolayer water adsorption by HS and the formation of an additional water layer due to a high relative humidity. HS enhanced moisture retentions of heavy clay soil (Liem et al. 2003a and 2003b). When combined with nutrients, soil OM (including HS) would form slowly available forms of nutrients (especially N) with a minimum loss through the soil. It would also stabilize soil structure, enhance the root habitat, and increase soil water holding capacity (NPK 1983).

For soil microorganisms, HS would be utilized as a source of organic carbon (C) and might serve as a functional group of the biosynthesis chains. The decomposition of HS included non-enzymatic, enzymatic, and enzyme-initiated reactions, in which C being consumed was converted into microorganism biomass (Filip et al. 2000 and Filip and Berthelin 2001). These microorganisms included numerous bacteria and fungi (Gramss et al. 1999). Beneficial soil organisms increased the supply of mineral nutrients to plants, indirectly stimulated plant growth by suppressing the growth of plant pathogens, and directly stimulated plant growth through the production of phytohormones (Davison 1988, Lynch 1990, Lugtenberg et al. 1991, and Lazarovits and Nowak 1997). It was hypothesized that HS influenced metabolic processes, nucleic acid syntheses, and ion uptakes (Ahmed and Ismail 2016).

Soil enzymes came from plants and microorganisms (Sollins et al. 1996), while HS functioned indirectly as catalysts to biochemical reactions, via the formation of humic enzyme complexes or modified humic substrates from the enzymatic reaction. Soil enzymes like urease, protease, phosphatase, and others had been found in complex forms with both inorganic and organic colloids, in particular with HS (Masciandaro and Ceccanti 1999 and Benitez et al. 2000).

HS contained carboxyl (COOH), carbonyl (C=O), phenol-hydroxyl (OH), and amide groups performing as carriers to soil enzymes, improving their long-term stability (Rosa et al. 2000). A direct reaction of HS and soil enzymes might result in the modification of HS, like the formation of FA or the enhancement of humic molecular size over time (Hofrichter et al. 1999 and Cozzolino and Piccolo 2002).

HS enhanced the structure of soil and formation of enzymes useful for the health of soil and plants; the most importantly being in the root zone (Walker et al. 2003). HS promoted the growth of roots and increased soil C inputs (Olk et al. 2018). HA improved root exudate yields on maize seedlings, with the release of chemical compounds to the root interface, benefiting the microbial population size and community (Canellas et al. 2019). The sorption of more hydrophobic organic exudates on the soil matrix was enhanced with the presence of HS (Rao 1990), making them more available to the organisms. Microbial activity resulted in the alteration of root exudates and the release of microbial metabolites, which in turn stimulated root exudation (Meharg and Kilham 1991 and 1995, and Phillips et al. 2004). Prikryl and Vancura (1980) showed that the release of root exudates was positively correlated with root growth.

The quality of soil would also be affected with the presence of toxic organic and inorganic compounds. In this case, HS had a unique structure of irregular polymers containing aromatic polyhydroxyl polycarbonic acids, able to bind both polar and hydrophobic toxic organic compounds, as well as inorganic ions. The binding of HS resulted in a change in speciation of the toxic contaminants followed by a change in their toxicity and bioaccumulation (Clapp et al. 1993). The activity of anaerobic Fe-reducing bacterium (*G. metallireducens*) increased significantly with the incorporation of HS, which was important with respect to the enhanced ability of anaerobic organisms in oxidizing organic contaminants, like in the case of benzene (Lovley et al. 1996a and 1996b). HS also facilitated the degradation of some toxic organic compounds through chemical hydrolysis followed by degradation by soil microorganisms, such as atrazine (Gamble and Khan 1988). HS accelerated the microbial degradation of diesel in soil (Liem et al. 2003a and 2003b). HS also affected the photo degradation of polyaromatic hydrocarbons, chlorinated aromatic hydrocarbons, chlorinated phenols, and other pesticides. This was possible due the ability of HS to form an excited state O molecule that attacked the organic compound that later decayed due to indirect photolysis (Kulikova et al. 2005). With respect to toxic inorganic ions, HS increased sand macro-aggregation significantly, maintaining soil pores for proper aeration and water movement. Due to their binding with HS, water-soluble metal components (the most toxic ones) were significantly reduced, while the strongly complex (and the least toxic) components increased significantly. HS adsorbed some toxic metals from their water phase, reducing their toxicity (Ibrahim and Goh 2002 and Stewart and Janin 2014).

Another factor influencing the quality of soil was salinity. It created abiotic stresses, affecting cellular ionic homeostasis, balance, and distribution (Niu et al. 1995 and Wang et al. 2012). Salinity also disrupted photosynthesis and N assimilation (Parida and Das 2005 and Munns and Tester 2008). While small amounts of salt might stimulate the growth of crops and uptake of nutrients, at higher rates, its toxic effect appeared (Khan et al. 2000). Plants growing in saline soil experienced an increase of osmotic pressure, resulting in decreasing water potential of soil and accumulation of salt ions, resulting in reduced nutrients uptake (Grattan and Grieve 1999) and low soil microbial activity (Yan et al. 2015). Salinity might happen due to irregular irrigation, inadequate drainage, and an incorrect fertilization program (George et al. 1997 and Wang et al. 2003). An excessive usage of fertilizers was one of the main causes of salinity increase in soil, severely damaging soil's micro-floral structure and composition (Nahid and Gomah 1991). There were various approaches to ameliorate this problem, including developing salinity-resistant cultivars, leaching excess soluble salts from plant roots, harvesting aerial parts of salt-accumulating plants, and incorporating treatments into the soil (Esringu et al. 2011). The use of plant growth-promoting rhizobacteria had been found to facilitate plant growth by reducing plant pathogens, increasing the uptake of nutrients, or influencing plant phytohormone production (Grichko and Glick 2001). The addition of OM overcame the negative effect of salinity (Walker and Bernal 2005 and 2008). HS were found to convert components in soil to more beneficial forms and to make them more effective (similar to plant growth-promoting rhizobacteria), increasing soil productivity (Sharif et al. 2002).

With respect to the benefits of HS to crops, this could be attributed to their hormonal-like activity (resembling auxin, specifically) due to the presence of polyamines within the structure, although this activity occurred at a much lower rate than that of real plant hormones (Young and Chen 1997, Phuong and Tichy 1976, and O'Donnell 1973). In addition, HS might also serve as a storage for hormone-like substances, transferring them to the plant's cell membrane (Nardi et al. 2002). It had been hypothesized that this growth regulator acted directly on genetic components of the living cell, affecting its duplication, transpiration, and translation processes (Syltie 1985). HA enhanced endogenous phytohormones (indole acetic acid and gibberellins) in bean plant shoots, as well as photosynthesis total pigments. The contents of carbohydrate, protein, and some nutrients in shoots and roots were also enhanced with the application of HA (Abbas 2013). HS promoted visible maize root elongation and proliferation, doubling its capability to take up nitrate (Vujinovic et al. 2020).

HS improved plant growth by enhancing nutrient supply, which occurred best when the nutrients were limited. They contained both hydrophilic components

(COOH and phenol-OH groups) as well as alkyl and aromatic functional groups. HS had an amphiphilic character, behaving as natural surfactants with an ability to adsorp on different surfaces including biological membranes (Samson and Visser 1989 and Vigneault et al. 2000). Varanini et al. (1993) stated that the presence of HS enhanced nutrients uptake due to their interaction with the phosphohydrolytic activity of PM H+-ATPase enzyme. HS changed the structure and fluidity of the membrane, increasing its permeability, thus increasing uptake of nutrients by living cells (Visser 1982 and Samson and Visser 1989). HS influenced nutrients uptake by helping to solubilize some insoluble nutrients and/or forming complexes reducing their losses during application (Govindasmy and Chandrasekaran 1992). It was hypothesized that humic products supplied S to plants and helped the uptake of nutrients (Akinremi et al. 2000).

It was hypothesized that HS improved plant polyelectrolyte nature and chelating properties (Patti et al. 2013). HS also contained C=O group with an ability to support electron transport, increasing photosynthesis efficiency (Rea and Pierandrei 1994). HS, due to their C double bonds (C=C) within the aromatic groups, had a scavenging activity by reacting with phenol-OH and ozone (O_3) radicals. HS showed a high antioxidant activity by dis-mutating O superoxide radicals to di-oxygens. They therefore played protective functions in ecosystems by interrupting radical reactions and preventing damages of living cell membranes and biological macromolecules (Lipski et al. 1999, Westerhoff et al. 1999, and Danilov and Ekelund 2001).

Synthetic chelators would tend to enhance micronutrients' availability by increasing solubility, however their excess might result in micronutrient deficiencies by lowering metal activities (Norvell 1991). HS, on the other hand, had lower stability constants for six different metals (Takahashi et al. 1997, Hering and Morel 1988, and Mackowiak et al. 2001), thus providing greater metal activities in the solution. They enhanced bioavailability of some micronutrients (Mackowiak et al. 2001) and the uptake of some macro- and micronutrients due to the metal complex formation (Rauthan and Schnitzer 1981). Humic metal complexes had also been found to reduce P fixation in soil, increasing P uptake and plant growth (Urrutia et al. 2013a and 2013b). HS made nutrients in soil more available to plants, thus benefiting their growth (Merwad 2017). HS, due to their negatively charged compounds, reduced the loss of NH_3 on urea applied to sandy loam soil (Reeza et al. 2009). A combination of Fe and humate was able to supply Fe to plants, transporting it from roots to plant tops. The process was slow but continuous, providing a long-term benefit that the Fe components were available for different growth stages (Cieschi et al. 2019). Fe humic complexes provided a readily available Fe form in rhizosphere and directly affected the plant physiology (Zanin et al. 2019). Humic improved the efficiency of K fertilizer. At reduced fertilizer rates with the incorporation of humic, the same or even enhanced productivity of plants was observed compared to the control (Khan et al. 2019a). Humic enhanced assimilation, evapotranspiration, gas exchange, chlorophyll contents, and some macronutrients uptake in plants. Product application methods were critical (i.e. fertigation vs. foliar), and the higher rates were not always better (Ameri and Tehranifar 2012).

HS accelerated seed germination and enhanced rhizome growth (Nardi et al. 2002 and Zancani et al. 2009 and 2011). HA and FA improved dry matter weights of shoots, roots, and nodules of soybean, peanut, and clover. Higher rates did not always give the best responses (Tan and Tantiwiramanond 1983). HA improved NO_3 reductase activity and net assimilation rates in leaves, as well as total N (stover and grain) uptake of corn (Khan et al. 2019b). HA enhanced the yield of tea and its quality, including the contents of polyphenols as well as some theine and soluble sugar (Huang et al. 2016). HA and FA at their optimum rates resulted in the enumeration of physiological groups of soil microorganisms. In certain organisms, a change in their metabolism happened, allowing them to grow on substrates which were previously un-utilizable. At higher rates, however, reverse results were observed (Visser 1985b).

With respect to its application in livestock, Alberta low-rank coal (humalite), with the exception of Fe and silica (Si), was found to contain very few nutrients (as defined by Church 1991). Based on one sample as shown in Table 4.3, this material contained only 6.4% crude protein and 0.7% crude fat. Carbohydrates and crude fiber were high at 17.2% and 52.9%, respectively. Ash (at 22.8%) represented inorganic components, most likely sand and clay, as well as trace elements. Humalite contained significant amounts of OM (77.2%), indicating a high content of HS. All analyses were completed according to Method 990.03; NITR; NT (2006), Method 954.02; LIPD; FA (2006), Method 978.10; CHO; FIB (2006), and Method 942.05; MIN; ASH (2006), except carbohydrate (by difference). The sample was also analyzed for major minerals, including Ca, Cl, Mg, P, K, Na, and S; as well as trace minerals including chromium (Cr), cobalt (Co), Cu, fluoride (F), Fe, iodide (I), Mn, Mo, nickel (Ni), selenium (Se), Si, and Zn. All analyses were completed according to ASTM D2013/D2013M-18 (2013), ASTM D6357-11 (2011), and ASTM D3176-15 (2015). The relatively high amounts of Fe and Si came from the nature of the material. Si indicated the presence of ash (sand and clay). Based on this nutritional analysis, humalite and humic products would be classed as feed supplements, not feeds. When applied to livestock, they would best be applied together with feeds so that their benefits could be clearly observed.

Humate promoted the electrolyte balance and immune potency of animals (Parks et al. 1996). Lignite (brown coal) had been used to treat diarrhea, dyspepsia, and acute intoxications, as they exerted a protective action on intestine mucous with antiphlogistic (anti-inflammatory), adsorptive, antitoxic, and antimicrobial properties (The European Agency for the Evaluation of Medicinal Products 1999).

Table 4.3 Livestock nutritional analysis for humalite.

Parameter	# of data	Content	Unit
Nutrients			
Crude protein	1	6.4	%
Crude fat	1	0.7	%
Carbohydrates	1	17.2	%
Crude fiber	1	52.9	%
Major minerals			
Ca	1	1.5	%
Cl	1	< 0.1	%
Mg	1	0.3	%
P	1	< 0.1	%
K	1	0.3	%
Na	1	0.3	%
S	1	1.1	%
Trace minerals			
Cr	1	12	ppm
Co	1	5	ppm
Cu	1	8	ppm
F	1	251	ppm
Fe	1	0.7	%
I	1	9	ppm
Mn	1	207	ppm
Mo	1	3	ppm
Ni	1	18	ppm
Se	1	2	ppm
Si	1	6.1	%
Zn	1	18	ppm
Others			
Ash	1	22.8	%
OM	1	77.2	%

Source: Canadian Humalite International Inc. (2020a).

The presence of phenol-OH, COOH, and C=O groups within FA provided various benefits, including antioxidant activity, stress management, and immune stimulation, as well as anti-inflammatory and antiviral properties (Yang et al. 1996 and Huck et al. 1991). FA showed a scavenging activity towards superoxide, hypochlorous acid, hydrogen peroxide (H_2O_2), and OH radical, peroxynitrite, and singlet O (Rodriguez et al. 2011). HA and FA absorbed different mutagens (Cozzi et al. 1993) and pathogens (Moura et al. 2007), reducing their toxicity. HA improved the performance and immune status of birds under low-nutrient density antibiotic free diets (Nagaraju et al. 2014). Humate had a high potential as an aflatoxin adsorbent to prevent its adsorption from the animal gastrointestinal tract (Ye et al. 2009).

Organic acids influenced gastrointestinal microbial ecosystem with respect to their consumption and growth on available nutrients (Freter et al. 1983). HA increased the growth of denitrifiers, pectin decomposers, aerobic and anaerobic cellulose decomposers, starch decomposers, and protein decomposers (Visser 1985a). Humic-containing products enhanced gut and small intestine morphometry of birds (Aristimunha et al. 2019). HS showed a gut-protecting and an immuno-stimulatory effect on broiler chickens (Mudronova et al. 2020). HA enhanced caecal amino acid levels, selected metabolite concentrations, and microbial populations of birds. At higher rates, however, the responses were reduced and in some cases were even negative (Shermer et al. 1998). FA increased lipid metabolism serum concentrations (low-density lipoprotein, leptin, growth hormone, insulin, and tri-iodothyronine) and serum metabolite (Ca, hemoglobin, and bilirubin) levels of pigs and cattle (Chang et al. 2014 and Chirase et al. 2000).

Feed disappearances had been observed with the addition of humate to rumen microorganisms indicating their enhanced activity (Hufstedler et al. 1993 and 1994). The presence of some nutrients in the intestinal tissue of living animals increased significantly over the control with the addition of HA to the feed (Visser 1973). HA and FA at their optimum rates stimulated respiration in rat liver mitochondria and increased the efficiency of the process of in-vitro oxidative phosphorylation (Visser 1987). HS improved the feed conversion ratio of birds (Rath et al. 2006) and enhanced body growth performance, serum biochemical indexes, and egg quality (Abdel-Mageed 2012). The use of peat as source of OM and HS resulted in better blood elements and serums, as well as the growth of pigs (Fuchs et al. 1995). HA was observed to reduce blood cholesterol, high-density lipoprotein, and white blood cells of finishing pigs (Kim et al. 2019). HA enhanced the number of rumen protozoa of lambs without negatively influencing blood parameters (Budag and Kara 2016). In the case of cow productivity, humic increased milk production at its optimum rate, while higher rates did not perform better (Tomassen and Faust 2000).

Included under livestock production would be the reduction of air pollutants, in which a special attention should be paid to their mitigation. Air pollutants under consideration were NH_3, hydrogen sulfide (H_2S), particulate matter, volatile matter, and odor (National Research Council of the National Academies 2003). The presence of these pollutants might result in health issues to animals, affecting their overall productivity (Robertson and Galbraith 1971 and Murphy and Cargill 2004), as well as to humans and the surrounding community (Zhang et al. 1998 and Hile 2015). Among other strategies, manure management and diet manipulation had been considered (Phillipe et al. 2011). To be specific, the use of lignite had been recommended since it contained a large percentage of labile C, which might suppress gaseous volatilization from manure (Chen et al. 2015), and that humate when fed to steers reduced rumen NH_3-N, lowering N loss to the environment (McMurphy et al. 2011). Terry et al. (2018a and 2018b) found that HS reduced ruminal NH_3 production in cattle, while other ruminal parameters were not changed. This might indicate the potential of humate in N metabolism and crude protein digestibility.

4.3 Disadvantages of Humic Substances

As summarized in Section 4.2, HS had unique physical and chemical/biochemical properties that influenced living cell metabolism and enzyme activity. HS interacted with both beneficial and harmful components in soil, affecting its quality. This would also apply in the case of livestock, where HS influenced the activities of the digestive system. These attributes would only be beneficial for soil, crops, and livestock if the amounts of HS applied were in moderation, or best at its optimum. Ineffective or even detrimental results had been observed for an excessive amount of HS.

As observed by Greene and Steelink (1962) and Morrison (1963), the degradation of HS might produce phenolic acids, including p-hydroxybenzoic, vanillic acids, and others. In excessive amounts, these compounds were found to have negative effects on seeds grown on germination paper (Krogmier and Bremner 1989).

HS bound certain enzymes through a cation-exchange mechanism. But at very high rates of HS, the formation of enzyme-humic precipitates was observed in reducing their activity (Ladd and Butler 1971). Ladd and Butler (1969) suggested that the COOH group within HS be the responsible component. Pospisil and Hrubcova (1974) as cited in Malcolm and Vaughan (1979) proposed that the molecular weight of HS was the one inhibiting the activity of certain enzymes. Haworth (1970) claimed that the whole aromatic core of HS was the responsible component. Dawes (1962) mentioned that some fractions within soil OM (at large amounts) inhibited the activity of certain enzymes due to metal chelation.

O'Donnell (1973), Cacco and Dell'Agnola (1984), and De Sanfilippo et al. (1990) observed that HS had the auxin-like properties of a growth hormone, promoting cell division and elongation for the growth of stems and roots. At very high rates, however, auxin might produce ethylene, resulting in an inhibitory action on root growth (Chadwick and Burg 1967), while Eliasson et al. (1989) found that auxin (by itself), in high amounts, resulted in an inhibitory action. The influence on plant growth from the synergic and antagonistic interactions between the two was summarized in great detail by Muday et al. (2012).

HS would best be used as soil amendments, not as a source of nutrients. While commercial humic products might contain some nutrients, applying too much of them with the intention of meeting nutrient requirements would result in an excessive amount of material being applied. While low-rank coal as the base material could be considered non-hazardous when used at reasonable rates, having too much of it might introduce unwanted components to soil, including ash and metals (Canadian Humalite International Inc. 2020a).

As soil amendments, HS would enhance the availability and uptake of nutrients by crops. Most fertilizers contain significant amounts of salts that, when combined with HS, would become more available to crops; meaning that more salts would also be taken up, resulting in detrimental effects to the crops. Some fertilizers might contain NH_3, which would be toxic to crops if taken up in excessive amounts (Follett et al. 1981). The same principle would also apply in the case of feed supplements (Church 1991). Excessive amounts of humic products might not be beneficial and could result in negative responses.

In addition, while this study was not directly related to low-rank coal and considering that the amount was excessive, this material might introduce toxic COOH and naphthenic acids (originated from the material itself or through microbial degradation) to soil and natural water (Bataineh et al. 2006).

4.4 Efficacy Results for Soil and Crops

A raw humic product with 4.0 pH and containing 85% HA was evaluated on Black Chernozem/Oxbow Association soil of silty clay loam texture. The soil was contaminated with brine spill back in the 1960s. The soil analysis found pH of 7.8, electrical conductivity (EC) of 14.81 mS/cm, soluble Na of 18.72 meq/100 g, and exchangeable Na of 6.16 meq/100 g. There was no vegetative growth in the area. Paiute weed and inland salt grass seedlings were then planted, and the soil was treated with the humic product as well as gypsum. The rates used were 0 (control), 0.2 t/ha (for the humic), and 50 t/ha (for the gypsum). Each treatment was repeated three times and replicated two times in the time. After 90 days, it was found that the gypsum treatment reduced soluble and exchangeable Na by 10.5%

and 25.8% compared to those of the control, respectively; while the humic treatment reduced soluble and exchangeable Na by 5.0 and 10.9%, respectively. These results were achieved through vegetative absorption and soil interaction. This study concluded that gypsum performed better, but at a much higher application rate. Humic achieved about 50% of the gypsum's performance with 250 times less product being applied (Gerrard 2006).

A raw humic product (pH = 4.1 and HA = 83%) and a processed humic product (pH = 10.1 and HA = 11.9%) were evaluated on a compacted field containing canola crop residues (chem-fallow) from the previous growing season. For the control area, no products were applied. For treated area 1, the raw product was broadcasted at a rate of 560 kg/ha. For treated area 2, the processed product was broadcasted at a rate of 37 l/ha. A cocktail cover crop mix containing proso millet, field peas, faba bean, oats, triticale, and tillage radish was subsequently seeded in August 2014 over the control, treated area 1, and treated area 2. Multiple soil readings using a penetrometer at 1380 and 2070 kPa were made in May 2015 (during dry weather) and August 2015 (during wet weather). Deeper readings indicated a reduced soil compaction. Humic products were found to reduce soil compaction or at least to maintain it. The best result for the dry product (12.5% improvement over the control) was observed during the dry weather at 1380 kPa. The best result for the processed product (42.9% improvement over the control) was observed during the wet weather at 2070 kPa (Schmidt 2015).

A raw humic product (5.17 pH, 70.0% HA, and 4.80 mS/cm EC) and a processed humic product (10.66 pH, 15% HA, and 28.8 mS/cm EC) were evaluated on sandy clay soil. The soil had 8.1 pH, 0.25 mS/cm EC, 0.15% OM, 55.0 cmol/kg CEC, 3.2% P, 0.25% $CaCO_3$, 0.01% N, and small amounts of other nutrients. The soil was spiked with sodium chloride (NaCl) to reach up to 60 mM salinity. The mixture was homogenized and transferred to pots and incubated for 30 days. N, P, K, and Zn nutrients were added at rates of 100, 80, 100, and 0.5 mg/kg, respectively. Hagein, Fardy10 cultivar of corn was grown and the soil was irrigated with deionized water throughout the study. For the raw humic application, the product was applied once to the soil at rates of 0.0 (control) up to 0.4% by weight. For the processed humic application, the product was applied twice to seedlings, at rates of 0.0 (control) up to 0.2% by weight. Each product application was replicated three times. The plants were harvested after a two month vegetation period. This study found that high salinity negatively affected plant growths and dry weights, while the humic products improved them. At 60 mM salinity, 0.4% raw humic product improved plant dry weights and uptake of N, P, K, Ca, Mg, Na, Fe, Cu, Mn, and Zn by 3.6, 2.6, 2.3, 2.6, 2.3, 2.4, 2.0, 5.5, 2.3, 4.7, and 3.4 times over the control, respectively. At 60 mM salinity, 0.1% processed humic product enhanced plant dry weights and N, P, K, Ca, Mg, Na, Fe, Cu, Mn, and Zn uptake by 1.7, 1.4, 1.4, 1.7, 1.4, 1.6, 1.4, 2.2, 1.0, 1.3, and 1.4 times over the control, respectively. Other results were not better than the control (Khaled and Fawy 2011).

A processed HA product (OM = 50.7 g/kg, available N = 903.11 mg/kg, available P = 500.21 mg/kg, available K = 2002.36 mg/kg, and pH = 6.8) was evaluated on moderately saline soil (clay:silt:sand = 45:18:37, OM = 8.54 g/kg, available N = 54.33 mg/kg, available P = 43.12 mg/kg, available K = 102.45 mg/kg, pH = 8.6, and total salt = 11.02 g/kg). The study area was located in an extremely warm desert climate with a −28.4 to 40.0 °C temperature range, 133.7 to 146.3 kcal/cm^2 of mean annual solar radiation, 2739.2 to 2991.8 hours of average annual sunshine duration, and 1876.6 to 2558.9 mm of average annual precipitation. Russian box thorn seedlings were raised for seven months for salt tolerance training, then transplanted to the soil. Thirty plants were grown in each test plot within an area of 10.8 m^2, then this was replicated 15 times for a total area of 162 m^2. The HA product was applied at rates of 0 (control) and 19.4 t/ha (treated) and the plants were harvested in five months. The humic product increased leaf, stem and branch, root, and above-ground tissue dry weights by 3.1, 5.1, 1.4, and 5.4 times over the control, respectively. This product increased the total Na plant absorption and soil OM by 2.6 and 1.1 times over the control, respectively (Feng et al. 2020).

A raw humic product (containing 78% HA and 10.0 pH) combined with 5% Ca was evaluated on soil with a very high Na content. Both laboratory and field projects were completed, in which the product was compared to gypsum (containing 20% Ca and 16% S) and Epsom salt (containing 10% Mg and 13% S); typical soil amendments used to treat high salinity soil. Parameters to be evaluated included sodium absorption ratio (SAR) and EC. The hypothesis was that due to the negative charge and large molecular weight of HS, Na ions would be bound into rather water-insoluble components, not affecting the soil and both SAR and EC calculations. Ca and Mg also reduced SAR, but they would add water-soluble Ca, Mg, and SO_4 to soil. All of these compounds would contribute to EC. It was found in the laboratory that the humic product at 1.0% rate by weight significantly reduced SAR from 213 to 60 (71.8% lower), while gypsum and Epsom salt at the same rate lowered it to 43 (79.8% lower) and 92 (56.8% lower), respectively. However, as expected, the humic product slightly reduced EC from 10.1 to 9.3 (7.9% lower) mS/cm, while gypsum and Epsom salt elevated it to 13.9 (37.6% higher) and 12.1 (19.8% higher) mS/cm, respectively. Similar results were observed for soil with SAR of 92. At the same product rate of 1.0% by weight, the humic product improved it to 19 (79.7% lower), while gypsum and Epsom salt to 19 (79.7% lower) and 14 (84.6% lower), respectively. EC was slightly reduced from 6.6 to 5.9 (10.6% lower) mS/cm with the addition of the humic product, while it was elevated to 14.6 (2.2 times higher) and 16.1 (2.4 times higher) mS/cm with the addition of gypsum and Epsom salt, respectively. This study found that the humic product, with much less Ca content than that of gypsum, performed better due to the formation of rather water-insoluble components within the soil. Subsequently, three soil replications prepared in the laboratory (mean SAR = 93 and EC = 6.1 mS/cm) and four soil replications completed on the field (mean SAR = 149 and EC = 10.3 mS/cm) were treated with the same humic product at rates of 1.0 and 1.35% by weight, respectively. In the laboratory, SAR was significantly reduced to 22 (76.3% lower) while EC slightly increased to 6.3 mS/cm (3.3% higher). On the field, SAR was significantly reduced to 84 (43.6% lower) while EC increased to 12.6 mS/cm (22.3% higher). These findings showed the importance of a complete blending between the product and soil, which was frequently challenging when working in the field (Orcutt et al. 2009, McDonald et al. 2010, Orcutt et al. 2010, and Liem 2016).

A processed HA product containing 15% total organic C, 1.6 mmol/g of acid functional groups, 4.8% K, 0.5% Fe, 0.15% N, 0.15% Si, 0.10% S, 0.09% Na, 0.07% Cl, 0.05% Ca, and 0.04% P was evaluated on tomato (cv. Field Busm) grown inside a greenhouse in pots filled with soil or sand at 21 to 28 °C under natural daylight and night cycles. The medium was treated with a nutrient stock (Hoagland's) solution plus 48 mg K/kg. In addition to the control, the humic product was either foliar applied (at a rate of 0.1 ml painted to shoots) or applied to the medium (at rates equivalent to 8 and 80 l/ha). The experiment was arranged in a split plot design with four replications. The foliar application improved fresh biomass weights by 3.8 times over the control for the sand medium and 2.7 times for the soil medium. The soil application performed similarly, i.e. 2.8 times over the control for the sand medium, and 2.4 times for the soil medium. The uptake of nutrients was also enhanced in the case of N: 2.1 to 2.4 times over the control for both foliar and soil applications. A processed FA product containing 5% FA, 2.5% total organic C, and 0.5% K was later evaluated the same way and at the same application rates. The foliar application enhanced fresh biomass weights by 3.8 times over the control for the sand medium and 2.3 times for the soil medium. The soil application performed similarly, i.e. 2.8 times over the control for the sand medium, and 2.6 times for the soil medium. Uptake of nutrients was also enhanced in the case of N: 1.3 to 1.7 times over the control for both foliar and soil applications. Results from both products indicated that foliar application was overall the better option, while soil application at the higher rate might not be economical. Better results were observed for the sand culture, which was the medium with less nutrients (Patti et al. 2013).

The same products were evaluated on Sudan grass, grown in saline loamy soil with 7.89 to 8.02 pH, EC of 3.01 to 12.33 mS/cm, $CaCO_3$ of 55.32 to 62.71 g/kg, OM of 6.42 to 8.63 g/kg, and NPK of less than 0.1 g/kg. The HA product was applied to the soil at a rate equivalent to 5 kg/ha, together with N, P, K, and S at 150, 15, 40, and 188 mg/kg soil, respectively. The FA product was applied to the soil at the same rate. The grass was planted in soil inside a greenhouse irrigated with water, then cut after 60, 120, and 180 days of growth. The HA product increased cumulative fresh and dry weights by 9.3 and 9.0% over the control, respectively. Cumulative plant heights, total chlorophyll, and proline were all increased by 22.1, 17.5, and 30.9%, respectively. Uptake of N, P, and K was up by 19.2, 40.3, and 23.3%, respectively. The FA product increased cumulative fresh and dry weights by 7.7 and 7.7% over the control, respectively. Cumulative plant heights, total chlorophyll, and proline were all increased by 17.0, 11.1, and 23.3%, respectively. Uptake of N, P, and K each increased by 35.3, 64.6, and 41.7%, respectively. These results showed that the HA product performed better in terms of plant weight and height, and the contents of total chlorophyll and proline. The FA product was better with respect to the uptake of N, P, and K (Merwad 2017).

A processed humic product containing 12% HA and 3% FA, as well as a raw humic product containing 68% HA, were evaluated by Verlinden et al. (2009) on different types of soil and crops. The first two experiments were completed for grass grown on loamy sand fields, containing 0.8 and 1.1% OM, with 5.2 and 5.4 pH in 24 plots of 1080 m^2 in total. The third experiment was completed on grass grown in a sandy field (OM = 1.8% and pH = 4.9) in 8 plots of 816 m^2 in total. Another two experiments were completed on corn grown in loam and sandy loam fields (OM = 0.9% and pH = 6.2 for each) in 24 plots of 1440 m^2 in total. The sixth experiment was completed on potato grown in a sandy loam field (OM = 0.6% and pH = 4.9) in 20 plots of 1128 m^2 in total. The last two experiments were completed in 24 l pots of 1 l each within a growth chamber on grass and spinach. The soil had 0.9% OM and 5.2 pH. Each soil differed somewhat in nutrient and salt contents: P = 13 to 49, K = 8 to 25, Mg = 6 to 20, Ca = 59 to 123, and Na = 1.0 to 2.4 (all units in mg/100 g dry soil). Fertilizers were applied to each case, according to the recommended application, which represented the control in these experiments. In addition to fertilizers, the processed humic product was applied at a rate of 50 l/ha for each experiment. In another set of experiments, the raw humic product was applied by weight at rates of 0.5% for the sixth experiment, 1.5% for the second, fourth, seventh, and eight experiments, and 3.5% for the first, third, and fifth experiments. The best result was observed for potato (the sixth experiment), in which 13 and 17% improvements on dry matter yields over the control were achieved with the use of the processed and raw humic products, respectively. Spinach grown in pots (the eight experiment using the raw humic product), grass grown in sand (the third experiment using the raw humic product), and grass grown in pots (the seventh experiment using the processed humic product) showed 13, 10, and 6% dry matter yield improvements, respectively. The other experiments resulted in less than 5% improvements over the control.

Four potato cultivars of Van Gogh, Milva, Lady Olympia, and Agata were grown in loamy soil of 8.2 pH containing 0.21% N, 18.5 mg/kg P, 166 mg/kg K, and 1.3% OM. A processed HA product of 6.8 pH containing 54.5% OM, 50.5% combined

HA + FA, 1.79% N, 24 mg/kg P, 135 mg/kg K, and below threshold values of heavy metals was broadcasted at rates from 0 (control) up to 600 kg/ha during planting. NP of 200 and 100 kg/ha were then applied to the soil. The experiment was conducted in a research farm within a period of two years. The plants were grown in May and harvested in September each year. The average daily temperature for the same period was 19.7 and 18.6 °C for the first and second years of experiment, and the total precipitation was 127 and 178 mm, respectively. The experiment was completed in a factorial design with two factors in randomized complete block design with three replications. For all four cultivars (all results were averaged), HA rate of 400 kg/ha was found to be optimum, in which plant heights, tuber numbers per plant, marketable tuber yields, total tuber yields, protein contents, vitamin C contents, and specific gravity were all increased by 4.9, 20.8, 31.4, 13.7, 7.2, 14.1, and 0.3% over control, respectively (Sanli et al. 2013).

Potato of the Caspar variety was sown on soil with a fall plow and spring cultivate system. The soil contained 47% clay/37% silt/16% sand, 7.48 pH, 1.16 mS/cm EC, 6.8% $CaCO_3$, 1.59% OM, and nutrients as follows (all in mg/kg): 150 N, 7.95 P, 196 K, 3.30 Mn, 1.44 Zn, 5.85 Fe, and 0.59 Cu. Additional nutrients were added to the soil at rates from 0 (control) up to 200N-100P-100K kg/ha as well as a raw humic product (containing 85% HA, 75% total organic C, 3.5 to 5.5 pH, 3.0% Ca, 0.5% Fe, 0.5% Mg, and other trace nutrients) at rates from 0 (control) up to 600 kg/ha. The experiment was conducted from May until October in a randomized complete block design with three replications per treatment. The experiment was repeated the following year. Mean air temperatures, precipitations, and relative humidity were 17.7 and 16.6 °C, 165.6 and 140.6 mm, and 49.1 and 52.1%, respectively, for both years. HA (with 0 NPK) enhanced tuber weights, tuber total yields, dry matters, starch contents, and protein contents by 40.9, 75.6, 11.8, 15.8, and 31.1% over the control (0 NPK), respectively, at 600 kg/ha humic rate. HA (with 0 NPK) enhanced tuber weights, tuber total yields, dry matters, starch contents, and protein contents by 6.9, 19.7, 9.2, 11.9, and 10.2% over the control (100N-50P-50K kg/ha), respectively, at 600 kg/ha humic rate. Only when 200N-100P-100K kg/ha were applied to the soil, HA (with 0 NPK) did not perform better than the control. HA (with 0 NPK) enhanced P, K, Ca, Mg, Fe, and Mn contents within potato tubers both at 0 NPK and 100N-50P-50K kg/ha, best for Fe at 41.4% over the control at 600 kg/ha humic rate. When 200N-100P-100K kg/ha nutrients were applied to the soil, HA (with 0 NPK) did not perform better than the control, except for K, in which the response was maintained. This study showed that when adequate nutrients were present, the benefit of HA was minimum (Ekin 2019).

A processed liquid product (HA = 6%) was evaluated on potato (cv. Russet Burbank) grown in a semi-arid region with an annual rainfall of 152.4 to 203.2 mm. The soil was calcareous, of 8.0 to 8.2 pH, containing 5 to 7% free lime, and 0.9 to 1.0% OM, and was watered by means of irrigation. Potato seeds were planted on agricultural field plots and the HA product was applied side-dressed at rates from 0 (control) up to 150 l/ha. The harvest was made five months later. Four replications were completed for each treatment. It was found that yields of potato were enhanced over the control up to product rates of 37 l/ha (up to 14.7% higher), maintained at 75 l/ha, then dropped to 3.2% at its highest rate. The same product was applied for Norkotah potato grown in sandy loam of 7.9 pH with 1.4% OM. Potato seeds were grown in agricultural field plots. The HA product rates were from 0 (control) up to 112 l/ha, with foliar application on seedlings. Four replications were completed for each treatment. During harvest, it was found that potato yields were improved by 71.7% over the control at a product rate of 37 l/ha. At higher rates, although they were still better than the control, yield improvements were reduced (to as low as 15.8% over the control). Both studies showed a non-linear relationship of HA rates on potato yields (Seyedbagheri 2010 and 2015).

A processed humic product of 7.63 pH, containing 72.0% OM, 1.98% N, 0.36% P, 3.40% K, 395 mg/kg Fe, 249 mg/kg Mn, and 32.18 mg/kg Zn was evaluated on fodder beet grown in two similar fields, i.e. clay and clay loam soil containing 0.58 and 0.55% OM, 41.08 and 31.38 cmol/kg CEC, 12.49 and 10.66 mS/cm EC, with 8.14 and 8.10 pH, respectively. Diphosphorus pentoxide (P_2O_5) at 77 kg/ha was added before planting; N (247 kg/ha) was added three times at 30, 55, and 75 days after planting; and dipotassium oxide (K_2O) at 185 kg/ha was added two times at 30 and 55 days after planting. The humic product was foliar applied at rates from 0 (control) up to 5.9 l/ha, three times at 30, 55, and 75 days from sowing. Both fields were irrigated using canal and drain water with 7.98 and 8.03 pH, 1.75 and 3.31 mS/cm EC, and 4.35 and 5.18 SAR, respectively. The study was completed in a split plot design with three replications. The plants were harvested seven months after sowing. Soil OM levels were found to increase with the addition of humic by 32.7% over the control, while EC levels decreased to 40.4%, all at the highest humic rate. Humic also improved total stable aggregates and soil moisture contents by 34.6 and 63.5% over the control, respectively, all at the highest humic rate. Most importantly, humic enhanced root and top yields by 51.5 and 51.0%, respectively, all at the highest humic rates. Overall, the best performance was observed for soil with higher salinity (Ahmed and Ismail 2016).

A greenhouse study was completed for clayey Oxisol and medium texture Oxisol types of soil. The former had 4.8 pH and contained 4.6% OM, 56.5% clay, 1.3 mg/dm^3 P, and 69 mg/dm^3 K. It also contained 0.8, 0.4, 1.0, and 11.2 Ca, Mg, Al, and CEC, respectively (all in cmol/dm^3). The latter had 5.1 pH and contained 0.7% OM, 68.0% sand, 0.3 mg/dm^3 P, and 42.7 mg/dm^3 K. It also contained 0.1, 0.1, 0.1, and 2.0 Ca, Mg, Al, and CEC, respectively (all in cmol/dm^3). The soil was incubated for 15 days with super phosphate and HA. P fertilizer was added to the soil at a rate of 400 mg/kg. A commercial Sigma-Aldrich humic product (containing 76.3% HA + FA, see Sections 3.2.2 and 3.2.3) was used as the source of HA. It had

9.8 pH and 376 g/kg C, as well as the following nutrients: 7.6, 0.3, 6.3, 4.4, 5.9, 0.3, and 1.1 of N, P, K, S, Ca, Mg, and Fe, respectively (all in g/kg), as well as traces of Cu, Mn, and Zn. This product was applied at rates from 0 (control) up to 100 mg C/kg. Soybean (cv. CD 250) seeds were sown in pots, each containing clayey Oxisol soil and medium texture Oxisol soil, respectively, maintained near its 100% water holding capacity. Additional nutrients were added: 100 mg/kg N, 100 mg/kg K, 30 mg/kg S, 1 mg/kg B, 2 mg/kg Zn, 6 mg/kg Mn, and 1.5 mg/kg Cu. A randomized block experimental design was used with four replications. Ten days after sowing, thinning was completed and top dress fertilizers were added (including 100N and 100K, all in mg/kg). The cultivation was completed after 38 days. Positive results were observed mainly for clayey Oxisol soil: up to 10.0% increase in total dry matters (achieved at 100 mg C/kg HA rate), up to 13.5% increase in plant P uptake (achieved at 50 mg C/kg HA rate), and up to 13.6% increase in plant N uptake over the control (achieved at 50 mg C/kg HA rate). In the case of medium texture Oxisol soil, the only improvement found was an increase of up to 45.5% in plant N uptake over the control (observed at 50 mg C/kg HA rate). This study also showed that high HA rates did not necessarily result in better responses (Rosa et al. 2019).

Barley (cv. Radegast) grains were germinated in Petri dishes on wet filter paper for 72 hours. The seedlings were placed in 1/10 strength of Hoagland solution. After seven days, they were exposed to 100 mM NaCl with the application of Sigma-Aldrich humic product. It was applied in the solution at rates from 0 (control) up to 6.0 mg/l. In another treatment, the same strength was foliar applied daily at a rate of 5 ml per shoot. The plants were harvested seven days after treatment. There were ten replications for the length measurement and six replications for other parameters. Compared to those of the control, humic treatment increased plant (root + shoot) lengths by 23.7 and 34.3%, when the product was added to the solution and foliar, respectively. Soluble proteins in the plant also increased by 8.2% (with the product applied to the solution) and 10.1% (with the product foliar applied). Total plant nutrients (K, Ca, Mg, Fe, and Zn) were enhanced by 10.4 and 2.2% over the control when the product was applied to the solution and foliar, respectively. Free amino acids, phenolic acids, and organic acids were up by 15.8, 46.4, and 145.1% (with the product applied to the solution) and by 22.4, 39.5, and 165.8% (with the product foliar applied), respectively. This study found that foliar was in general the better choice for product application (Jarosova et al. 2016).

Barley seeds were germinated in a hydroponic solution of 5.3 pH containing 2.12 mM NH_4NO_3, 0.77 mM K_2SO_4/KCl, 1.25 mM calcium chloride ($CaCl_2$), 0.82 mM $MgSO_4$, and other trace nutrients. Five days after germination, the seedlings were moved to pots filled with the culture solution and cultivated for a further ten days. A processed HA product (molecular weight = 8163, 7.62 mmol/g C of COOH group, 8.56 mmol/g C of phenol-OH group, 46.6% C, 4.37% H, 2.98% N, 39.2% O, 0.16% S, and 6.66% ash) was added to a solution at rates from 0 (control) up to 25 mg C/l. Fifteen days after cultivation, lengths and fresh weights of barley roots were measured. The best improvements were 26.0 and 84.6% over the control, respectively; each was achieved at the highest product rate. It was also found that HA contained an auxin-like structure resulting in higher (or at least maintaining) antioxidant enzymatic activities, best observed for thermostability of catalase and ascorbate peroxidase at more than 40% over the control (Nagasawa et al. 2016).

A raw HA product of 3.5 pH, containing 86% OM, 85% polymeric polyhydroxy acid, 1% N, 0.9% K, 0.57% Mg, 2.3% S, 0.88% Fe, and 0.02% Mn was evaluated on 15-day pepper (cv. Demre) seedlings planted in pots inside a growth chamber (22 °C with a day/night lighting system). The soil contained 7.64 pH, 1.12 mS/cm EC, 0.035% total salt, 3.05% total N, 18.5 mg/kg P, and 0.71 cmol/kg exchangeable K. It was spiked with 8 mM/kg NaCl resulting in 2.12 mS/cm EC. The soil was irrigated with distilled water and treated with P at rates from 0 (control) up to 150 mg/kg, and humic at rates from 0 (control) up to 1500 mg/kg. The experiment was conducted for ten weeks and found that plant total (shoot + root) dry weights at 0 P were enhanced by 18.5% over the control with the addition of HA at its highest rate. At the medium P rate (50 mg/kg), the weights were enhanced by 24.5% with the addition of HA at its highest rate. At the highest P rate, the addition of HA both at its medium (750 mg/kg) and highest (1500 mg/kg) rates, while improving plant dry weights by up to 27.9% over the control, reduced the actual weights to 12.2% lower than that of its best. These results showed that higher rates of combined nutrients and HA might result in negative responses (Cimrin et al. 2010).

Seeds of three hybrid red sweet pepper cultivars (Barbero, Ferrari, and Imperio) were germinated in six weeks, then transplanted to sandy soil 43 days after sowing, all inside a greenhouse at 75% relative humidity and 26/19 °C (day/night). The soil had 7:4:89 of clay:silt:sand, 8.18 pH, and 1.96 mS/cm EC. It contained 12.71, 3.93, 2.17, 0.96, 6.86, 4.16, and 8.7 meq/l Ca, Mg, Na, K, Cl, bicarbonate (HCO_3), and SO_4, respectively. The soil was drip irrigated for 173 days, in which the following nutrients were applied: 245 N, 122 P, and 664 K kg/ha. A processed humic product (containing 76% HA, 4% FA, and 12% K_2O) was applied foliar at rates from 0 (control) up to 1.5 g/l on leaves at 20, 40, 60 days after transplanting. Harvest started 85 days after transplanting and the crop was harvested 12 times. Plant fresh and dry weights enhanced best at the highest humic rate by 4.9 and 20.6% over the control, respectively; both happened for Barbero cultivar. The same enhancements were observed for total yields, fruit fresh weights, fruit dry weights, vitamin C contents, % total soluble solids, and % total sugar (best at 15.7, 8.5, 13.2, 10.6, 17.8, and 24.8% over the control, respectively) at a humic rate of 1.5 g/l. The other two cultivars had lower responses, but all were still better than their control (Ibrahim et al. 2019).

A processed HA product with less than 7 pH, 56.3% C, 4.7% H, 36.3% O, 2.7% N, 2430 meq/kg COOH, and 3010 meq/kg phenol group was evaluated on sugarcane grown in sandy loam soil of 8.0 pH and 0.61 mS/cm EC. The size for each plot was 40 m^2. NPK nutrients were applied to the soil at rates of 27.5, 2.75, and 9.34 g/m^2. The HA product was then applied to the soil at rates from 0 (control) up to 6 g/m^2 at planting and on the 35th day after planting. The study was completed in a randomized block design, in which each treatment was replicated four times. The plants were harvested 330 days after planting. It was found that the single cane weights, cane yields, Brix Index, yields of sugar, and total nutrient (N, P, K, Ca, Mg, Fe, Mn, Zn, Cu) contents of sugarcane were all enhanced by 17.0, 9.3, 6.4, 23.8, and 23.1% over the control, respectively, with the application of 6 g/m^2 humic product on the 35th day after planting (Govindasmy and Chandrasekaran 1992).

Bean plants were grown in pots inside a greenhouse in a loamy soil with 7.20 pH, 0.85 mS/cm conductivity, 0.62% $CaCO_3$, 385.2 mmol/kg P_2O_5, and 455.3 mmol/kg K_2O. The soil was spiked with Na, K, Ca, and Mg based salts from 0 (control) up to 120 mM and a processed humic product at rates from 0 (control) up to 0.01% by weight. The product contained 75% (HA + FA) and 20% K_2O. After 45 days of soil acclimatization, seeds were sown, then 100, 30, and 130 mg/kg/week N, P, and K, respectively, were added starting 10 days after sowing until harvest 60 weeks after sowing. The growing conditions were 25 to 30/10 °C day/night temperature at 30 to 40% relative humidity. It was found that the salts reduced all plant growing parameters, while HA enhanced all of them. Compared to those of the control at the highest salt application, the highest HA rate alleviated soil EC by up to 9.1%, enhanced shoot dry matters by up to 55.4%, enhanced root dry matters by up to 111.0%, reduced proline contents (a stress indicator) by up to 26.1%, reduced electrolyte leakages by up to 14.3%, enhanced chlorophyll contents by up to 29.7%, increased plant NO_3 by up to 20.5%, and increased N and P in roots and shoots by up to 14.0% (Aydin et al. 2012).

Vertisol soil (pH = 7.24, organic C = 0.75%, EC = 0.83 mS/cm, $CaCO_3$ = 0.22%, N = 0.08%, available P = 7.96 mg/kg, Na = 39.1 mg/kg, K = 175.5 mg/kg, Ca = 3860 mg/kg, Mg = 282.0 mg/kg, Fe = 5.56 mg/kg, Cu = 1.3 mg/kg, Zn = 0.2 mg/kg, and Mn = 10.4 mg/kg) was chosen as a growing medium for this study. The experiment was conducted inside a greenhouse in a completely randomized factorial design with the following treatments. NaCl salt was mixed into the soil at rates from 0 (control) up to 60 mM to increase its salinity. It was then treated with 100, 80, 100, and 0.5 mg/kg of N, P, K, and Zn, respectively. In the first experiment, a raw HA product (containing 65% HA, 4.87 pH, and 5.80 mS/cm EC) was applied at rates from 0 (control) up to 2 g/kg. Durum wheat of the Salihli cultivar was then grown in the soil. In the second experiment, no raw HA product was applied. Instead, after 20 and 35 days, the seedlings were sprayed with a processed HA product (HA = 12%, pH = 12.86, and EC = 32.8 mS/cm) at rates from 0 (control) up to 0.2% each time. For each experiment, the plants were harvested after a two-month vegetation period. In all cases, the presence of NaCl in the soil reduced the uptake of all nutrients and dry weights of plants. The raw HA product at its optimum rate (2 g/kg) enhanced all parameters. The best improvements were observed at the highest salt rate, mainly for Fe uptake (at 5.1 times higher than that of the control) and plant dry weights (at 3.4 times higher). The processed HA product at its optimum rate (0.1%) also enhanced the uptake of all nutrients. The best improvements were observed at the highest salt rate, mainly for Fe uptake (at 2.2 times higher than that of control) and for plant dry weights (at 1.5 times higher). Most interestingly, the optimum rates for both HA products were not necessarily the highest ones. Excessive rates of HA sometimes reduced its performance. Overall, this study showed that HA performed best when the growing condition was unfavorable (Asik et al. 2009).

A processed humic product containing 55.8% C, 1.8% H, 40.9% O, 1.2% N, and 0.3% S; with 9.4 meq/l total acidity, 8.45 meq/l COOH group, 0.95 meq/l phenol-OH group, 2.24 meq/l C=O group, and < 0.01 meq/l C-metoxyl group was evaluated on wheat of the Zheng Yin-1 variety. The plants were grown in pots inside a shade house filled with soil and irrigated to maintain 13 to 30% moisture contents. When the main stems were close to meiosis stage, the humic product was foliar applied at rates of 0 (control) and 0.05%. For a period of 13 days, the product reduced the stomatal conductance by up to 72.7% compared to that of the control. The experiment was repeated with the soil irrigated to maintain 18% water content, followed by 9 days of drought. The product was applied at rates of 0 (control), 0.01, and 0.05%. From days 5 to 9 after application, the product maintained the stomatal conductance at between 0.2 and 0.4 cm/second (0.05% product rate) and between 0.1 and 0.2 cm/second (0.01% product rate), compared to almost 0 and 0.2 cm/second in the case of the control. Nine days after the drought simulation, the relative water contents at 0.05% product rate were up by 8.8% over the control. Chlorophyll and root total P contents were enhanced by up to 35.7 and 40.8% over the control (at 0.05% product rate), respectively. Under the drought condition, yields were down by 30% compared to that of the non-drought condition. When the humic product was applied, yields were down only by 10.0 and 3.0% compared to those of the control, at rates of 0.01 and 0.05%, respectively. The number of grains per ear and percentage fertile spikelet increased by 29.4 and 11.8% over those of the drought control, respectively. The same plants were grown in pots as before and sprayed with the product at the early grain-filling stage at a rate of 0.05%. After two weeks, the plants were exposed to a wind of 3.8 m/second at 32 to 34 °C with a relative humidity of 30% for 1 day. The humic product was found to enhance relative water contents of the leaf to 4.5% over the control and increased the uptake of P by plants to 61.0%. A field experiment on agricultural plots was then completed at 27 different sites during a drought and hot-dry condition. The average rainfall was 11 mm and the pan evaporation ranged from 184 to 240 mm.

The humic product at rates of 0 (control) and 0.04% was sprayed at the boot and grain-filling stages. Each treatment was replicated three to four times for each site. The product performed best when sprayed at the boot stage, i.e. 9.4% higher than that of the control, compared to 7.2% when sprayed at the early grain-filling stage; both for yield levels of over 4500 kg/ha (Xudan 1986).

Soil with 8.2 pH, 10.4 meq/100 g CEC, a very low OM content (0.48%), and very low nutrient contents (0.033% N and 0.077% P) was used as a medium to evaluate the fractionation of P. The soil was mixed with 1.0% by weight MAP and processed HA product (containing 1.15 pH, 5.2 meq/g COOH, 1.8 meq/g OH, and 3.38 meq/g C=O) at rates of 0 (control) and 2.0% by weight. The soil was incubated for 15 days at room temperature in the laboratory and maintained at its field moisture capacity. It was observed that HA increased the water-soluble P component in soil two times of the control. The same product was then evaluated on the fractionation of P in the soil and on yields when wheat was planted in test pots inside a laboratory. The same soil was treated with 0.01% MAP and the humic product at rates of 0 (control) and 0.01% by weight. There were six replications completed for each treatment, including control. During harvest, it was found that HA increased the water-soluble P component by 27.8%, available P by 54.5%, P uptake by 98.2%, and yields by 53.3% over the control. Lastly, the same humic product was evaluated on yields of wheat grown in test fields using the same soil. Urea at a rate of 50 kg N/ha was broadcasted to the soil three times during the growing season, together with DAP at rates of 0 (control) and 150 g/21 m^2. The HA product was broadcasted at rates of 0 (control) and 150 g/21 m^2. There were three replications completed for each treatment, including control. After 100 days of growing, HA increased yields by 27.9% (with P) and 25.4% (no P) over the control (Wang et al. 1995).

A greenhouse study was completed, in which Hungarian vetch was grown in soil of mostly sand (54%) medium with 7.20 pH, 0.62 mS/cm EC, 16.20 cmol/kg CEC, 0.81% OM, 2.55 g/kg $CaCO_3$, 0.1 g/kg total N, 4.86 mg/kg P_2O_5, and exchangeable Ca, Mg, K, and Na at 9.52, 3.10, 2.65, and 0.25 cmol/kg, respectively. It also contained available Fe, Mn, Zn, Cu, and B at 1.36, 2.25, 1.17, 1.15, and 0.12 mg/kg, respectively. The soil was spiked with up to 60 mM NaCl (or equivalent to 8.02 mS/cm of EC). A processed HA product (containing 15% HA, 10% OM, and 1% K_2O) was applied to the soil at rates from 0 (control) up to 90 mg/kg by weight. The study was completed in a period of two years, using a completely randomized design with three replications per treatment, containing ten plants in each replication. Plant amino acids increased with the increase in salinity stress. HA alleviated the negative effects of salinity stress and decreased amino acid contents. For tryptophan and phenylalanine, the best improvements (2.2 and 1.9 times lower than those of the control, respectively) were observed at the highest salinity, when the HA product was applied at its highest rate. At the highest salinity, plant hormones (gibberellic and salicylic acids) were improved with HA (with the best improvements at 26.6 and 52.3% over the control, respectively) when the product was applied at 90 and 60 mg/kg by weight, respectively. Nutrient compositions in soil (including N, K, Ca, Mg, P, S, Fe, Cu, Mn, and Zn) were all lower with the increase of salinity. The application of HA helped increase these parameters. The best improvements were mainly observed at the highest salinity when the HA product was applied at its highest rate (i.e. 37.6 times higher than that of the control in the case of N). The application of HA product also improved soil properties. For OM, the best improvement (82.4% over the control) was observed at the highest salinity, when the HA product was applied at its highest rate (Esringu et al. 2016).

Canola was grown in 3 kg pots of sandy loam soil (with 6.9 pH, 7.6 mg/g C, 0.8 mg/g N, 0.1 mg/g S, and 9.8 cmol/kg CEC). A raw humic product (3.5 pH, 384 mg/g C, 12.7 mg/g N, 5.2 mg/g S, and 197.2 cmol/kg CEC) was applied to the soil at rates from 0 (control) up to 10 g/pot. Seeds of canola were sown in each pot placed inside a greenhouse, in a randomly randomized factorial design with three replications per treatment. Nutrients (N, P, K, S, Zn, Fe, Mn, Cu, B, and Mo) were added to the soil at 100, 50, 100, 20, 4, 4, 2, 1, 1, and 0.4 mg/kg soil. After six weeks, the plants were harvested, and it was observed that a low humic rate of 1 g/pot resulted the best dry matter yields, S uptake, N uptake, P uptake, and K uptake at 14.6, 8.7, 7.7, 9.6, and 14.2% over the control, respectively. The results became more significant when S was excluded from the treatment, i.e. 27.0, 47.8, 19.8, 37.8, and 55.9% over the control, respectively (achieved at the highest humic rate). A subsequent study was completed by applying higher rates of the humic product from 0 (control) up to 200 g/pot to the same crop using the same arrangement but without the addition of S, and only one half of other nutrients. Dry matter yields were improved by 26.2% at the highest humic rate. Uptake of S, N, P, and K was individually improved by 139.8, 60.8, 39.9, and 47.7% over the control, respectively (Akinremi et al. 2000).

Three-day old maize (var. Polaris) seedlings were incubated with Hoagland's solution for 48 hours. A processed humic product containing 55.26% C, 4.45% H, 7.53% N, 5.52% ash, 20.45% aliphatic C, 36.3% peptidic and carbohydratic C, 5.95% aromatic C, 7.32% phenolic C, and 29.98% C=O C was applied at rates of 0 (control) and 0.1 mg C/l. Results from 15 random plants (repeated five times) showed that the uptake of NO_3 was enhanced by 88.5% over the control when the root seedlings were pre-incubated for 16 hours with the humic product. The gibberellin-like activity of the product was evaluated on 15 seedling roots (repeated five times) after an incubation period of 48 hours. The HA product enhanced the invertase activity by 25.0% over the control. Most importantly, the fresh root weight was increased by 2.6 times compared to that of the control (Nardi et al. 2000).

Lee and Bartlett (1976) evaluated a processed HA product containing 1.0 mg C/ml, 135 ppm Ca, 52.1 ppm N, 11.4 ppm Mg, and trace nutrients on corn of the single cross variety 902 JX. Corn seeds were germinated in perlite, then transplanted after the first leaf appeared. The first experiment used a loam type of soil inside a greenhouse. The soil was treated with nutrients as follows: 5, 5, 2, 1, 2 mM of $Ca(NO_3)_2$, KCl, $MgSO_4$, MKP, and NH_4NO_3, respectively, as well as trace micronutrients including Fe, B, Mn, Zn, Cu, and Mo. Corn seedlings were transplanted in jars and treated with the humic product at rates from 0 (control) up to 50 ppm C, and grown at a temperature of 25 °C with 18 hours light per day with air bubbled into the medium. The nutrients were replaced every four days and the seedlings were harvested in 12 days. Each treatment was replicated three times. It was found that plant top and root yields were enhanced best at a humic product rate of 5 ppm C, i.e. 92.3 and 63.2% over the control, respectively. At higher rates, the responses were not better. In the second experiment, the seeds were transplanted in pots inside a greenhouse containing two types of soil of pH 6.53 and 5.25, with 5.05 and 0.65% OM each. The first soil (loamy fine sand) had more nutrients at 0.18%, 55.5 ppm, 160 ppm, 990 ppm, and 105 ppm of N, P, K, Ca, and Mg, respectively, compared to 0.02%, 8.0 ppm, 44 ppm, 8 ppm, and 3 ppm, respectively for the other (loamy sand). Each soil was fertilized with the following nutrients: 10, 10, and 2.5 μM/g of NH_4NO_3, MKP, and $MgSO_4$, respectively. In addition, the low OM soil was treated with 4 meq/100 g $CaCO_3$. The HA product was applied at rates of 0 (control) and 5 ppm C; each treatment was replicated four times. Corn plants were harvested one month after transplanting and it was found that HA enhanced total (top and root) mass yields by 35.1% over the control for the second soil, but reduced them by 1.7% for the first soil. This indicated that HA was beneficial at its optimum rate for crops grown in a less favorable environment. It was not effective and could even be detrimental for crops grown in a more than adequate growing environment.

The uptake of B and shoot dry weights of maize (cv. Euralis Es Armandi FAO 640) were studied in a greenhouse experiment. The soil used was Eutric Vertisol, containing a balance texture of sand, silt, and clay. The pH of the soil was 7.48, its EC was 0.45 mS/cm, and its OM was 2.53%. N, P, K, S, Ca, Mg, Na, Fe, Cu, Zn, Mn, and B contents were 1400, 15.15, 228, 12.75, 9262, 988, 96, 6.76, 1.48, 3.52, 92.88, and 0.56 mg/kg, respectively. B was applied to the soil at rates from 0 (control) up to 10 mg/kg. Two humic products were evaluated, i.e. a processed product (12% HA, 12.86 pH, and 32.80 mS/cm EC) and a raw product (65% HA, 4.87 pH, and 5.80 mS/cm EC). Each product was applied at rates from 0 (control) up to 500 μg HA/kg soil. Five seeds were planted in pots, each containing soil with 70% field capacity. After 37 days, maize shoots were analyzed for shoot dry weights and B uptake. The processed humic product enhanced shoot dry weights by 6.6% over the control at 10.0 mg/kg B and 250 μg HA/kg soil. B uptake was improved best at

16.7% at 10.0 mg/kg B and 500 μg HA/kg soil. The raw product enhanced shoot dry weights by 10.9% over the control at 10.0 B mg/kg and 250 μg HA/kg soil. B uptake was improved best at 11.8% at 10.0 mg/kg B and 500 μg HA/kg soil. This experiment indicated that the processed humic product was more efficient than its counterpart, considering it had a lower HA content but resulted in a similar overall performance (Celik et al. 2019).

A raw humic product of leonardite originating from North Dakota (containing 79.3% HA, see Section 3.2.4) was evaluated on corn grown in loamy sand (containing 4.3 pH, 4.8% OM, as well as P, K, Ca, and Mg at 18, 10, 125, and 10×10^{-6} kg/kg soil, respectively) and clay soil (containing 5.9 pH, 6.4% OM, as well as P, K, Ca, Mg at 74, 10, 218, and 2070×10^{-6} kg/kg soil, respectively) filled in 3 kg pots. P was applied in bands at a rate of 55 kg $P_2O_5/10^4$ m^2 in four applications below surface. N and K were applied two times to each pot at rates of 0.2 g/pot, with the second application four weeks after seeding. Seeds were then grown in each pot. The HA product was applied to the soil at rates from 0 (control) up to 10×10^3 kg/10^4 m^2. Each experiment was replicated four times. Harvests were made seven weeks after seeding on the loamy sand and eight weeks after seeding on the clay soil. On the loamy sand, HA was found to increase yields, P uptake, N uptake, and K uptake by 19.1, 21.0, 24.8, and 21.1% over the control, respectively (all at the highest humic rate). The use of HA on the clay soil, however, had no effects on either yields or uptake. These findings showed that HA was efficient when applied to the soil with low OM and nutrient contents, meaning it was effective for a less favorable growing environment (Duplessis and Mackenzie 1983).

Raw leonardite was evaluated on winter lentil (cv. Sazak 91). The experiment was completed outdoor in two consecutive years on 6.3 m^2 plots with sandy loam soil with 8.7 and 8.3 pH, containing 16.3 and 18.6% clay, 15.6 and 17.4% $CaCO_3$, 4.84 and 5.11 ppm P, 0.019 and 0.020% total salt, and 1.84 and 1.79% OM, respectively for each year. The humic product was applied in bands close to rows of lentil sown on soil, at rates from 0 (control) up to 600 kg/ha. The experiment was completed in a factorial randomized complete block design replicated three times per treatment. Harvests were completed at the end of each season over the two years of the experiment. It was found that HA enhanced the number of seeds per plant by 24.2% over the control at 300 kg/ha humic rate. Seed yields and weights were improved best (by 24.2 and 2.7%, respectively) at a humic rate of 600 kg/ha (Dogan et al. 2014).

A raw and a processed humic product (each containing approximately 88% HA) were evaluated on geranium cuttings and Alaska pea seeds placed in an aqueous solution. The solution contained each product at rates from 0 (control) up to 0.5%. The seeds were subject to a 12-hour light cycle at a temperature of 28 °C and 25% relative humidity. After 13 days of growing, the raw product at a rate of 0.5% was found to approximately quadruple the number of root initials over the control. For the processed product at a rate of 0.05%, the increase was approximately 2.5 times. After 13 days of growing, the raw product at a rate of 0.5% enhanced total root lengths by approximately 17 times over the control, while in the case of the processed product at a rate of 0.05%, the increase was approximately 7 times. At a rate of 1.0%, the raw product increased the pea root tip elongation by 8.2% over the control, while in the case of the processed product at rates 0.025 to 0.05%, the increase was 6.7%. This study showed the difference between raw (water-insoluble) and processed (water-soluble) HA products in terms of their effective rates and performances. The raw product performed better, but at a much higher rate (O' Donnell 1973).

A processed and purified FA product (with 6.2 pH and containing only 0.26% ash) was evaluated on the growth of cucumber of the Boston pickling variety. The cucumber was grown inside a controlled chamber (26/24 °C during day/night including controlled day/night lighting). Seeds were spread on filter paper. Five-day old seedlings were grown in distilled water for two days for acclimatization, and later in Hoagland's solution (without Ca to avoid precipitation). The FA product was applied at rates from 0 (control) up to 2000 ppm and was refreshed once every six days. Each experiment was replicated three times. The plants were harvested in six weeks and it was found that 100 to 300 ppm FA rates were the best. Shoot and root lengths were enhanced by up to 80.7 and 39.5%, shoot and dry weights by up to 2.4 and 2.3 times, and the number of leaves and flowers by up to 43.0% and 2.5 times over the control, respectively. N, P, K, Ca, Mg, Cu, Fe, and Zn contents of the shoot were 1.6 to 3.9 times higher compared to those of the control. The root contained 2.3 times more N than that of the control (Rauthan and Schnitzer 1981).

Cucumber of the Beta-Alpha (HATIM F1) cultivar was sown in soil inside a greenhouse, in which a processed humic product (containing 80% HA) was sprayed one month later on the seedlings at rates from 0 (control) up to 300 mg/l. This experiment was run under a completely randomized block design, using four plants per treatment. Three months later during harvest, it was found that fruit weights and numbers per plant were enhanced with the incorporation of HA. The best results were 14.9 and 14.8% higher than those of the control, respectively, at 100 mg/l humic rate. At 300 mg/l rate, both responses were reduced and even lower than those of the control. The percentage of Brix (sugar content of an aqueous solution) was best achieved at the higher humic rate and was 5.2% better than that of the control (Al-Madhagi 2019).

Two processed humic products were evaluated on lettuce (S. Anna local variety) and tomato (S. Marzano local variety) seeds. The first product (pH = 7) contained 20.9 C:H, while the second (pH = 7) contained 16.5 C:H. Each contained 900 cmol/kg total acidity. Each product was added to the seeds that were placed on paper inside a laboratory. The products were added at rates from 0 (control) up to

5000 ppm (first product) and 200 ppm (second product). The seeds were then evaluated after six days. Each treatment was replicated five times using a completely randomized design. The first product performed best at 1000 ppm rate (for lettuce), at which mean germination rates and dry seedling weights were enhanced by 3.2 and 3.5% over the control, respectively; and at 5000 ppm rate (for tomato), at which both parameters were enhanced by 63.0 and 30.6% over the control, respectively. The second product performed the best at its highest rate, at which mean germination rates and dry seedling weights were enhanced by 1.6 and 1.2% (for lettuce) and by 12.4 and 6.1% (for tomato) over the control, respectively (Piccolo et al. 1993).

A processed humic product containing 30.4% HA, 17.5% FA, 875 mmol/100 g total acidity, 5.1% N, 2.3% P, and 4.5% K was evaluated on lettuce (cvs. Dark Green and Big Bell) grown in an agricultural field containing clay soil. It had 7.97 pH, 0.83 mS/cm EC, and the following ions (all in meq/l): 6.3 Na, 4.2 SO_4, 3.1 Cl, 1.3 Ca, 1.2 HCO_3, 0.5 K, 0.4 Mg, and 0.14 CO_3. The soil was irrigated and fertilized using NPK. For the control, their rates were 136, 56, and 59 kg/ha, respectively. For "Treated 1," the NPK rates were 50% of those of the control and the HA product was applied to soil besides the plants three times up to nine weeks after growing, at a rate of 3.5 l/ha each time. For "Treated 2," the NPK rates were 50% of those of the control and the HA product was applied to plants (foliar) three times up to nine weeks after growing, at a rate of 1.7 l/ha each. The plants were harvested three months after growing in a two-year growing season. The experiment design was a split plot with three replications. Results presented were mean values for the whole growing season. For Dark Green cultivar, 50% NPK + HA application to soil increased crop yields by 11.5% over the control at 100% NPK. For Big Bell, crop yields were slightly reduced by 1.8%. For Dark Green cultivar, 50% NPK + HA application to plants (foliar) slightly reduced crop yields by 3.3% compared to those of the control at 100% NPK. For Big Bell, crop yields were increased by 29.7%. This study showed that at 50% reduced NPK, HA improved or slightly reduced crop yields compared to those of the control at 100% NPK (Shahein et al. 2015).

A three-year study was completed on an outdoor field with sandy loam soil (6.7 pH, 0.53% organic C, 0.31 mS/cm EC, 8.4 to 24.8% water contents, 93.6 to 109.5 mg/kg N, 4.2 to 18.5 mg/kg P, and 96.3 to 130.4 mg/kg K). The field received 118.2 to 154.2 mm effective rain at temperature ranges of 13.0 to 37.7 °C, and 1.7 to 6.8 mm evaporation. Tomato (cv. Heemsohna) seedlings that were 20 days old were planted, and several harvests were made in approximately 50 days, completed in approximately four months per growing season. Additional moisture was created via a drip irrigation system to maintain soil moisture, with a total water requirement of between 328.6 to 354.2 mm. Four fertigation levels were completed: 100, 80, 60, and 40% of the recommended rates of NPK, where the 100% rate was equal to 311, 85, and 497 kg/ha N, P_2O_5, and K_2O, respectively. In addition, a processed humic product (12% HA) was applied at rates of 0 (control) and 0.5 l/ha per fertigation. The study was completed in a randomized block design, replicated four times, or eight treatment combinations in total. The incorporation of humic enhanced all results over their control (NPK only, without humic), where the least improvement was found for the highest rate of NPK (at 100%). The fertilizer use that produced the best improvements was with NPK rates of 80 and 40%, respectively. N uptake, P uptake, K uptake, leaf chlorophyll contents, total yields, total soluble solids, firmness, and ascorbic acid contents were all increased, best at 10.7, 14.9, 12.1, 8.9, 12.6, 0.8, 6.6, and 6.9% over the control at 40, 80, 80, 40, 80, 40, 40, and 40% NPK, respectively. This study showed that humic worked well in reduced nutrient growing conditions (Suman et al. 2017).

A processed HA product of 7 pH, containing 3 to 13, 0.5 to 12, and 0.5 to 9 meq/g total acidity, COOH group, and phenol-OH group, respectively, was evaluated on cantaloupe, lettuce, and onion seedlings. All seeds were grown on paper inside a laboratory, then transferred to Petri dishes. The HA product, dissolved in either water or a half strength Hoagland's nutrient solution, was applied at rates from 0 (control) up to 20 000 mg/l. Each treatment was replicated four times in a completely random design. Roots were measured after three days (for cantaloupe and lettuce) and four days (for onion). In the case of onion, 500 mg/l HA enhanced root lengths by 3.3 times (when water was used) and 84.1% higher (when Hoagland's solution was used) compared to those of the control. At 1000 mg/l HA rate, similar results were observed. Beyond these rates, lower performances were observed. At the highest rate, root lengths were in fact reduced by 19.0% (both when water and Hoagland's solution were used) compared to those of the control. Similar trends were observed for those of lettuce (at 62.2 and 32.5% higher for water and Hoagland's nutrient solutions, respectively, at an optimum HA rate of 500 mg/l) and cantaloupe (at 81.8% and 2.1 times higher for water and Hoagland's nutrient solutions, respectively, at an optimum HA rate of 1000 mg/l). The highest HA rate resulted in negative responses compared to those of the control. The same experiment for onion was repeated and the seeds were sterilized to create an axenic growing condition. The optimum HA rate was found at 500 mg/l, where root lengths were enhanced by 2.1 and 2.5 times compared to those of the control for water and Hoagland's nutrient solutions, respectively (Van de Venter et al. 1991).

Chardonnay grape vines were planted on coarse sand inside a greenhouse treated with two HA products. A raw HA product (55% HA) was mixed with the soil at rates from 0 (control) up to 0.5% active ingredients by weight. A liquid HA product (15% HA) was mixed with the soil at rates from 0 (control) up to 0.2% active ingredients by weight. The vines were watered twice weekly with 400 ml water containing 20N-20P-20K plus trace elements of mainly Fe (1000 ppm) and

Mn (500 ppm). The experiment was arranged as a randomized complete block design with six blocks and three vine treatment replications, and was concluded 11 months after planting. It was found that the dry HA product best enhanced leaf areas per vine (at its higher rate) and plant dry weights (at 0.2% active ingredients by weight) by 35.0 and 29.3% over the control, respectively. At its highest rate, the liquid HA product resulted in a decrease of 23.9% for leaf areas per vine and another decrease of 17.7% for plant dry weights compared to those of the control. This indicated that the liquid HA product was much more available than the dry product; and at the applied rate (which was too high), the effect of HA was detrimental to the growth of plants. In terms of nutrient concentrations in plants, the dry HA product (at its highest rate) enhanced total macronutrients and total micronutrients by 3.8 and 17.1% over the control, respectively. In the case of the liquid HA product (at its highest rate), total macronutrients were enhanced by 20.9%, but total micronutrients were reduced by 2.8% over the control, respectively. Due to a high HA content, the raw product increased soil OM best at 2.6 times over the control at its highest rate. The liquid product at its highest rate did not change the soil OM (Reynolds et al. 1995).

Seventy two Italia table grape vines were grown in sandy clay loam with the following properties: pH = 7.84, EC = 0.32 mS/cm, Na = 0.82 cmol/kg, Cl = 0.19 g/kg, OM = 8.74 g/kg, N = 0.61 g/kg, P_2O_5 = 78.28 mg/kg, K = 1.17 cmol/kg, Ca = 12.40 cmol/kg, Mg = 3.75 cmol/kg, Mn = 11.07 mg/kg, B = 0.67 mg/kg, Cu = 4.22 mg/kg, Zn = 1.72 mg/kg, and Fe = 4.60 mg/kg. The soil was drip irrigated and treated with 140 kg N/ha, 60 P kg/ha, and 170 kg K/ha. A processed humic product (containing 52.46% C, 5.77% H, 5.40% N, 0.70% S, 35.66% O, 6.9% ash, total acidity = 5.0, COOH group = 2.7, phenol-COOH group = 2.2, and E_4/E_6 = 7.4) was applied foliar at a rate of 100 mg/l during pre-bloom, full-bloom, fruit set, and ripening. Water was used as the control. A randomized block design was applied with three blocks, and each treatment in the block consisted of three grapevines. Harvest was made after four months, where 27 bunches per treatment were picked, and a mean of nine bunches were considered as a replication. A significant increase in fruit weights (at approx. 10% over the control) was observed when the humic product was applied during full bloom, while chlorophyll and Brix contents increased slightly at other treatment periods (Ferrara and Brunetti 2010).

Tobacco (var. Coker 254) seeds were planted on paper/Petri dish soaked with a Hoagland's solution (without Ca) and either a processed HA or FA product at rates from 0 (control) up to 1000 and 2000 ppm, respectively. The HA product of 6 to 7 pH contained 493 to 593 meq/100g CEC, 2.46% N, 0.45% P, and trace nutrients including 950 ppm Fe. The FA product of the same pH contained 828 to 909 meq/100g CEC, 1.70% N, 1.05% P, and trace nutrients, including 115 ppm Fe. The growing was completed inside a laboratory, in an environment of 28 °C (day) and 22 °C (night) with a 12-hour day and night lighting system. Seedlings were evaluated up to 17 days after growing. The treatments were arranged in a randomized complete block design with five replications. On the 17th day, the HA product increased both the number and the length of roots by 39.4 and 56.6% over the control, respectively, at a product rate of 75 ppm. The FA product increased both the number and the length of roots by 9.2 and 38.5%, respectively, at a product rate of 100 ppm. Higher rates of each product resulted in reduced performances (Mylonas and McCants 1980).

A processed and purified HA product with 7.0 pH, containing < 2% ash, 2.0% Mg, 1.73% N, and trace nutrients was evaluated on teak seedlings grown in clay soil of 4.7 pH (containing 7.8 % OM, 4.3% extractable HS, 0.38% total N, 23.13 mg/kg available P, 30.40 meq/100 g Ca, 7.02 meq/100 g Mg, 6.64 meq/100 g K, and 0.32 meq/100 g Na) and in sandy soil with 6.2 pH (containing 1.0% OM, 0.8% extractable HS, 0.15% total N, 23.75 mg/kg available P, 2.10 meq/100 g Ca, 0.61 meq/100 g Mg, 0.06 meq/100 g K, and 0.08 meq/100 g Na). The HA product was mixed in the soil at rates from 0 (control) up to 1000 mg/kg by weight. Teak seeds were planted in pots inside a greenhouse and harvested four months later. The experimental design was a completely randomized block, in which each treatment was replicated three times for each type of soil. HA at its highest rate improved plant heights grown in high and low OM soils by 2.0 and 2.6 times over the control, respectively. The same rate did not enhance plant stem diameters for the first soil but enhanced them by 66.7% for the second soil. At its highest HA rate, dry matter yields for plants grown in high and low OM soils increased by 2.4 and 5.4 times compared to those of the control, respectively. This study showed that HA performed better when plants were grown in a less favorable growing environment (Fagbenro and Agboola 1993).

Spinach (cv. Balady) plants were grown from seedings on sandy soil in an experimental station over two seasons and watered using a drip irrigation system. The soil had 1.7 mS/cm EC, 16.57% field capacity, 5.25% wilting point, and contained 0.982 meq/l Na, 7.020 meq/l Ca, 0.527 meq/l Mg, and 0.310 meq/l K. The plants were treated with $(NH_4)_2SO_4$ and a processed humic product of 9 to 10 pH, containing 80% HA, 10 to 12% K_2O, and 100 ppm micronutrients (including Fe, Mn, and Zn). The humic product was thoroughly mixed with water. Three rates of N fertilizer (up to 247 units N/ha) and three rates of the humic product from 0 (control) up to 10 l/ha HA were evaluated in split plots with three replications. $(NH_4)_2SO_4$ was applied in two equal amounts, one during soil preparation and another one month after sowing. The humic product was applied in two equal amounts, one at 21 days and another at 36 days after sowing. For each case, P_2O_5 was added during soil preparation at 79 units P_2O_5/ha, and K_2SO4 was added at one month after seeding at a rate of 119 units K_2O/ha. After 50 days, the plants were harvested. The humic product was found to enhance total yields of spinach, with the best result at 247 units N/ha and 10 l/ha HA, resulting in 7.4 and 9.4%

increases over the control during the first and second seasons, respectively. Protein values on spinach leaves were all enhanced due to the application of this humic product. Its best results were recorded at 247 units N/ha and 10 l/ha HA, resulting in 16.3 and 20.1% increases during the first and second seasons, respectively (Hafez et al. 2015).

Noroozisharaf and Kaviani (2018) studied the effect of a raw HA product (containing 1.3% N., 62.2% C, 36.8% O, and 2.4% H) on garden thyme. Seedlings were purchased and transferred, then harvested after six months before flowering. The plantlets were cultured in plots filled with clay-loam soil and perlite (pH = 7.32, EC = 1.41 mS/cm, 0.076 mg/kg total N, 5.33 mg/kg available P, and 273 mg/kg available K) and continuously irrigated. Each plot contained nine plants with three replications for each treatment in a randomized complete block design. The HA product was mixed well with the soil at five different rates from 0 (control) up to 100 g/m^2. Samples were taken before the flowering stage. N, P, K, Mg, and Fe contents of the leaves were all higher compared to those of the control, with the best results at the maximum product rate of 100 g/m^2 (ranging from 2.3 times higher for Mg up to 5.4 times higher for P). Compared to those of the control, the essential oil contents (a sum of 32 different compounds) were enhanced by up to 2.5 times at a product rate of 75 g/m^2, and were slightly lower (2.3 times higher) at the maximum product rate.

Pea plants were grown in sandy loam soil under drip irrigation on agricultural fields with 7.8 pH, 1.8 mS/cm EC, 0.16% OM, and 19.2% soil moisture field capacity. Macronutrient contents were 15.0, 9.4, and 16.0 mg/100 g of N, P, and K, respectively; 5.30, 3.00, and 2.00 meq/l of SO_4, Ca, and Mg, respectively; and small amounts of micronutrients, including Cu, Fe, Mn, and Zn at 4.0, 7.8, 3.3, and 1.86 ppm, respectively. The plants were grown in plots, to which P fertilizer was added during soil preparation at rates from 0 (control) up to 120 kg P_2O_5/ha. A processed humic product, containing 60% HA and 15% FA (plus 0.7, 0.06, and 8.0% N, P, and K, as well as trace micronutrients) was applied after 35 days of sowing at rates from 0 (control) up to 10% of foliar application. The experiment was laid out in a randomized complete block design, consisting of four replications. At 75 days after sowing, the product enhanced dry plant weights, green pod yields, and protein contents by 62.7, 30.8, and 60.0% over the control, respectively. All were achieved at 120 kg P_2O_5/ha and 10% humic product. Results for 80 kg P_2O_5/ha and 10% humic product did not differ much, indicating that the best performance was between 80 and 120 kg P_2O_5/ha and 10% humic product. Macro and micronutrients within pea seeds were all enhanced with the addition of the humic product, with the best results observed at 93.3% (for P) and 2.0 times (for Zn) higher than those of the control at the highest rates of P and humic product, respectively (Kandil 2014).

A processed humic product containing 13.2% HA and 3.3% FA was evaluated on watermelon (cv. Crimson Sweet) grown in pots located inside a net house. The growing medium was soil (pH = 5.30, Ca = 0.56 cmol/dm^3, Mg = 5.90 cmol/dm^3, K = 0.01 cmol/dm^3, and OM = 2.33 g/kg) mixed with manure (pH = 9.09, Ca = 0.32 cmol/dm^3, Mg = 4.40 cmol/dm^3, K = 2.78 cmol/dm^3, and OM = 156.00 g/kg) and sand in a ratio of 1:2:1. A complete randomized block design with five treatments using the humic product was completed, involving five replications of ten seedlings each, with a total of 250 seedlings. The soil was irrigated daily, and the humic product was sprayed at rates from 0 (control) up to 30 ml/m^2 on seedlings at 10, 15, and 20 days after sowing. Harvest was made 25 days after sowing. It was found that the humic product enhanced shoot dry weights by approximately 30% over the control, at a rate of 22.5 ml/m^2. The leaf chlorophyll contents and root volumes were also improved (with the best result at approximately 17 and 9% higher than those of the control) at humic product rates of 22.5 and 7.5 ml/m^2, respectively. At higher rates, the responses were found to be lower (Silva-Matos et al. 2012).

Seedlings of watermelon plants were transplanted during their three to four true leaf stages in ditch rows and were drip irrigated. The soil was 90% sand with 8.20 pH, 1.50 mS/cm EC, and 0.82% OM. N, P, K, SO_4, Ca, Mg, and Cl contents were almost 0, 0.443, 18.50, 55.65, 2.65, 2.40, and 53.00 meq/l, respectively. Na was 4.35 meq/l. A processed humic product was mixed with water so as to contain 23 g/l HA and 0.5 g/l FA as well as 5.0N-0.5P-15.0K (all in %), and was injected into the drip irrigation system. The mixture was applied at rates from 0 (control) up to 14.8 l/ha. Three hybrids were evaluated in this experiment: Giza 1, Aswan, and Sugar Belle. The experiment was arranged in a split plot design, four treatments using humic as the main plots and three sub treatments for each hybrid, with three replications completed in two seasons. Harvests were made during the first and second seasons. The best marketable yields for all three hybrids were achieved at the highest humic rate during the second season: Sugar Belle by 52.6%, followed by Giza 1 by 47.7%, and Aswan by 33.7%, over the control. At this humic rate, N leaf contents were 11.5, 60.0, and 12.5% higher; P leaf contents were 73.2, 6.7 and 2.7% higher; and K leaf contents were 20.2, 19.1, and 17.7% higher than those of the control, respectively (Salman et al. 2005).

Seeds of papaya were grown in pots inside a canvassed shelter. The same growing medium and humic product as those of Silva-Matos et al. (2012) were utilized in this study. A complete randomized block design with five rates of the humic product from 0 (control) up to 30 ml/m^2, involving five replications and ten seedlings per pot (or 250 seedlings in total). The seedlings were sprayed at 15, 25, and 35 days after sowing and were irrigated daily. Forty-five days after sowing, the plants were harvested and evaluated. The shoot dry weight was best improved (at approximately 23% over the control) at a product rate of 15 ml/m^2, the leaf chlorophyll was best enhanced (at approximately 3% over the control) at a rate of 7.5 ml/m^2, and the root dry mass was also maximized (at approximately 22% over the control) at a rate of 7.5 ml/m^2. This experiment showed that there were optimum humic rates (i.e. 7.5 to 15 ml/m^2), beyond which the responses were lower (Cavalcante et al. 2011).

Arabica coffee (var. Catuai Vermelho I-144) seedlings of 28 weeks old were transplanted into pots inside a greenhouse filled with Oxisol soil (3 kg each). The soil had 4.1 pH, 0.32% OM, 12 mg/kg NH_4-N, 10 mg/kg NO_3-N, 1.9 mg/kg P, 30 mg/kg K, and other nutrients in small amounts. Thirteen g/kg CO_3 was added to the soil to reach its optimum pH of 6.2. P (300 mg/kg) was added during transplant. N (400 mg/kg), K (400 mg/kg), S (40 mg/kg), Cu (4 mg/kg), and Mn (0.15 mg/kg) were later added to the soil. B and Zn were added at 3 different rates: 0.3 and 0.6 mg/kg (foliar), 0.6 and 1.2 (foliar), and 1.2 and 6 mg/kg (soil), respectively. A processed FA product (19.1 mS/cm EC, 5.6 pH, 7.35 E_4/E_6, 39% C and 0.35% N) was foliar applied at rates from 0 (control) up to 1 g C/l at 30, 60, and 90 days after planting. A factorial design was used through the combination of three application modes of micronutrients with four rates of the FA product, with three replications. The best results were achieved at the lowest B and Zn rates applied foliar; the worst results were found at the highest B and Zn rates applied to soil. Leaf and stem dry matters were enhanced best at 0.5 g C/l product rate at 45.5 and 62.5% over the control, respectively. Higher rates resulted in lower responses. For the accumulation of B, the same rate gave the best result at 63.1% enhancement over the control. For the accumulation of Zn, the best improvement (10.5% higher than that of the control) was achieved at the highest product rate (1 g C/l). No difference was observed in the case of root dry matter (Justi et al. 2019).

Sari and Abdoellah (2017) evaluated the performance of a processed humic product containing 34.5% C and 28.7 C/N on the growth of coffee seedlings. Arabica coffee of Andungsari I was planted in a nursery. After two months, seedlings were transferred to a polybag and grown in Gley soil under a controlled environment inside a greenhouse. The humic product was blended into the soil at rates from 0 (control) up to 30 g/2 kg soil. Each treatment was replicated three times and was performed in a completely randomized design. It was found that root fresh weights were increased with the humic application. The best result was 1.5 times higher than that of the control (at the highest product rate). Stem fresh weights were also increased, with the best result 2.7 times higher than that of the control (at the highest product rate). Leaf fresh weights were also improved, with the best result 3.0 times higher than that of the control (at 10 g/2 kg soil of product rate), and remained the same at the highest humic rate.

A processed humic product containing 50% HA was evaluated on yields of rice (cv. Huadiao 5) grown in a costal saline field with the following soil parameters: pH = 8.58, organic C = 11.77 g C/kg, total N = 0.92 g N/kg, and Cl = 0.16%. The experiment was completed in plots using a randomized complete block design with eight treatments, replicated three times, involving 300 kg N/ha, humic product at 600 kg/ha, gypsum at 600 kg/ha, and a control. Before rice transplanting, 60 kg P_2O_5/ha and 120 kg K_2O/ha were applied. Moisture was provided by rain at 626.5 mm over three months of the growing season and the air temperature averaged 23.2 °C. The best yield was achieved using N fertilizer + humic product, which was 18.4% higher than that of N fertilizer only, and 7.2% higher than that of N fertilizer + humic product + gypsum (all were mean values). This indicated that using more nutrients and/or HS did not always generate better results (Sun et al. 2019).

A processed humic product (with 10.10 pH, 4.98% OM, 0.22% N, 0.69% K, 0.39% Na, 0.79% Ca, 0.06% Mg, 0.28% Mn, and traces of P, Cu, Fe, and Zn; as per Suntari et al. 2013) was evaluated on rice of the Inpari 6 Jete variety grown in Vertisol soil. The soil was fertilized using 92 kg N/ha (50% during the planting of seedlings, and 50% 21 days later), 72 kg P_2O_5/ha (during planting), and 120 kg K_2O/ha (during planting). Urea 46-0-0 was used as the N fertilizer. The humic product was applied during planting together with the N fertilizer at rates from 0 (control) up to 400 kg/ha. At 14 days after planting, the soil available NH_4-N and NO_3-N contents were improved with the addition of the humic product. The best results were observed at product rates of 150 kg/ha (for NH_4-N, i.e. 4.6 times higher than that of the control) and 250 kg/ha (for NO_3-N, i.e. 5.8 times higher than that of the control). Harvest was completed six months after planting. It was found that the 200 kg/ha product rate produced the most effective dry milled improvement, at 22.1% higher than that of the control. Higher product rates resulted in lower results on these three measured parameters (Suntari et al. 2015).

Da Rocha et al. (2019) evaluated one-year-old guava (cv. Paluma) plants grown in Ultisol soil in an experimental farm. At up to 20 cm depth, the soil was 78.8% clay with 5.02 pH, 0.73% OM, and 4.33 cmol/dm^3 CEC, containing P, K, Ca, and Mg at 5.31, 68.01, 0.98, and 0.50, respectively (all in cmol/dm^3). From 20 to 40 cm depth, the soil was 73.2% clay with 5.03 pH, 0.40% OM and 3.11 cmol/dm^3 CEC, containing P, K, Ca, and Mg at 1.69, 50.78, 0.24, and 0.76, respectively (all in cmol/dm^3). The soil was micro-sprinkled, irrigated, and all plants were fertilized with 1166 kg/ha urea (45-0-0), 467 kg/ha P fertilizer (0-18-0), and 1166 kg/ha potash (0-0-60), bimonthly. A processed humic product containing 13.2 HA and 3.3% FA was applied at rates from 0 (control) up to 40 ml/plant, every 60 days after pruning. The experiment lasted for 18 months with four replications of two plants each. The average values for monthly precipitation, air temperature, and air humidity ranged from 0 to 195 mm, 21.2 to 28.7 °C, and 45 to 92%, respectively. N, P, S, Ca, Mg, Cu, Fe, Zn, and Mn in leaves were increased by 13.6, 8.4, 10.8, 36.0, 26.3, 30.0, 39.1, 36.9, and 15.0% over the control, corresponding to product rates of 40, 20, 40, 30, 20, 40, 30, 40, and 40 ml/plant, respectively. This indicated that higher product rates were not always better. In the case of K and B, responses for all rates were somehow negative. The humic product resulted in the best fruit yield enhancement (10.9% over the control) at a rate 20 ml/plant.

British Isles seeds were grown in a greenhouse and the seedlings were transferred to pots containing 2:1:1 of soil:peat:sand (pH = 7.19, $CaCO_3$ = 1.85%, OM = 5.78%, and EC = 0.2 mS/cm). A processed humic product (containing 15% HA + FA and 1% K_2O) was evaluated at rates from 0 (control) up to 120 mg/kg by weight. Each pot was treated four times, in a ten-day interval, up to 40 days after planting. The plants were grown under natural light conditions, 27 °C (day) and 14 °C (night) temperatures, and 75% relative humidity. The experiment was undertaken based on a completely randomized design with three replications per treatment, and ten plants for each replication. The humic product at 40 mg/kg enhanced the number of buds, plant heights, the number of main shoots, plant diameters, the number of flowers, root lengths, fresh root weights, dry root weights, fresh leaf weights, and dry leaf weights best at 4.3, 1.2, 2.0, 1.4, 2.0, 1.2, 2.0, 2.0, 2.9, and 4.0 times those of the control, respectively. At higher rates, all responses were lower, and in some cases even worse than the control (Esringu et al. 2015).

A processed humic product containing 13.5% HA was evaluated on strawberry of the Paros cultivar grown within a greenhouse. The medium was 1:1 leaf mold and soil. All treatments commenced after all plants were established, producing five to six expanded leaves. The product rates were from 0 (control) up to 750 mg/l for the soil drenched application and up to 1200 mg/l for the foliar application. Shoot fresh, shoot dry, root fresh, and root dry masses, as well as leaf areas, were all increased with the addition of humic at rates of 600 to 750 mg/l (soil drench application) by up to 70.8, 64.6, 66.9, 78.4, and 44.8%, respectively, over the control. For foliar application at rates of 600 to 900 mg/l, the improvements were up to 92.3, 79.5, 81.1, 93.1, and 46.3%, respectively, over the control. For soil drench application at rates of 600 to 750 mg/l, vitamin C contents, total soluble solids, chlorophyll contents, and plant yields were all increased by up to 45.1, 40.0, 5.4, 77.9%, respectively, over the control. For foliar application at rates of 600 to 900 mg/l, the improvements were up to 59.5, 60.0, 6.8, and 102.2%, respectively, over the control. At other rates, including higher ones, the responses were lower. Overall, foliar was the more effective application method (Eshghi and Garazhian 2015).

A processed humic product containing 2 to 4% HA and 8.0 to 9.5 pH was applied to oyster mushroom grown in 5 cm chopped paddy straw. The medium was soaked in water for 16 hours, then soaked with the humic product at rates from 0 (control) up to 100%, followed by sterilization, cooling, blending with mushroom spawn, and incubation for 22 to 31 days. During harvest, at 60% product rate, it was found that mushroom yields, protein contents, and total fiber contents were all increased by 3.1 times, 47.7%, and 2.2 times over the control, respectively. At higher rates, the responses were not better, i.e. at 100% product rate, there were 28.8 and 16.9% improvements over the control for yields and protein contents, but a reduction of 17.1% for total fiber compared to that of the control. These findings showed that higher product rates were not always beneficial (Prakash et al. 2011 and Neyveli Lignite Corporation Ltd. 2008).

Cauliflower (var. Botrytis cv. Kartopu) seedlings with three to four true leaves were grown inside a greenhouse in sandy loam soil (81% sand, 11% clay, and 8% silt) containing 7.8 pH, 0.29 mS/cm EC, 1230 kg/m^3 bulk density, 25% field capacity, 1.4% OM, 20.56% $CaCO_3$, 0.56% N, 19.6 ppm P_2O_5, and 221.2 ppm K_2O. Nutrients were applied to the soil at rates of 40 kg/ha N, 60 kg/ha P_2O_5, and 5 kg/ha K_2O before planting, followed by 40 kg/ha N four weeks after planting. The soil was irrigated daily and had its temperature and humidity maintained at 24.2 °C and 63%, respectively. The soil was spiked with NaCl at concentrations of 0.3 (control) up to 8 mS/cm as the main factor, and a processed humic product (with 12.86 pH, 12% HA, and 32.8 mS/cm EC) at rates from 0 (control) up to 2 g/kg as the split plot. The experiment was a split plot arranged in a complete factorial with four repetitions. All crops were harvested after 90 days with curds cut at the soil surface. NaCl at rates of 0.3 and 2 mS/cm did not affect curd weights per plant compared to the control, while the addition of the humic product at 2 g/kg enhanced them by 7.0% over the control. At 4, 6, and 8 mS/cm NaCl rates, curd weights per plant were 15.6, 28.8, and 39.8% lower than those of control, respectively. At a NaCl rate of 4 mS/cm, the addition of HS at 1 and 2 g/kg improved on these results, with the outcome at 13.5 and 11.9% lower than those of the control. At higher rates of NaCl (6 and 8 mS/cm), the addition of HS at any rates could not help the curd weights. Note: the curd weight per plant corresponded directly to the marketable yield (Turhan 2019).

A processed humic product containing 86% HA and 6% K_2O (Abbas 2013) was evaluated on eight-year old Florida Prince peach trees budded on Nemaguard rootstock and grown in calcareous sandy loam soil under a drip irrigation system. The soil had the following properties: 72% sand/17% silt/9% clay, 14% $CaCO_3$, 2.00 mS/cm EC, 7.52 pH. The available ions (all in meq/l) were as follows: 0.74 K, 17.74 SO4, 7.5 Mg, 10 Ca, 4.0 Na, and 2.0 Cl. The trees were planted with a 4 × 5 m^2 spacing, and fertilized with OM, NPK, and mineral oil before being given a treatment using the product. Fruitlets were thinned to leave 300 fruits per tree, then the humic product was applied (to the soil and/or foliar) at rates from 0 (control) up to 0.50%. Each treatment was applied in December after fruit set and was repeated four times at 15-day intervals. Each treatment was replicated three times on trees arranged in randomized complete block design. Fruits from each tree were harvested in April of the following year. The whole experiment was repeated for another year, and both results were included during the evaluation. Fruit yields were enhanced best with the addition of HA at the highest product rates (0.50%

soil application and 0.50% foliar application) at 75.5 and 78.1% higher than those of the control for the first and second seasons, respectively. Total soluble solids were also improved by 51.3 and 41.0% at the second highest product rates (0.25% soil application and 0.25% foliar application) over the control for the first and second seasons, respectively. Ancothyanins (the color and brightness indicator) were also enhanced by 87.8 and 81.1% at the highest product rates over the control for the first and second seasons, respectively (Abd El-Razek et al. 2012).

The same humic product was evaluated on Washington navel orange trees budded on sour orange rootstock, grown in clay loamy soil. NPK nutrients were applied as recommended, then the product was applied (foliar and soil drenched) at rates of 0 (control) and 1.5 g/l (of 3 l/tree), six times in a season. The complete randomized block design with four replications was used in this experiment. The whole experiment was repeated in the following year and consistent responses were reported. At the end of the season (six months), the best improvements of fruit weights and yields per tree were 7.6 and 21.1% higher than those of the control, respectively, when the humic product was applied to the soil. The best improvements of total acidity contents and fruit juice volumes were 28.8 and 35.9% higher than those of the control, respectively, when the product was applied foliar. The best improvements of total sugar and vitamin C contents were 16.7 and 16.9% higher than those of the control, respectively when the product was applied to the soil. These results indicated that, with the exception of the total acidity and fruit juice volume, the humic product performed best when it was applied to the soil. Still, the foliar application resulted in better responses than those of the control (Sharaf et al. 2011).

Broccoli (var. Italica) seedlings were transplanted to sandy soil containing 87% sand/10% silt/3% clay, 18.6% field capacity, 20.7 cm/hour hydraulic conductivity, 8.2 pH, 7.98% $CaCO_3$, 0.23 mS/cm EC, 0.02% OM, 27 mg/kg N, 5.5 mg/kg P, and 96 mg/kg K. The soil was fertigated 15 days after transplanting and terminated 15 days prior to harvest (i.e. three months after transplanting). The soil was fertilized with 250N-75P-200K kg/ha (the recommended rate for this this type of crop and soil) and their reduced rates (i.e. 50 and 75% recommended rate), and treated using a process humic product at rates of 0 (control) and 120 l/ha. The product contained 7.7 pH, 14.8% HA, 3.5% FA, 3.9% N, 0.13% P, 3.22% K, 248 ppm Zn, 436 ppm Fe, and 216 ppm Mn. The experiment was set up in a completely randomized block design, in which each treatment was replicated three times. Each plot was 100 m^2 in area and was surrounded with border strips. During harvest, it was found that HA enhanced vegetable marketable yields by 19.7% over the control, at 75% NPK. Total soluble solids and protein contents were best enhanced at 50% NPK plus HA (i.e. 2.9 and 9.0% over the control, respectively). Vitamin C contents were enhanced by 3.9% over the control at 100% NPK plus HA. With the exception of vitamin C, all the best results were achieved at rates other than the recommended NPK rate. These results indicated that HA performed best when the growing conditions were not optimum (Selim and Mosa 2012).

Rady et al. (2016) evaluated Egyptian cotton grown in loamy sand soil at two different sites. Rainfall in the area was 7.5 mm/year with minimum/maximum temperatures of 14.5/31.0 °C and minimum/maximum evaporation rates of 1.9/7.3 mm/day. Soil in the first site had 7.84 pH, 79% sand/10% silt /11% clay, 1670 kg/m^3 bulk density, 25.33% field capacity, 3.46 mS/cm EC, 10.11 meq/100 g CEC, 5.03% $CaCO_3$, 1.30% OM, and available nutrients (all in mg/kg) including 50.00 N, 520.93 P, 55.36 K, 2.82 Fe, 6.6 Mn, 0.85 Zn, and 0.35 Cu, as well as soluble ions (all in meq/l) including 12.0 Ca, 8.5 Mg, 0.75 K, 16.32 Na, 20.59 Cl, and 14.73 SO_4. Soil in the second site was more saline. It had 7.66 pH, 78% sand/9% silt/13% clay, 1510 kg/m^3 bulk density, 23.31% field capacity, 12.86 mS/cm EC, 9.91 meq/100 g CEC, 3.30% $CaCO_3$, 1.14% OM, and available nutrients (all in mg/kg) including 40.00 N, 592.49 P, 44.11 K, 2.97 Fe, 6.52 Mn, 0.82 Zn, and 0.33 Cu, as well as soluble ions (all in meq/l) including 19.0 Ca, 12.5 Mg, 2.53 K, 115.84 Na, 80.11 Cl, and 66.47 SO_4. Cotton seeds were planted and thinned for two plants per hill. The soil was treated with 375 kg/ha of 15.5% P_2O_5 and 175 kg/ha of 33.5% N, and surface irrigated in 10-day intervals using water with 7.46 pH, 3.22 mS/cm EC, 5.38 SAR, 7.50 meq/l SO_4, and 19.50 meq/l Cl. A processed humic product (90.29% HA, 0.95% N, 1.04% P, 1.46% K, 2.81% Ca, 0.92% Mg, 0.48% S, and some micronutrients) was applied to the soil surface at rates of 0 (control) and 15 kg/ha. Treatments were replicated three times in a randomized complete block design. Harvest was completed five months after planting. For control (0 HA), it was found that the second soil (EC = 12.86 mS/cm) had 44.6% lower yield (seed + lint) than that of the first soil (EC = 3.46 mS/cm). HA helped improving total yields by 16.1 and 14.2% over the control for the first and second soil, respectively. While salinity reduced fiber strengths by 9.3%, HA enhanced them by 7.5% over the control for the second soil but did not significantly help for the first soil. These findings showed that in general, HA performed better in soil with a higher salinity (i.e. in a less favorable growing environment). This study also evaluated the micronaire value (i.e. its fineness and maturity). The salinity did not affect this parameter, which was 4.30 and 4.37 (a "base" reading) for both sites, while HA enhanced it to 3.93 and 4.07 (a "premium" reading) for both sites (Barnhardt Natural Fibers 2015).

A processed humic product containing 51 to 57% organic C, 4 to 6% N, and 0.2 to 1.0% P was evaluated on sunflowers (hybrid Hysun-33) grown in 8 kg pots. Soil of sandy loam with 8.03 pH, 18.12% $CaCO_3$, 1.34 mS/cm EC, 2.11% OM, 7.48 meq/100 g CEC, 0.15% N, 157.6 ppm K, and other trace nutrients was treated with nutrients at rates of 90N-60P-60K kg/ha and the humic product at rates from 0 (control) up to 25 g/l. Six seeds were then sown in each pot, and the germinated seeds were thinned to two plants per pot. The experiment was set in a completely randomized design, in which each treatment was replicated four times. Sixty-eight days later, before flowering, harvest was made, and the following responses were observed. HA at a rate of 1.0 g/l enhanced dry shoot and root weights by 37.3 and 21.2%, respectively over the control. Higher rates resulted in reduced responses, especially at the highest rate, at which both responses were lower than those of the control. Leaf nutrient concentrations were also improved best (up to 13.0% for P) at a humic rate of 1.0 g/l. At rates of 5 g/l and higher, the results were worse than those of the control. These findings showed that high HA rates were not always beneficial, and in fact could be detrimental (Sadiq et al. 2014).

Linseed or flax of the PKV-NL-260 cultivar was sown in 3.3 m^2 test plots located in Nagpur, India, in which the experimental design was a randomized block with three replications. A processed humic product (containing 1300 ppm HA, 50 ppm P, 168 ppm K, and 121 ppm Na) was sprayed two times at 35 and 55 days after sowing at rates of 0 (control) up to 500 ppm. During harvest (approximately 100 days after sowing), HA at a rate of 350 ppm enhanced seed oil contents and yields by 6.8 and 34.0% over the control, respectively. Higher rates resulted in lower responses, although they were still better than those of the control (Neware et al. 2017).

Strawberry (cv. Douglas) plants were grown within a greenhouse in calcareous loam/sandy clay loam soil. They were fed with 200 kg/ha N, 100 kg P_2O_5/ha, and 400 kg K_2O/ha as well as two humic products. The first one was a raw product containing 85% HA at rates from 0 (control) to 400 kg/ha before planting. The second one was a processed product containing 15% HA at rates from 0 to 10 l/ha per month, applied by drip irrigation after transplanting. The experiment was designed in a randomized block design with four replications for each treatment. This study found that the use of humic products for this specific application was not beneficial on the leaf nutrient contents. It was concluded that high humic rates applied in excessive calcareous soil had an inhibitory effect on the uptake of some nutrients (Pilanal and Kaplan 2003).

Hartz and Bottoms (2010) evaluated four different processed liquid humic products (6 to 11% HA) and a processed dry humic product (50% HA) applied to soil in two separate studies. The first study was completed in a greenhouse maintained at 24/20 °C day/night temperatures. The soil contained 7.3 to 7.8 pH, 8 to 11 g/kg OM, and 3 to 12 mg/kg HCO_3-P. Green Towers romaine lettuce seeds were sown in pots filled with soil treated with nutrients at an equivalent rate of 16N-24P kg/ha and each of the humic products at equivalent rates of 0 (control) and 2.2 kg HA/ha, respectively. Each pot was watered daily with water containing 100 mg/l NO_3-N. The whole plants were harvested 6.5 weeks after sowing. The experiment was completed in a randomized complete block design with five replications for each treatment. The second study was completed on the field, where Yolo silt loam soil was mixed with 12 mg/kg P. P (34 kg/ha) and N (34 kg/ha) nutrients were then applied to soil together with each of the humic products at rates from 0 (control) to 3.4 kg HA/ha. Heinz 9780 processing tomato plants were transplanted and harvested 6.5 weeks later. The experiment was completed in a randomized complete block design with five replications for each treatment. Both greenhouse and field studies found that HA did not benefit the lettuce dry weights and tomato fruit yields or quality. However, it should be noted that the timing for the application of the humic products (before sowing and transplanting) might not have been ideal, and that the soil might have contained more than adequate nutrients.

Neri et al. (2002) evaluated the performance of a processed humic product containing 55.0% combined HA + FA, 1.20% organic N, and 2.6% total N on strawberry (cv. Onda) grown in an agricultural field. The soil was sandy loam with 7.59 pH, 1.58% OM, 0.53% total N, 32.4 mg/kg available P, and 286.02 mg/kg exchangeable K. The soil was mixed with bovine manure, then strawberry plants were transplanted. Nutrients (containing N, B, Mn, Zn, and Fe) were sprayed eight times with and without the humic product (0.18 g humic product/l), which was sprayed foliar up to eight times until fruit maturity. The experiment was completed in a randomized block design with three replications for each treatment. It was found that the humic application at bloom were not beneficial to plant production and in fact decreased the fruit set. During fruit maturation and harvest, the product application resulted in greener leaves, as it enhanced the availability of nutrients. Humic was therefore best applied foliar for late applications, helping to sustain plant vigor during final fruit growth.

Mahoney et al. (2016) evaluated the performance of a fulvic product and a humic product, with an unknown FA or HA strength. The tested crops were dry beans of various types. The experiment was completed in a two-year period on 35 individual trials. The soil was described as silt loam to clay loam, containing over 3% soil OM, with good to excellent soil nutrients. Extra NPK were applied to soil during seeding. The fulvic product was applied pre-emergence, while the humic product was applied in seed furrows. The experiment was completed in a randomized block design with at least four replications for each treatment. Yield improvements were observed at high rates of the fulvic product, although they were not significantly different compared to the control. No effective responses were monitored in the case of the humic product when compared to those of the control. It was important to note that the fulvic product was originally recommended to be applied foliar to crops, and that nutrients in soil were more than adequate.

The non-positive responses to the application of humic products as shown by Pilanal and Kaplan (2003), Hartz and Bottoms (2010), Neri et al. (2002), and Mahoney et al. (2016) demonstrated critical issues with the application of humic products. These issues ranged over growing conditions, types of crops, effective rates, methods, and timing of application. In each case, the study might better be completed with a wider range of parameters, so that insufficient, optimum, and excessive growing conditions could be evaluated.

4.5 Efficacy Results for Livestock

A raw humic product containing 70% HA, 3.6% crude protein, 0.05% crude fat, and 9.0% Na was blended in feed at rates from 0 (control) up to 0.3% total feed by weight. It was found to be effective as an aflatoxin binder in broiler chickens exposed to the aflatoxin contaminated feed. There were 200 Ross broiler chicks included in this study (four replications per treatment with ten chicks per replication). The birds were fed with a blend of corn, soybean, and others, containing 2895.65 to 3155.90 kcal/kg metabolizable energy, 20.13 to 22.80% crude protein, 3.39 to 3.75% crude fat, 6.39 to 7.17% ash, and 0.40% P_2O_5. The feed was contaminated with aflatoxin B1 at a rate of 100 μg/kg. At the end of the study (after 42 days) at a humic product rate of 0.1%, the body weight gain was 13.7% higher than that of the control (100 μg/kg aflatoxin B1). At a product rate of 0.3%, the feed efficiency (feed to gain ratio) was enhanced by 7.4% and mortality rate was reduced by 50.0%. At the same product rate, the aflatoxin residue in liver was reduced by 83.0% and antibody titer was increased by 46.5% compared to those of the control (Arafat et al. 2017).

A processed humic product containing 5.4, 2.2, and 3.2 meq/g of total acidity, COOH group, and phenol-OH group, respectively (Dekker et al. 2003), was evaluated for its ability as an aflatoxin binder on broiler chickens. Based on an in-vitro study, adsorptions of 10.3, 7.4, and 11.9 mg of aflatoxin B_1/g of this product were observed at pH 3, 5, and 7, respectively. When blended with the feed at a rate of 0.35% by weight, the product was able to bind 92.9 and 87.5% of 2 ppm aflatoxin B_1 at pH of 3 and 7, respectively. An in-vivo study was then completed on 7-day-old male Ross 788 chicks. There were 420 chicks of similar weight assigned into seven treatment groups, with four replications per treatment, and 15 birds per replication. Aflatoxin B_1 rates were from 0 (control) up to 2 mg/kg feed. The diet contained feed, water, and the humic product at rates of 0 (control) and 0.35% total feed by weight. At the end of the study (after 42 days of age), the application of humic at 1 mg/kg aflatoxin B1 maintained body weight over the control, while liver lesions were reduced by 8.2%. At 2 mg/kg aflatoxin B_1, body weight was improved by 7.6% over the control with the incorporation of humic, and liver lesions were reduced by 34.3%. This indicated that the humic product performed best under a more toxic environment (Van Rensburg et al. 2006).

A processed humic product containing 65.6% organic acid (combined HA and FA), 1.04 to 1.20% N, 0.08% P_2O_5, 0.30% K_2O, 0.22 to 0.24% S, 0.37 to 0.74% Ca, 0.10 to 0.23% Mg, 6.7 to 8.9% Na, 0.15% Cl, 700 to 2940 mg/kg Fe, 45 to 77 mg/kg Cu,

34 to 54 mg/kg Zn, 34 to 52 mg/kg Mn, 20 to 27 mg/kg Se, and 0.5 to 1.9 mg/kg Cr was evaluated on cattle. There were 2000 mixed-breed beef cattle of about the same weight (328 kg) that were equally distributed in 16 pens. They were fed with the humic product at rates of 0 (control) and 0.006% body weight/day, plus a feed-lot diet containing a blend of sorghum, barley silage, and others. The feed had 2964 kcal/kg metabolizable energy, 14.8% crude protein, 3.0% fat, 8.4% crude fiber, 15.9% neutral detergent fiber, 0.81% Ca, 0.36% P, 0.16% S, 0.69% K, 25.6 mg/kg monensin, and 20.5 mg/kg virginlamycin. There was an adaptation period for the cattle of 16 days before the introduction of the feed. The experiment was ended after market specifications on the animal body weight (approx. 444 kg) and fat depth (approx. 9 mm) were met. The humic product increased average daily weight gains by 12.8%, enhanced the feed conversion ratio by 11.1%, and reduced days in feed (from 81.4 down to 69.0 days, that was a 15.2% decrease) compared to those of the control. Initial cases of bovine respiratory disease and mortalities were unchanged (Cusack 2008).

One hundred and eighty commercial layer hybrid Hysex Brown hens at 36 weeks of age were evaluated in three groups consisting of one control and two treatment groups. Each treatment group was replicated four times and consisted of 15 birds. They were fed with a blend of wheat, maize, soybean meal, and others containing 17.05% crude protein, 3.66% crude cellulose, 11.11% crude ash, 2.35% ether extract, 3.41% Ca, 0.58% P, 0.43% methionine, 1.32% lysine, and 2802 kcal/kg metabolizable energy. A processed humic product containing 85% polymeric poly-hydroxy acid was added to the feed at rates of up to 0.006% of total feed by weight. The animals were fed for 16 weeks and egg production records were taken daily. The egg production was best improved (2.8% over the control) at 0.003% humic product rate. The feed efficiency was enhanced best by 6.4% over the control at 0.006% humic product rate. At the same rate, the egg weight was improved by 3.1%. Egg shell thickness, egg yolk index, egg albumen index, and egg Haugh unit were all improved by up to 5.5% over the control at the higher humic rate (Kucukersan et al. 2005).

A processed humic product containing 0.016% combined humic, fulvic, ulmic, and humatomelanic acids, 0.07% silica oxide (SiO_2), 50 ppm Mn, 60 ppm Fe, 60 ppm Zn, 5 ppm Cu, 0.2 ppm Co, 1.1 ppm I, 0.5 ppm Se, and other trace minerals was evaluated on Isa × Isa straight-run broilers. There were 400 birds randomly distributed in 20 floor pens, and each pen consisted of 20 chicks. They were fed with a blend of ground corn, soybean meal, and others. It contained 3106 to 3256 kcal/kg metabolized energy, 21.03 to 23.51% crude protein, 1.17 to 1.41% lysine, 0.49 to 0.56% methionine, 0.98 to 1.10% methionine + cysteine, 0.98 to 1.10% Ca, and 0.40 to 0.50% P. The birds were fed with the feed and the humic product at rates of 0 (control) and 0.25% total feed by weight. Each broiler was weighed at hatch, 21, and 42 days. It was found that from hatch to 42 days, body weights increased by 3.5%, mortality decreased by 12.5%, and the feed to gain ratio was improved by 3.5%, compared to the control. The hot carcass weight at 42 days was improved by 6.4% and abdominal fat pad by 11.1% over the control (Kocabagli et al. 2002).

The same humic product was evaluated on Hisex Brown layers. Three hundred birds of 54 weeks of age and uniform body weight were fed with the product at rates of 0 (control) up to 0.2% total feed by weight as well as a commercial feed. The feed was a blend of corn, soybean meal, and others containing 2530 kcal/kg metabolizable energy, 15.53 to 15.66% crude protein, 3.86% Ca, and 0.61% P. The study was completed in 75 days. The humic product at 0.1% rate reduced chicken mortality by 46.9%, improved the feed to egg produced ratio by 9.1%, increased egg production by 9.1%, and improved egg weights by 2.2% over the control. At the higher rate of humic product, lower responses were observed, but all were still better compared to those of the control (Yoruk et al. 2004).

A processed humic product containing 16% HA with application rates from 0 (control) up to 0.06% total feed by weight was applied to nine-week-old female Japanese quails. There were 140 of them, assigned to one control and three replications of seven quails (each group with 35 birds). In addition to the humic product, each bird was fed with a blend of corn, soybean, and others, containing 19.98 to 20.03% crude protein, 2.50 to 2.51 % Ca, 0.34 to 0.36% P, 2890 to 2900 kcal/kg metabolizable energy, 1.11% lysine, and 0.70% methionine + cystine. At the end of the experiment (five weeks), the humic product at 0.036% improved blood parameters by 7.7% (red blood cells), 13.4% (hemoglobin), and 10.0% (hematocrit value) over the control, while maintaining its oxidative stress. Higher rates increased the oxidative stress and therefore were not recommended (Ipek et al. 2008).

Twenty late pregnant Barki goats (each 35 months old and with a body weight of approximately 45.4 kg) were allocated two treatments in a completely randomized design. For 70 days, they were fed with a blend of maize silage, berseem hay, cracked maize, and others (931.1 g/kg OM, 143.5 g/kg crude protein, 39.1 g/kg ether extract, 363.1 g/kg non-fiber carbohydrate, 385.4 g/kg neutral detergent fiber, 172.8 g/kg acid detergent fiber, 199.2 g/kg hemicellulose, 140.2 g/kg cellulose, 32.6 g/kg lignin, and 11.55 g/kg acid insoluble ash) and a processed humic product (containing 90% HA). The humic rates were 0 (control) and 2 g/day animal (treated). Compared to those of the control, when treated with humic, ruminal NH_3-N was reduced by 16.5%, total short chain fatty acids were up by 7.3%, and protozoa was lowered by 9.3%. The total protein in blood profile was enhanced by 7.9%, milk production was improved by 18.4%, and the feed efficiency was up by 20.7%. With respect to weaning goat kids, humic enhanced their weights after 60 days by 21.6% and their average daily weight gains by 31.3% over the control (El-Zaiat et al. 2018).

A raw humic product containing 56.2% C and with a molecular weight of 1127 was fed to sauvage fish. Four hundred and fifty uniform-sized juvenile fish were placed in 15 fully controlled-environment tanks and, after 14 days of acclimation, were fed with a control diet and the humic product for 60 days. The diet contained fish meal, corn, wheat meal, soybean meal, minerals, and vitamins. It had 34.6 to 35.1% crude protein, 5.2 to 5.6% crude lipid, 6.3 to 6.6% ash, 1.2 to 1.5% PO_4, and 1.5 to 1.8% Ca. The humic product rates were from 0 (control) up to 2.0% of total feed by weight. It was found that 1.5% product rate was the best: final average weights were enhanced by 14.1%, weight gains by 34.5%, specific growth rates by 9.3%, feed conversion rates by 22.3%, and survival rates by 1.2% over the control. At other rates, all responses were lower, but were still better or the same as those of the control. Activities of intestinal digestive enzymes were improved by approximately 74% at 2.0% product rate (for protease) and approximately 6% at 1.5% product rate (for lipase). Activities of intestinal antioxidant parameters were generally improved, with the best result observed at 1.0% product rate for superoxide dismutase, at approximately 50% over the control. Activities of intestinal non-specific immune enzymes also improved, with the best result observed at 2.0% product rate for Complement 3, at approximately 71% over the control. The intestinal microflora analysis also found that the product enhanced its diversity. The most obvious difference was found to be on the Simpson Index, where 1.5% product rate enhanced diversification by approximately four times (Gao et al. 2017).

A raw humic product with 3.5 to 5.0 pH and 25 to 35% HA (DTG 2019) was evaluated on crossbred piglets (Topigs Norsvin, Senden Deutschland; TN70 × Pietrain). Fifteen animals were weaned on day 21 and then housed in three boxes. They were fed for four weeks with a commercial feed containing 3274 kcal/kg metabolizable energy, 176 to 180 g/kg crude protein, 38.0 to 45.4 g/kg crude fat, 402 to 407 g/kg starch, 47.4 to 58.5 g/kg acid detergent fiber, 132 to 140 g/kg neutral detergent fiber, 13.3 to 13.4 g/kg lysine, 3.47 to 3.61 g/kg methionine, and minerals (mainly Fe and Zn at 468 to 505 and 128 to 130 mg/kg, respectively); and the humic product at rates from 0 (control) up to 3.0% total feed by weight. At the end of the study, the feed conversion ratio, animal body weights, and average daily weight gains were found to be enhanced best at 1.5% product rate (i.e. 4.9, 9.2 and 14.0% higher than those of the control, respectively). At the higher product rate, all responses were lower (Visscher et al. 2019).

A processed humic product with 7 pH and 12.05% FA was evaluated on crossbred (Landrace × Yorkshire) growing-finishing pigs, each with a body weight of approximately 30 kg. Two hundred and sixteen castrated male pigs were allocated to 36 pens, six pigs per pen. The animals were fed with a blend of maize, soybean meal, wheat bran, and others. The feed contained 3236 to 3248 kcal/kg metabolizable energy, 14.52 to 16.40% crude protein, 0.52 to 0.56% Ca, 0.19 to 0.22% P, and 0.72 to 0.85% lysine. Together with the feed, the animals were fed with the product at rates from 0 (control) up to 0.8% total feed by weight for 87 days. The study ended when each pig reached approximately 90 kg of body weight. It was found that 0.6% product rate gave the best gain to feed ratio, i.e. 9.1% over the control. With the humic product at its highest rate, both slaughter and hot carcass weights were improved by 6.7 and 8.6% respectively over the control. At rates of 0.6 to 0.8% total feed by weight, the incorporation of this product resulted in the reduction of backfat by 13.6%, increased longissimus muscle areas by 13.2%, and increased marbling by 34.8% (Bai et al. 2013).

A raw humic product containing 39.4% HA, 27.8% FA, 3.6% crude protein, 1.9% crude ash, and 0.05% crude fat was evaluated on the feed efficiency, blood characteristics, and meat quality of crossbred (Landrace × Yorkshire × Duroc) finishing pigs. There were 48 pigs with an average initial body weight of approximately 56.5 kg, fed with a commercial product and the humic product for eight weeks. The commercial product contained a blend of corn, soybean meal, wheat bran, and others. The feed had 3341 to 3353 kcal/kg metabolizable energy, 19.01 to 19.06% crude protein, 0.98 to 1.01% lysine, 0.38% methionine, 0.78 to 0.79% Ca, and 0.64 to 0.67% P. The humic product rates were from 0 (control) up to 10% total feed by weight. At its highest rate, the product improved average daily weight gains and the feed efficiency by 12.1 and 13.7%, respectively. The blood analysis showed a 34.4% increase in lymphocytes over the control. The meat analysis showed that the product improved marbling (by 67.9%), firmness (by 15.7%), and color (by 18.7%), as well as reducing backfat (by 17.8%) compared to those of the control (Wang et al. 2008).

A raw humic product containing 42.6% HA, 5.0% FA, 3.4% crude protein, 31.8% crude ash, 20.5% moisture, and < 0.1% crude fat was evaluated on (Camborough-22 × PIC boar) pigs. A commercial feed containing a blend of ground corn, soybean meal, dried whey, and others was applied together with the humic product. The feed contained 789 kcal/kg metabolizable energy, 22.9% crude protein, 1.55% lysine, 0.82% methionine + cysteine, 0.31% tryptophan, 1.11% threonine, 1.18% Ca, and 0.85% P. One hundred and twenty pigs (weaned at approximately 21 days, each with 6.45 kg body weight) were fed with the commercial feed and the humic product at rates of 0 (control), 0.5, and 1.0% total feed by weight. Each treatment had eight pens; each pen had five pigs. They were fed for 98 days until each animal reached 60 kg body weight. At 0.5% product rate, average daily weight gains were enhanced by 4.4% and the feed efficiency was increased by 6.8% over the control. All responses were lower at 1.0% product rate. One hundred and ninety-two barrow (male) and gilt (female) pigs (weaned at approximately 25 days, each with 8.21 and 7.92 kg body weight, respectively) were fed with the commercial feed and the humic product at rates of 0 (control) and 0.5% total feed by weight. Each treatment had four barrow pens and four gilt pens, that is, each had a total of eight pigs. The pigs were fed for 149 days until each animal reached 110 kg body weight.

For barrow and gilt, the humic product enhanced average daily weight gains by 11.6 and 23.4%, and the feed efficiency by 10.7 and 15.8% over the control, respectively. The product also improved hot carcass weights by 1.8%, reduced backfat thickness by 13.9%, and enhanced lean percentage by 3.4% compared to those of the control. Leg muscle depths and Japanese pass rates were all enhanced by 6.1 and 7.0% over the control, respectively. Ninety-six pigs (weaned at approximately 22 days) were allocated to six groups (16 pigs per pen), receiving the humic product at rates of 0 (control) and 0.5%. Each treatment had six pens, with eight pigs each in it. They were fed for 52 days and measured for aerial NH_3. Based on the average NH_3 concentration in 24 hours, the humic product reduced its level by 18.1% compared to the control (Ji et al. 2006).

A processed humic product containing 12.02% HA, 1.44% K, 0.61% Na, 0.13% P, 0.11% N, and 0.004% Fe was evaluated on the performance of an anaerobic lagoon. The lagoon received flushed water from swine barns close by. The water contained fresh feces and spilled feeds in the lagoon supernatant with the following properties: pH = 7.5, chemical O demand = 756 mg/l, total Kjeldahl N = 142 mg/l, fixed suspended solids = 295 mg/l, and total P = 46 mg/l (all mean values). The evaluation was completed in pipes of approximately 25 l capacity each, placed in the lagoon. Each represented an isolated lagoon mesocosm. The evaluation was completed in one month. Among the parameters evaluated, the humic product reduced methane (CH_4) gas and volatile organic compounds (tetrahydrofuran) in the pipe headspace by 34.0 and 57.2%, respectively, compared to those of the control at an optimum rate of 0.06% by volume of flush water. For the supernatant (top layer), the same rate reduced chemical O demand by 22.2%, fixed suspended solids by 30.4%, and total P by 13.0% compared to those of the control. The pH and total Kjeldahl N remained practically unchanged. For sludge (bottom layer), the optimum rate reduced chemical O demand by 2.4%, total Kjeldahl N by 1.8%, and total P by 0.9% compared to those of the control. The pH remained practically unchanged, while fixed suspended solids were not analyzed (Shah and Kolar 2012).

The same product was evaluated on two identical barns, tunnel-ventilated with deep pits, housing 1125 pigs each. Each barn had an equal footprint and pit capacity of approximately 1542 and 2270 m^3, respectively. The manure contained up to 6.4% total solids and had 8 pH. The product (525 l in total) was applied to the first barn periodically in several stages over 18 weeks. The second barn received none and functioned as the control. On week 20, each barn was monitored for odor flux using both field and laboratory oflactometry analyses and it was found that the first barn released 20.5% less odor compared to that of the control. Manure from each barn was then removed and surface applied to two separate fields, each with a 60 m diameter, and measured for odor concentration using both field and laboratory oflactometry analyses. A pile treated with humic was found to have 60.3% less odor strength than that of the one without the product (Fabian-Wheeler et al. 2015).

A raw product (containing 55% HA and FA; HumaTech 2018) was evaluated for its performance at reducing NH_3 emission from an open beef cattle feedlot at rates of 0 (control) and 9000 kg/ha. It was thoroughly blended with the mixture of soil, feces, and urine. The work was completed in a laboratory on 1950 g samples, which were connected to an NH_3 collection trap and monitored over 21 days. The product did not significantly change pH and moisture of the mixture; however, it reduced the NH_3 emission by 67.6% compared to that of the control at the end of the observation. It was hypothesized that N could be conserved, and N/NH_4 emissions were decreased by increasing C (i.e. HS) and N (i.e. manure) ratio (Shi et al. 2001).

4.6 Product Registration

Product efficacy data from a range of literature have shown both scientifically and practically that HS (including HA and FA), when applied at reasonable rates, are beneficial for soil, crops, and livestock. Still, limitations exist that must be taken into consideration to avoid any misconception and misunderstanding on the use of HS in the agricultural sector. These include the following:

1) crop growing conditions or livestock health conditions;
2) types and varieties of crops or livestock;
3) intended parameters to be evaluated;
4) characteristics of commercial humic products;
5) methods of product application;
6) timing of product application; and
7) product application rates (i.e. higher rates are not always better).

For a specific commercial humic product, all of these items can be summarized in a product application label. This is the main document accompanying the product during its marketing and referred to during field application. It does not contain detailed science on how humic products benefit agriculture, other than a short summary on its mechanism and intended use for a specific application. It is essential that the label is not only cleared in terms of usage but, most importantly, is precise, is not misleading, and meets all of the requirements specified under a given jurisdiction.

A jurisdiction in this context may relate to:

1) a region, such as a country, province, or state; or
2) a certain method or practice for how the products are manufactured or applied, which refers specifically to inputs used to produce certified organic products.

A regional jurisdiction is represented by an approved government agency. An organic product jurisdiction is represented by one (or more) approved private or government agencies. In order to meet all requirements within a certain jurisdiction, each product has to be registered. The product application label is the main document. It can be supported by other documents if this is requested by the agency.

Firstly, the applicant needs to decide whether a candidate product is intended for soil and crops or livestock. The same product may be applicable for both, yet still they have to be registered separately. Regulations for crops are way ahead of those for livestock, at least in North America (note: this may reflect the fact that humic applications for crops are more common than those for livestock). The applicant then needs to determine which region(s) this product is to be marketed and sold in, and whether it is to be marketed and sold as an organic product or not. Any approved organic products still have to be registered under regional jurisdictions to allow them to be marketed and sold there. Registrations may be imposed on the company, processing facilities, and/or finished products. Each product brand requires a unique registration.

Secondly, the applicant should contact the relevant jurisdiction(s) to obtain a product application form. The form contains detailed instructions, including requests for: the main information about the product, any supporting information, and applicable fees. The main information about the product is presented in the form of a product application label, containing the following:

1) identification: origin and name;
2) contents: HS, HA, FA, and/or nutrients contained within;
3) claims: purpose and efficacy;
4) applications: recommended rates, timing, and methods;
5) packaging; and
6) compatibility and safety.

Supporting information may include the following:

1) laboratory results for representative samples of the product;
2) product manufacturing, storage, and packaging;
3) field data on product efficacy, compatibility, and safety; and
4) existing registrations with other jurisdictions (if any).

Thirdly, the applicant is to provide all required information requested by the agency as well as any applicable fees. The agency reviews all information and may request that more be provided. The whole process may take anything from one week to a couple of months. The decision outcome is either that the product is registered (if approved) or it is rejected. Appeals are common and involve the provision of additional information about the candidate product and/or modifications on the product to meet requirements.

Fourthly, after the product has been approved for registration, the applicant has to maintain its status by completing the following tasks:

1) Periodically (usually annually) renew its registration, including paying the fees. Renewing a product is usually faster and simpler than submitting a new application. Any modifications to the product (name, formulation, claim, and packaging) and/or its intended application must be reported. The status of the product will then be determined, either as a new or modified product.
2) Periodically submit tonnage reports and fees for the product currently sold within a region. These can be monthly, quarterly, biannually, or annually. Organic product jurisdictions do not control sales within any regions; therefore, they do not have these requirements.
3) Submit additional information as requested by the agency. Regional agencies usually request such information when there are questions outstanding for registered products being marketed and sold there. Organic product agencies sometimes randomly select products currently being registered and request that updated information on them be provided.

No products are allowed to be marketed and sold prior to their approvals. Any violations may result in the products being red-tagged, or disallowed from being marketed and sold, until all steps have been completed. Some jurisdictions allow products in dispute to still be marketed and sold within a specific time allocated for the applicant to complete all the requested work. This usually happens for existing, modified, or renewed products, not for new products.

4.6.1 Regional Registration for Crops

A regional jurisdiction is represented by a designated government agency, i.e. the Canadian Food Inspection Agency (CFIA) for Canada, the California Department of Food and Agriculture (CDFA) for the state of California, the Colorado Department of Agriculture (CDA) for the state of Colorado, the Florida Department of Agriculture and Consumer Service (FDACS) for the state of Florida, the state of Hawaii Department of Agriculture (SHDA) for the state of Hawaii, the Idaho state Department of Agriculture (ISDA) for the state of Idaho, the Oregon Department of Agriculture (ODA) for the state of Oregon, the South Dakota Department of Agriculture (SDDA) for the state of South Dakota, the Washington state Department of Agriculture (WSDA) for the state of Washington, and so on. No agencies related to fertilizers and soil amendments exist at the federal level of the US government; all registrations are handled at the state level. Note: with respect to the use of pesticides and biostimulants in agriculture, all products must be registered at the federal level, which creates a unique position for HS, HA, and FA, as discussed later in this section. The main information about the product is to be included in the product application label, which is ideally applicable for one jurisdiction only. The agency will review the label as well as other information provided in the application form. When necessary, the agency may request supporting information from the applicant. With the exception of CFIA, all reviews are completed on paper. Site visits or inspections are not required.

The origin of the product indicates where the manufacturing facilities are. This may create challenges, as products that are registered out of country, province, or state may not meet requirements set by the local jurisdiction. The differences here can be as simple as approved terminology or wording, formatting of more detailed information requested on product efficacy and/or safety, and, most critically, the humic methods of analysis. Each product claims HA and/or FA as percentages of the product. No jurisdictions consider HA and FA to be "fertilizers," since they do not contain significant amounts of nutrients. Instead, they are considered "soil amendments" or other similar terms in this category (such as "supplements"). CDFA defines "auxiliary soil and plant substance" to be any chemical or biological substance or mixture of substances or device distributed in this state to be applied to soil, plants, or seeds for soil corrective purposes. This does not include commercial fertilizers, agricultural minerals, pesticides, or soil amendments, except biochar or manures. "Packaged soil amendment" means any substance distributed for the purpose of promoting plant growth or improving the quality of crops by conditioning soils solely through physical means. Processed humic products fit more in the first category, while raw humic products belong to the second category (California Department of Food and Agriculture 2017a). CDA defines "plant amendments" as any device or substance supplied to soil, plants, or seeds which is intended to improve germination, growth, yield, product quality, reproduction, flavor, or other desirable characteristics of plants, except commercial fertilizers, soil amendments, agricultural liming materials, animal and vegetable manures, pesticides, plant regulators, and other materials which may be exempted by regulations promulgated by the commissioner. CDA defines "soil conditioner" as a substance, defined by regulation of the commissioner, intended to improve the chemical or physical characteristics of the soil. It does not include commercial fertilizers, plant amendments, agricultural liming materials, untreated animal or untreated vegetable manures, untreated peat moss, untreated peat humus, or any other materials that may be exempted by regulation of the commissioner. In this regard, processed humic products are considered "plant amendments," while raw humic products are "soil conditioners" (Colorado Department of Agriculture 2005). Similar terms and definitions are provided by other jurisdictions, which are not presented here.

Several jurisdictions, including SHDA (State of Hawaii Department of Agriculture 2020), FDACS (Florida Department of Agriculture and Consumer Service 2019), and WSDA (Washington State Department of Agriculture 2020) do not require humic products to be registered. Such products may be directly advertised and sold without such procedures being completed. Other jurisdictions require product samples to be taken (which may include specific procedures to follow) and analyzed for HA and/or FA. The challenge is deciding what method to use to determine HA and/or FA contents within the sample. The most common ones are the colorimetric and CDFA methods. The newly established ISO 19822 is gaining popularity due to its accuracy. The colorimetric is the most popular method to use due to its practicality. This method also gives the highest HA results compared to others. Several jurisdictions, including CDFA, ODA, WSDA, and ISDA (as included in the "Oregon Fertilizer Program Guide" document), recognize the status and position of HA and hydrophobic FA. The CDFA method and ISO 19822 are the only recognized methods for each, respectively (Oregon Department of Agriculture 2019). Note: 1) while the state of Washington has been included in this document, it does not require any commercial humic products to be registered within its jurisdiction (as noted above); 2) ISO 19822 is sometimes identified as "the analytical method of Lamar et al. 2014, Journal of AOAC International, 97 (3): 721-730." For the analysis of HA, the state of California has recently been considering switching to ISO 19822 (Humic Products Trade Association 2017). Lately, the CDFA has approved the use of the HPTA (or ISO 19822) test method seal for all HA products sold in California (Humic Products Trade Association 2020).

Another thing to watch carefully is the moisture content. "As is analysis" includes all components within the product: HS, ash, and moisture. "Dry matter" analysis only includes HS and ash, and therefore shows higher humic values. When nutrients are also present, it is important to watch what forms being claimed. CFIA recognized total N but only the water-soluble components for P, K, and other nutrients. Other jurisdictions may ask only for portions of N being claimed, i.e. NH_4, NO_3, or nitrites (NO_2). These discrepancies may result in confusion at the fact that the total of the product ingredients does not add up to 100%.

If the humic product contains sufficient nutrients, it may be registered as a fertilizer. Claims here have to be made based on the presence of nutrients. How much nutrients are adequate for a product to be considered a fertilizer depends on the jurisdiction. The Fertilizers Act and Regulations in Canada, for example, defines "fertilizers" as plant nutrients, while "supplements" (used as the term here instead of "amendments") as substances other than nutrients that improve plant growth and crop yields. To be precise, "fertilizer" means any substance or mixture of substances, containing N, P, K, or other plant food, manufactured, sold, or represented for use as a plant nutrient (including the one derived through biotechnology), while "supplement" (or "amendment") means any substance or mixture of substances, other than a fertilizer (including the one derived through biotechnology) that is manufactured, sold, or represented for use in the improvement of the physical condition of soils or to aid plant growth or crop yields (Minister of Justice 2015 and 2020). The amount of major plant nutrients (N, P, and K) contained in every "mixed fertilizer" cannot be less than 24% of the total content. For a "fertilizer," it must contain at least 50% that is of animal or vegetable origin that supplies 25% N and at least 18% that is major plant nutrients. Some exceptions apply, including customer-formula, specialty, and other fertilizers registered under the Act (Canadian Food Inspection Agency 2009). This of course may not apply for other jurisdictions. However, it provides an indication that if a product contains HS and significant (or in that proximate) nutrients, then it may be considered to be a fertilizer. The most practical way to proceed is to contact the agency, address the fact that the product contains both HS and nutrients (including macro and micronutrients), and seek advice on the status of the product.

Depending on the jurisdiction, proof of efficacy may be required for HS, HA, and FA claims, including Canada. For Canada, there is a list of supplements, in which their registrations are exempt, as their efficacy has been accepted. Schedule II Class 5 includes OM supplements like compost, humus or leaf mould, and peat. However, HS, HA, FA, and their materials of origin are not included and therefore proof of efficacy has to be provided for their registrations (Minister of Justice 2015). Proof of efficacy, among other things, includes actual field trials on relevant soil and growing environment appropriate to the region, in which the products are to be marketed and sold. If the humic product is not claimed to be based on their main function, then it may be exempt from registration. HS products of raw leonardite have approximately 4 pH (Canadian Humalite International Inc. 2020a). If it is claimed that they alter soil pH, then their registrations are not required. For other states, no proof of efficacy is required. HS, in general, are recognized to improve soil texture and aid nutrients uptake by plants. While most jurisdictions do not ask for proof of efficacy for HS, a statement is still needed that the intention of the product is to enhance soil texture and nutrients uptake by plants.

Recently, a document has been released by CFIA with the intention of replacing the role of Schedule II in the current Fertilizer Regulations when the proposed modernized regulations come into force. The document is titled "Primary Fertilizer and Supplement Materials," and is part of the Guidance Document Repository. Among other things, the document lists primary supplement materials, which now include FA, HA, HS, and humus. FA are defined as the extracted portion of HS that are soluble in both alkali and acid aqueous solutions. HA are defined as the portion of the alkali-extracted HS that are insoluble in strongly acidic solution. They will precipitate from the alkali extraction in acid solutions of

2 pH or less. HS are defined as the major organic constituents of soil OM and the aquatic environment, consisting of complex heterogeneous mixtures of C-based substances formed by biochemical reactions during the decay and transformation of plant and microbial remains. They are primarily composed of three main fractions: HA, FA, and humin, which are operationally defined by their differential solubility in dilute alkali and acid solutions. High concentrations of HS can be derived from terrestrial deposits of leonardite, oxidized lignite, oxidized subbituminous coals, humalite, carbonaceous shale, peat, and sapropel. Humus is defined as a homogenous and friable mixture of biochemically stable partially decomposed OM (Canadian Food Inspection Agency 2020). It is interesting to notice similarities between these CFIA definitions and inclusions and the ones determined by the Association of American Plant Food Control Officials (AAPFCO). Should this document be officially accepted, FA, HA, HS, and humus containing products become "exempt from registration," in the sense that an actual registration is still required but proof of efficacy is not.

Similar to CFIA, SDDA previously would not issue any registrations or sanction the distribution of any products containing HA and FA as ingredients in the state of South Dakota. Therefore, any products containing HA and/FA would require proof of claims of usefulness and value resulting from three years of data in the state or similar regions. Claiming that such products contain OM (the parent group of HA and FA) would not be acceptable, and the submitted label would be considered misbranded. Recently, however, the state has recognized HA and FA, and any commercial humic products can be claimed to be soil amendments (South Dakota Department of Agriculture 2020).

Attitude changes in Canada and the state of South Dakota are two good examples showing that more and more jurisdictions are recognizing the efficacy of HS (as well as HA and FA) for soil and crops. Scientific information on the benefits of HS for the agricultural sector has currently been accepted as valid, not only in the laboratory, but most importantly in agricultural fields. Still, the basic understanding of the material has not changed. HS are soil amendments or supplements, not fertilizers, and have to be represented accordingly.

Claims on FA contents are more challenging since no standard methods are officially adopted by most jurisdictions. Neither the colorimetric nor CDFA methods quantify FA. ISO 19822 is the only accurate one available for this purpose. Some use the mass balance method, which is only an approximate measure of how much OM is contained within the product, and relies of the fact that the product usually has low pH (thus other humic components, i.e. HA and humin, are not present). Others estimate the amount of FA as being the difference between HA + FA (the colorimetric method) and HA (the CDFA method). As discussed in Section 3.2.5, this approach is not acceptable, considering that combining two analytical methods of different principles is inherently incorrect. A slightly better approach is to estimate the amount of FA as being the difference between HA + FA (the colorimetric method) and HA (the modified colorimetric method, as suggested by Lamar and Talbolt 2009). This approach is acceptable, considering that both results come from the same colorimetric principle. However, it is less accurate than ISO 19822, which has the ability to directly and accurately quantify FA.

Several jurisdictions, including Canada, California, and Oregon, require product samples to be taken and analyzed for heavy metals and/or pathogens (Canadian Food Inspection Agency 2017a, California Department of Food and Agriculture 2017a, and Oregon Department of Agriculture 2019). Safety may apply to target crops and/or soil or environment. Others do not need such information, assuming that product efficacy guarantees its safety.

Recommended rates (in kg/ha for dry products and l/ha for liquid products) are ideally determined for a specific or regional growing environment with some variation in soil type, climate, and other growing environmental factors. This approach, however, will result in a huge cost and long timeline for any manufacturer needing to complete field trials for each application. In many cases, the recommended rates are provided in ranges, which include growing variations within the region. It is still recommended that the producers consult with agricultural coaches before applying them to the field, where soil samples are typically required and reviewed. When the products also contain nutrients, their recommended rates are usually crop specific.

Recommended rates depend on the amounts of HA and/or FA contained in the product as well as product solubility in water. Raw and pelletized humic products have the highest rates due to their low water solubility, followed by liquid humic products, and lastly dry soluble humic products. In most cases, dry soluble humic products are to be made into liquid products before field application. Recommended rates are also related to their field application methods. A raw or pelletized product can be broadcasted (at high application rates) or side- or mid-row banded (at low application rates). A liquid product can be broadcasted or sprinkled, boom-sprayed, or dripped. Broadcasting or sprinkling requires the highest recommended rate, followed by boom-sprayed, then dripped with the lowest application rate. Lastly, frequency and timing of the field application determines recommended rates. Larger amounts of a raw humic product can be applied only once in fall, allowing it to break down ready for the following spring growing season. Smaller amounts of a liquid or dry soluble humic product can be applied once during seeding or on seedlings, or several times on mature crops at several growing stages, typically before flowering.

Packaging also has to be declared for a product to be marketed and sold under a certain jurisdiction. This includes product weight or volume, as well as its density or specific gravity. Usually it can be categorized as either large or small packaging, which may determine its registration requirements. Any modifications on packaging have to be reported to the relevant agency and are usually considered to be a minor change.

Product compatibility with nutrients and/or agricultural chemicals needs to be carefully addressed considering that HA are not soluble at low pH. Precautions must be taken when mixing the product with low pH nutrients. HA and FA may also react with certain types of N (NH_4 and NO_3) and metals (Ca, Mg, Fe, and others). Dilution with water is critical to ensure that the final blend can be applied on the field using the typical applicators. Most importantly, a small laboratory-scale trial is always recommended prior to any applications to the field.

A regional agency provides a registration number or certificate for an approved product indicating that the product is allowed to be marketed and sold within the jurisdiction until its expiry date. An organic product agency provides a registration number or certificate for an approved product indicating that the product can be marketed and sold as an organic product within the jurisdiction until its expiry date. This certificate must be presented together with the product at all times during its marketing, storage, and field application, together with its application label and safety data sheet (SDS).

In addition to state jurisdictions, the US federal government, through the Federal Insecticide, Fungicide, and Rodenticide Act (FIFRA), requires the registration of all pesticide products intended for use within the country before they can be marketed and sold at the federal level. This creates a unique status and position for HS, HA, and FA in the agricultural sector. Pesticides are defined as any substances that are intended for a pesticidal purpose, i.e. for use for the purpose of preventing, destroying, repelling, or mitigating any pests or use as a plant regulator, defoliant, or desiccant. With respect to plant regulators, they are defined as products other than (US Environmental Protection Agency 2019):

1) plant nutrients and trace elements (also known as fertilizers, containing one or more macronutrients or micronutrient trace elements necessary to the normal growth of plants in a form readily usable by plants);
2) plant inoculants (containing microorganisms to be applied to the plant or soil for the purpose of enhancing the availability or uptake of plant nutrients through the root system);
3) soil amendments (including soil additives and conditioners, containing a substance or substances intended for the purpose of improving soil characteristics favorable for plant growth); and
4) vitamin-hormone products (containing a mixture of plant hormones, plant nutrients, inoculants, or soil amendments).

Based on these definitions, raw and pelletized humic products, as long as they are claimed for the improvement of soil, can be classed as soil amendments. Enhanced (nutrients enriched) liquid humic, liquid fulvic, pelletized humic, and dry soluble humic products, as long as they contain significant amounts of nutrients and claim to provide nutrients necessary to the normal growth of plants, may be classed as plant nutrients and trace elements. In this case, they do not require any registration. Liquid humic, liquid fulvic, and dry soluble humic products without any significant nutrients within them, are applied to crops and/or soil in relatively small amounts and do not therefore qualify to be classed as plant nutrients and trace elements, nor as soil amendments. They are classed as plant regulators, which fall under the FIFRA, and together with defoliants, desiccant, and N stabilizers, are considered pesticides. These products require registration in this regard.

In addition, the US federal government, through the US Environmental Protection Agency (USEPA), recognizes new but growing types of products known as plant biostimulants (PBS), which are discussed in a draft guidance document. They are defined as products containing naturally occurring substances and microbes that are used either by themselves or in combination with other naturally occurring substances or microbes for the purpose of stimulating natural processes in plants or in the soil in order to, among other things, improve nutrient and/or water use efficiency by plants, help plants tolerate abiotic stress, or improve the physical, chemical, and/or biological characteristics of the soil as a medium for plant growth. While there is no applicable regulatory definition of PBS under FIFRA, this draft guidance has been issued under FIFRA, with the main intention of providing guidance to USEPA personnel and decision-makers, as well as pesticide registrants on identifying product label claims. Under this document, HA, FA, and organic acids from leonardite are considered plant regulators, and therefore fall under FIFRA as pesticides (US Environmental Protection Agency 2019). They require registration in this regard.

To be specific, under this document, HA and FA are classed as complex polymeric polyhydroxy acids (CPPA). CPPA are defined as a complex mixture of naturally occurring organic substances found in dead plant materials. The components of CPPA are widespread in nature, being found in soils, freshwater, and saltwater environments as a result of decaying plant materials, and are used to condition agricultural soils. Its major components are humic acid, fulvic acid, and tannins, and their relative concentrations in soil and water systems are influenced by environmental conditions, such as climate, soil types, vegetation, and hydrology. CPPA is made by concentrating the organic substances from water leached through forest soil using a proprietary manufacturing process (US Environmental Protection Agency 2013). As discussed by Humic Product Trade Association (2019), the definitions of HA and FA by USEPA may not meet the ones understood as per ISO 19822. The Humic Products Trade Association (HPTA) therefore suggests the USEPA apply good scientific testing as a basis for regulating products that may not be classed as plant regulators by FIFRA.

4.6.2 Organic Registration for Crops

One of the most important applications of humic products (and also an excellent marketing line) is in the production of organic certified products for people. The most recognized organic jurisdiction is the US Department of Agriculture-National Organic Program (USDA-NOP), which is represented by private agencies including Organic Materials Review Institute (OMRI), ProCert, and EcoCert. Other jurisdictions include the Washington Organic Standard (represented by itself, WOS), the California Organic Standard (represented by itself, COS), the Canada Organic Regime (COR, represented by OMRI Canada, ProCert, and EcoCert), and the Japanese Agricultural Standard (JAS, represented by ProCert). However, it is important to note that humic products themselves are not considered to be "products" by these jurisdictions, and therefore cannot be certified accordingly. Instead, they are considered as "inputs" that may be reviewed and approved as being in compliance with the standards for the production of certified organic products.

OMRI reviews the application label and other requested information submitted by an applicant according to the USDA-NOP standard as per the NOP regulations at 7 CFR Part 205, the USDA-NOP Program Handbook, and the OMRI Generic Materials List (Organic Materials Review Institute 2017a and 2017b). The NOP regulations differentiate between synthetic and non-synthetic substances, agricultural and non-agricultural substances, and organic and non-organic (genetically modified) inputs.

With respect to crop production materials, OMRI uses three classes of materials: crop fertilizers and soil amendments (CF); crop pest, weed, and disease control (CP); and crop management tools and production aids (CT). Crop fertilizers contain plant nutrients that can be applied to soil or plants. Soil amendments include all materials that can be applied as a soil conditioner. These substances must meet 7 CFR Part 205.203 of the NOP regulations. Crop pest, weed, and disease control substances are used as pest control, weed control, and plant growth regulators; and must meet 7 CFR Part 205.206 of the NOP regulations. Most of them must be registered with the USEPA (see Section 4.6.2). Crop management tools and production aids do not provide plant nutrients, soil conditioning, or crop production function, intended for post-harvest handling, adjuvants, equipment cleaners, compost inoculants, and plant extracts. In general, 7 CFR Part 205.105 of the NOP regulations applies for these substances.

Each material has one of the following designations: allowed (A), allowed with restrictions (R), and prohibited (P). Any approved materials are then listed under OMRI Generic Materials List as crop production materials. Without going into too much detail, the following are critical items to discuss. Allowed substances include non-synthetic materials not prohibited by 7 CFR Part 205.602 and synthetic materials allowed by 7 CFR Part 205.601 of the NOP regulations. Allowed with restrictions substances include soil fertility, crop nutrition, and crop pest/weed/disease management practice standards (as per 7 CFR Parts 205.203 and 205.206 of the NOP regulations), as well as some annotations and exceptions (as per 7 CFR Parts 205.601 and 205.602 of the NOP regulations). Prohibited substances include synthetic materials not listed under 7 CFR Part 205.601 and non-synthetic materials prohibited under 7 CFR Part 205.602, which are generally defined under 7 CFR Part 205.105 of the NOP regulations.

In addition to crop production materials, OMRI also lists processing and handling materials, including processing agricultural/non-agricultural ingredients and aids, pest controls, sanitizers and cleaners, and packaging and containers. These materials must also meet 7 CFR Part 205 of the NOP regulations.

USDA-NOP allows humates (defined as stable decomposed OM), which include leonardite, lignite, or coal. When processed or altered, humates become humic acids, which can be non-synthetic (naturally occurring deposits of humic acids and water-extracted humates are both allowed), alkali-extracted (allowed), or derivatives fortified with prohibited synthetic fertilizers (prohibited). Fulvic acids, defined as fractions of humates soluble at neutral or acid pH, are allowed.

Potassium hydroxide (KOH) or caustic potash is allowed with restrictions to manufacturing alkali-extracted humic products. In other words, it may be used to raise the pH of humic solutions, as raw humates are typically only soluble in water at high pH. Ammonium hydroxide (NH_4OH) is not listed, however together with KOH, is mentioned under "humic acids-alkali extracted material" and is classed as materials allowed for extracting humates. No claims on nutrients (i.e. K and N) are allowed, however. Sodium hydroxide (NaOH) or caustic soda is actually the most effective extraction chemical, but it is prohibited and the presence of Na in the product is not beneficial for soil and crops. In addition to (or even instead of) KOH, H_2O_2 (allowed with restrictions), O_3 (allowed), and non-synthetic or restricted microbial products or microorganisms (allowed) may be used as processing aids that may help to break down humates in water. Note: unlike USDA-NOP, JAS does not allow any chemicals to extract humates. This may indicate that either mechanical or microbial extraction is the only allowable method.

Adding nutrients to fortify humic products can be a challenge since the use of most synthetic materials, such as urea, NH_4 products, KOH, phosphoric acid (H_3PO_4), and sulfuric acid (H_2SO_4) are prohibited. The following are selected materials that are allowed or allowed with restrictions, containing a significant percentage of macro and micronutrients. Some synthetic forms are still allowed, yet the product list has to carefully be consulted. The values are approximated as they may vary from one product or analysis to another. Moisture and impurities contained within each product are considered (Canadian Humalite International Inc. 2020b and Card et al. 2015):

1) blood meal, fish products, and Chilean nitrate ($NaNO_3$) contain 10 to 15% N;
2) guano (high P), bone meal, and phosphate rock contain 10 to 25% P_2O_5;
3) K_2SO_4 and potash contain 50 to 60% K_2O;
4) gypsum, manganese sulfate ($MnSO_4$), and K_2SO_4 contain 50 to 55% SO_4;
5) gypsum, phosphate rock, and limestone contain 20 to 40% Ca; quick lime (CaO) and hydrated lime ($Ca(OH)_2$) contain 65 to 70% Ca;
6) Epsom salt, dolomite, and magnesium chloride ($MgCl_2$) contain 10 to 25% Mg;
7) boric acid (H_3BO_3) contains 55% B_2O_3;
8) KCl and $MgCl_2$ contain 45 to 70% Cl;
9) copper oxide (CuO) and copper sulfate ($CuSO_4$) contain 25 to 75% Cu;
10) ferric oxide (Fe_2O_3), ferrous sulfate ($FeSO_4$), and ferric sulfate ($Fe_2(SO_4)_3$) contain 30 to 65% Fe;
11) manganous oxide (MnO) and $MnSO_4$ contain 30 to 75% Mn;
12) molybdic oxide (MoO_3) contains 65% Mo; and
13) zinc sulfate ($ZnSO_4$) and zinc oxide (ZnO) contain 45 to 75% Zn.

Note: salts like gypsum, potash, and others contain certain amounts of water and impurities that must be taken into consideration when calculating their effective (nutritional) components.

Non-synthetic acetic acid ($C_2H_4O_2$, or vinegar, not glacial acetic acid), ascorbic acid ($C_8H_8O_6$), and citric acid ($C_6H_8O_7$) are allowed to be used to lower the pH of liquid humic products. They are, however, weak acids that, when used in conjunction with KOH, must be used in large amounts to be effective. Low-rank coal is also a weak acid that can be used to lower the solution's pH. H_3PO_4 and H_2SO_4 are much stronger. However, they are only allowed with restrictions for stabilizing liquid fish products, not humic products. Defoamers containing organic agricultural and restricted ingredients are allowed with restrictions, including silicone-based products. Their addition is sometimes required as reactions between humates and certain chemicals may produce excessive foams. Cl materials including chlorine dioxide (ClO_2) and sodium hypochlorite (NaOCl, or bleach) are allowed with restrictions to maximum labeled rates for disinfecting and sanitizing equipment or tools.

Lignin sulfonates and calcium lignosulfonate ($C_{20}H_{24}CaO_{10}S_2$), as long as no claims on nutrients are made, as well as bentonite and clay, are allowed. They are typical fillers used to manufacture some dry humic products. Non-synthetic dust suppressants are allowed, which are commonly required when handling materials containing very small particulates. Calcium stearate ($C_{36}H_{70}CaO_4$), which is a common flowing or anti-caking agent, is, however, prohibited.

In Canada, the Canadian General Standards Board provides guidelines for organic production systems on general principles and management standards (Canadian General Standards Board 2015a) and permitted substances lists

(Canadian General Standards Board 2015b). Among other things, these guidelines include the following:

1) organic product composition: product composition (solid and liquid products for crop and livestock productions) and categorization of organic products (based on percentages of organic content);
2) humates, humic acid, and fulvic acid are permitted as organic products under soil amendments and crop nutrition criteria, as long as they are extracted using non-synthetic substances, microbial fermentation, or KOH (not to exceed the amount required for extraction), and should not have arsenic (As), cadmium (Cd), Cr, Cu, lead (Pb), and mercury (Hg) levels exceeding their limits as set by Developpement Durable, Environnement at Parcs (2008) at 13, 3, 210, 400, 150, and 0.8 mg/kg (or ppm), respectively;
3) other permitted organic products under soil amendments and crop nutrition criteria include blood meal, fish meal, Chilean nitrate, guano, bone meal, phosphate rock, K_2SO_4, potash, gypsum, $MnSO_4$, limestone, Epsom salt, $MgCl_2$, CuO, CuSO4, Fe_2O_3, $Fe_2(SO4)_3$, MnO, $MnSO_4$, $ZnSO_4$, and ZnO, which are also accepted under USDA-NOP (see the previous page); and
4) permitted organic products under crop production aids include acetic acid (vinegar, not glacial acetic acid), ascorbic acid, citric acid, lignin sulfonates, calcium lignosulfonate, and bentonite, which are also accepted under USDA-NOP.

These guidelines do not have any legal bearing for any organic products to be marketed and sold in Canada or any other jurisdiction. However, they provide important information on the principles and standards of the organic agricultural practice, which are frequently requested by end users (Canadian Humalite International Inc. 2020b).

4.6.3 Registration for Livestock

Similar to the crop application, commercial humic products for livestock have to be registered before they can be marketed and sold under a certain jurisdiction. The most recognized organic jurisdiction is the USDA-NOP, which is represented by private agencies, most importantly OMRI. Organic Materials Review Institute (2017a) and (2017b), and Cornell Law School 2020) are the main reference.

Under livestock production materials, there are four use classes, i.e. livestock feed ingredients (LF), livestock health care (LH), livestock external parasiticides and pesticides (LP), and livestock management tools and production aids (LT). Livestock feed ingredients are added to livestock feed in small quantities, identified as feed additives and feed supplements, to fulfill a specific nutritional need. While "feeds" contain significant amounts of nutrients, feed "additives" include vitamins and minerals, and feed "supplements" include substances to improve the nutritional balance or performance of a total feed ration as per the Association of American Feed Control Officials (AAFCO). They can be fed blended with other feeds or as a free choice. Animal by-products are not allowed for mammals and for poultry applications, while agricultural products must be from certified organic sources, with their uses to meet livestock feed standards as per 7 CFR Part 205.237 of the NOP regulations. Carriers are allowed by AAFCO as long as they meet the same regulation. A carrier is defined as "an edible material to which ingredients are added to facilitate uniform incorporation of the latter into feeds, in which active particles are absorbed, impregnated, or coated into or onto the edible material in such way as to physically carry the active ingredient." Livestock health care materials include drugs, parasiticides, health care substances, non-synthetic and synthetic medications, and biologics (as per 7 CFR Parts 205.238, 205.603, and 205.604 of the NOP regulations). Livestock external parasiticides and pesticides include pesticides used to manage parasite and pests in livestock facilities as per 7 CFR Parts 205.603 and 205.604, and must meet 205.238 of the NOP regulations. Livestock management tools and production aids are the ones used for cleaning, grooming, and other non-direct functioning on animals, as per 7 CFR Parts 205.603, 205.604, and 205.105, and 205.200 of the NOP regulations.

Each material for the livestock production has one of the following designations: allowed (A), allowed with restrictions (R), and prohibited (P). Any approved materials are then listed under OMRI Generic Materials List as livestock production materials or processing and handling materials. Allowed substances are non-synthetic materials not prohibited by 7 CFR Part 205.604 and synthetic materials permitted under 7 CFR Part 205.603 of the NOP regulations. Allowed with restrictions substances include livestock feed standards, health care practice standards, pest and parasite management standards, and specific annotations as per 7 CFR Parts 205.237, 205.238, 205.603 of the NOP regulations. Prohibited substances are defined in 7 CFR Part 205.105 with additional requirements under Parts 205.603 (synthetic materials that are not specifically listed here) and 205.604 (non-synthetic materials that are specifically prohibited here) of the NOP regulations.

In addition to livestock production materials, OMRI also lists processing and handling materials, including processing agricultural/non-agricultural ingredients and aids, pest controls, sanitizers and cleaners, and packaging and containers. These materials must also meet 7 CFR Part 205 of the NOP regulations.

Unfortunately, HS, HA, FA, and their related products are not found under the livestock listing. The closest candidates for them are carriers and minerals (both are livestock feed ingredients). In order for a humic product to be considered a carrier, it has to be claimed that it physically carries the active ingredient. In this case, the product has to contain an optimum texture, particulate sizing, and moisture content benefitting the overall blend of the feed ration. They have to be used in accordance with the Food and Drug Administration (FDA) and AAFCO. Minerals are allowed by FDA and listed by AAFCO, as long as they are not derived from mammalian and poultry slaughter by-products. They cannot be fed in amounts above those needed for adequate nutrition and health maintenance for the animal at its specific stage of life (Organic Materials Review Institute 2017a). Under AAFCO, a carrier for premixes (including for trace minerals) should constitute at least 1% of the total ration. No specific lists are provided for carriers, however. Many products are listed under "additives," including reed-sedge peat, which has some similarity in origin to that of humic-containing materials. While there are many mineral products, none of them are directly related to humic-containing products. However, their definitions do not preclude humic materials like low-rank coal (lignite, leonardite, or humalite), i.e. as "ingredients that come from mined and processed rock and ore deposits, chemically manufactured salts, recovered natural salts, residue or remains of living organisms, and organic salts or organically bound elements as well other similar ingredients."

Since low-rank coal usually contains a relatively high Fe content (mean = 0.57% from 35 data, Canadian Humalite International Inc. 2020a), its content may be claimed as Fe_2O_3, which is listed under mineral products. HPTA has petitioned FDA for new regulations that allow for the safe use of humate, HS, and FA as a source of Fe in animal feed. HS are defined as natural OM found in soil, contributing to the brown or black color in surface soil. FA are soluble compounds that are extracted from HS (Wheeler 2015).

Another category included in AAFCO is the "special purpose products." Among other products being listed, manure odor control may fit with humic-containing products (Association of American Feed Control Officials Inc. 2016). As shown earlier in this chapter, humic products have been proven to mitigate odor generated from manure pile stocks effectively.

In Canada, the Feeds Act and Regulations are the main legislative authorities with respect to livestock feeds marketed and sold within the country. Others include the Health of Animals Act, the Health of Animals Regulations, and the Organic Products Regulations (Canadian Food Inspection Agency 2017b). "Feed" is defined as any substance or mixture of substances containing amino acids, antioxidants, carbohydrates, condiments, enzymes, fats, minerals, non-protein N products, proteins or vitamins, or pelletizing, coloring, foaming, or flavoring agents and any other substance manufactured, sold, or represented for use for consumption by livestock, providing the nutritional requirements of livestock, or the purpose of preventing or correcting nutritional disorders of livestock, or any substance for use in any such substance or mixture of substances (Minister of Justice 2019b). "Supplement" is defined as a feed that is used with another feed to improve the nutritive balance of the total and that is intended to be fed undiluted as a supplement to other feeds, offered free choice with other parts of the ration separately available, or further diluted and mixed to produce a complete feed that is acceptable for registration (Minister of Justice 2019a). Since low-rank coal does not contain significant amounts of nutrients, it is best classed as a supplement, which is still considered to be a feed under this jurisdiction. Low rank-coal, humic, HS, HA, and FA are not included in Part II of the Feeds Regulations Schedules IV and V. They list substances that are exempt from registration as they are already recognized by the Act and Regulations, while other substances have to be registered (their safety and efficacy need to be proven). Fe_2O_3 is listed under Schedule IV, Miscellaneous Products. Since low-rank coal usually contains a relatively high amount of Fe, it may be classed under this category.

In the state of California, "commercial feed" is regulated and defined as all materials which are intended for use as feed or mixing in feed except preparations which are manufactured and distributed for feeding to domestic pets, such as dogs, cats, and birds. Further, "mineral feed" is defined as a commercial feed intended to supply primarily mineral elements or inorganic nutrients for animal nutrition. "Inert material" is defined as ingredients that are not assimilated in the digestive process, including sand, granite grit, charcoal, and clay. Different forms of Fe are listed under "mineral feeds." As already noted, since low-rank coal contains a relatively high amount of Fe, it may be classed under this category (California Department of Food and Agriculture 2017b). A "drug" is defined as any substance intended or represented for use in the diagnosis, cure, mitigation, treatment, or prevention of any disease in any animal, and any other substance, except feed, which is intended to affect the structure or any function of the body of any animal (and is regulated). "Aflatoxins," including their different varieties, are also recognized under this regulation (California Department of Food and Agriculture 2017b). For the state of Texas, commercial feeds include "minerals," substances used as an ingredient or compound of a mixture of materials used as a feed of an animal. Customer-formula feed includes, among other things, "an individual mineral substance not mixed with another material" (Texas A&M University System 2005 and 2019). In this regard, low-rank coal may be registered as a drug that prevents the toxicity of aflatoxins in animals (as shown earlier in this chapter).

The state of Wisconsin defines commercial feed, custom-mixed feed, feed ingredient, and feed product primarily by their uses and not by what materials are specifically included in them (Justia US Law 2012). The state of Oregon does not define either commercial or custom feeds in any detail (Oregon Department of Agriculture 2020). The state of Washington defines commercial feed, customer-formula feed, and feed ingredient by their basic composition and uses, but with much detail (Washington State Department of Agriculture 2005). There are also definitions and terms relating to feed ingredients and other items adopted from AAFCO (Washington State Department of Agriculture 2018).

No standard methods have been set for HS, HA, and FA for the livestock application. The ones applicable for soil and crops may then be used in this regard. The analysis can be completed "as is" or for "dry matter," and the presence of nutrients (if any) can be claimed accordingly. During product registration, detailed information and documents may be requested by the agency with some applicable fees imposed. After the product has been registered, the applicant must maintain its status by renewing it periodically, providing a routine reporting (typically on tonnage, as well as other requested information), and by submitting applicable fees (Canadian Humalite International Inc. 2020b).

4.7 Closure

While the science of HS, HA, and FA is solid, and each has been proven effective for agricultural application, cautions and limitations must be taken into consideration. One of the most critical factors here is that applying the product at high rates does not always result in better responses. In fact, negative responses have often been observed, indicating that humic applications in agriculture can be disadvantageous. The question is: how much is too much? Processed products, which are soluble in water, are more effective than raw products, and therefore have much lower optimum application rates. Applying HS using the surface-broadcast method over the field requires a large quantity of product per hectare in order to achieve effective results. However, applying the same product at the same rate using the band-application method (close to seeds) will definitely be too much and will likely result in detrimental effects for the crops. For a good quality of soil (in terms of the contents of OM and nutrients), less HS is required to produce positive responses. In this case, the typical recommended rates of application have to be reduced in order to avoid negative results. Also, it is important to evaluate the humic product carefully, since they may contain significant amounts of nutrients already. The same concept applies in the case of livestock: HS will likely not have dramatic effects on healthy animals or on those who are fed with more than adequate amounts of nutrients. It is also important to understand that different nutrients (and different types of nutrients) are required for crops or animals depending on their growth phase. Therefore, the incorporation of HS into nutrients produces the most observable results when applied at the right time. The type and variety of crops or animals also makes differences to their response; all of these factors must be carefully evaluated.

The science on the use of HS in agriculture can be summarized practically in the form of product application labels, and each product requires one. The emphasis on such labels is on the recommended rates and the timing or methods of application for the intended crops or livestock. This should be accompanied by a short summary of the mechanism for how the product works. The challenge here is that each label has to be registered before the product can be marketed and sold, and each jurisdiction has its own requirement, terms, and definitions that have to be carefully studied in advance. The position and status of humic-containing products is rather unique, as humic-containing products do not contain significant amounts of nutrients. They cannot, then, be considered fertilizers or feeds but should be registered as soil amendments or feed supplements (if these categories exist within a specific jurisdiction). If the products contain some nutrients, their status may change.

The regulations for the crop application are more advanced than those for livestock; other than the regional jurisdiction, there also exists the organic product jurisdiction. The most practical strategy is to find out whether HS, HA, FA, or their materials of origin (i.e. low-rank coal, humates, lignite, leonardite, or humalite) are listed within the jurisdiction. If so, this means that no proof of efficacy is required, since their effectiveness for either crops or livestock has already been well documented and recognized. Proof of safety may still be required since each commercial product is most likely different from another, and each manufacturing process has to meet the designated requirements. Things become challenging when the jurisdiction does not have HS, HA, FA, or their materials of origin on the list. In this case, proof of efficacy and safety for a specific crop or livestock application is mandatory. Detailed information is typically required, including numerous field results with replications on different sites and seasons for a specific growing region and a specific type of crop or livestock. This must be accompanied by a sound statistical analysis.

Last but not least, it is important to consider the analytical method used in making a claim concerning the content(s) of HS, HA, and/or FA on the application label. One method may yield higher HA and/or FA results than another. Thus, careful research is required before comparing one product to another in order to be sure that specified rates are commensurable. If the method is not mentioned in the application label, it is most likely to be the colorimetric method, as this is the most practical method that yields the highest results.

Reference

Abbas, S.M. (2013). The influence of biostimulants on the growth and on the biochemical composition of *Vicia faba* cv. Giza 3 beans. *Romanian Biotechnological Letters* 18(2): 8061–8068.

Abd El-Razek, E., Abd-Allah, A.S.E., and Saleh, M.M.S. (2012). Yield and fruit quality of Florida Prince peach trees as affected by foliar and soil applications of humic acid. *Journal of Applied Sciences Research* 8(12): 5724–5729.

Abdel-Mageed, M.A.A. (2012). Effect of dietary humic substances supplementation on performance and immunity of Japanese quail. *Egyptian Poultry Science Journal* 32(III): 654–660.

Ahmed, H.M.R.M and Ismail, F.S.H. (2016). Effectiveness of humic acid application in improving saline soil properties and fodder beet production. *Journal of Soil Sciences and Agriculture Engineering* 7(9): 623–634.

Akinremi, O.O., Janzen, H.H., Lemke, R.L., and Larney, F.J. (2000). Response of canola, wheat and green beans to leonardite additions. *Canadian Journal of Soil Science* 80(3): 437–443.

Al-Madhagi, I.A.H. (2019). Effect of humic acid and yeast on the yield of greenhouse cucumber. *Journal of Horticulture and Postharvest Research* 2(1): 67–82.

Ameri, A. and Tehranifar, A. (2012). Effect of humic acid on nutrient uptake and physiological characteristic *Fragaria ananassa* var: Camarosa. *Journal of Biological and Environmental Sciences* 6(16): 77–79.

Arafat, R.Y., Khan, S.H, and Saima. (2017). Evaluation of humic acid as an aflatoxin binder in broiler chickens. *Annals Animal Science* 17(1): 241–255.

Aristimunha, P.C., Mallheiros, R.D., Ferket, P.R., et al. (2019). Effect of dietary organic acids and humic substance supplementation on performance, immune response and gut morphology of boiler chickens. *Journal of Applied Poultry Research* 29: 85–94. doi: 10.3382/japr/pfz031.

Asik, B.B., Turan, M.A., Celik, H., and Katkat, A.V. (2009). Effects of humic substances on plant growth and mineral nutrients uptake of wheat (*Triticum durum* cv. Salihli) under conditions of salinity. *Asian Journal of Crop Science* 1(2): 87–95.

Association of American Feed Control Officials (2016). *2016 Official Publication* 355–486. Champaign, IL.

ASTM D2013/D2013M-18 (2013). *Standard practice for preparing coal samples for analysis*. West Conshohocken, PA: ASTM International.

ASTM D2492-02(2012) (2012). *Standard test method for forms of sulfur in coal*. West Conshohocken, PA: ASTM International.

ASTM D3174-12 (2012). *Standard test method for ash in the analysis sample of coal and coke from coal*. West Conshohocken, PA: ASTM International.

ASTM D3176-15 (2015). *Standard practice for ultimate analysis of coal and coke*. West Conshohocken, PA: ASTM International.

ASTM D5373-16 (2016). *Standard test methods for determination of carbon, hydrogen and nitrogen in analysis samples of coal and carbon in analysis samples of coal and coke*. West Conshohocken, PA: ASTM International.

ASTM D6357-11 (2011). *Test methods for determination of trace elements in coal, coke, & combustion residues from coal utilization processes by Inductively Coupled Plasma atomic emission, Inductively Coupled Plasma mass, & graphite furnace Atomic Absorption spectrometry*. West Conshohocken, PA: ASTM International.

Aydin, A., Kant, C. and Turan, M. (2012). Humic acid application alleviates salinity stress of bean (*Phaseolus vulgaris* L.) plants decreasing membrane leakage. *African Journal of Agricultural Research* 7(7): 1073–1086.

Bai, H.X, Chang, Q.F., Shi, B.M, and Shan, A.S. (2013). Effects of fulvic acid on growth performance and meat quality in growing-finishing pigs. *Livestock Science* 158: 118–123.

Barnhardt Natural Fibers (2015). What is a Micronaire in Cotton, and Why Does It Matter? http://www.barnhardtcotton.net/blog/what-is-a-micronaire-in-cotton-and-why-does-it-matter/?print=pdf (accessed 23 February 2018).

Bataineh, M., Scott, A.C., Fedorak, P.M., and Martin, J.W. (2006). Capillary HPLC/QTOF-MS for characterizing complex napthenic acid mixtures and their microbial transformation. *Analytical Chemistry* 78: 8354–8361.

Benitez, E., Melgar, R., Sainz, H., et al. (2000). Enzyme activities in the rhizosphere of pepper (*Capsicum annuum* L.) grown with olive cake mulches. *Soil Biology and Biochemistry* 32: 1829–1835.

Budag, C. and Kara, U. (2016). The effect of use humic acid in some blood parameters and rumen protozoa in Norduz lambs. *Igdir University Journal of the Institute of Science and Technology* 6(3): 185–190.

Cacco, G. and Dell'Agnola, G. (1984). Plant growth regulator activity of soluble humic complexes. *Canadian Journal of Soil Science* 64: 225–228.

California Department of Food and Agriculture (2017a). Food and agricultural code, division 7, agricultural chemicals, livestock remedies, and commercial feeds. *Ch. 5. Fertilizing Materials [14501-14682]* 4–9 and 36–38. Sacramento, CA.

California Department of Food and Agriculture (2017b). Inspection service division, food and agricultural code. *Ch. 6. Commercial Feed and Title 3, California Code of Regulations-Plant Industry* 1–14. Sacramento, CA.

Canadian Food Inspection Agency (2009). Registration triggers for NPK fertilizers. *Guidance Document Repository (GDR)* 1–4. Ottawa, ON.

Canadian Food Inspection Agency (2017a). Safety guidelines for fertilizers and supplements. *T-4-93* 1–11. Ottawa, ON.

Canadian Food Inspection Agency (2017b). Introduction. *Regulation of Livestock Feed in Canada* 1–7. Ottawa, ON.

Canadian Food Inspection Agency (2020). Primary fertilizer and supplement materials. *Guidance Document Repository (GDR)* 11–14. Ottawa, ON.

Canadian General Standards Board (2015a). Organic production systems. *General Principles and Management Standards (Amended March 2018)* 46–49. Gatineau, ON.

Canadian General Standards Board (2015b). Organic production systems. *Permitted Substances Lists (Amended March 2018)* 3–17. Gatineau, ON.

Canadian Humalite International Inc. (2020a). Raw Material and Products. *Laboratory Analyses and Technical Data* 1–656. Edmonton, AB.

Canadian Humalite International Inc. (2020b). Commercial products. *Registration and Application Labels* 1–286. Edmonton, AB.

Canellas, L.P., Olivares, F.L., Canellas, N.O.A., et al. (2019). Humic acids increase the maize seedlings exudation yield. *Chemical and Biological Technologies in Agriculture* 6: 1–14. doi: 10.1186/s40538-018-139-7.

Card, A., Whiting, D., Wilson, C., et al. (2015). Organic fertilizers. *CMG Garden Notes #234* 1–8. Fort Collins, CO: Master Gardener, Colorado State University Extension.

Cavalcante, I.H.L., Da Silva, R.R.S., Albano, F.G., et al. (2011). Foliar spray of humic substances on seedling production of papaya (pawpaw). *Journal of Agronomy* 10(4): 118–122.

Celik, H., Turan, M.A., Asik, B.B., et al. (2019). Effects of soil-applied materials on the dry weight and boron uptake of maize shoots (*Zea mays* L.) under high boron conditions. *Communications in Soil Science and Plant Analysis* 50(7): 811–826.

Chadwick, A.V. and Burg, S.P. (1967). An explanation of the inhibition of root growth caused by indole-3-acetic acid. *Plant Physiology* 42: 415–420.

Chang, Q., Lu, Z., He, M., et al. (2014). Effects of dietary supplementation of fulvic acid on lipid metabolism of finishing pigs. *American Society of Animal Science* 92: 4921–4926.

Chen, D., Sun, J., Bai, M., et al. (2015). A new cost-effective method to mitigate ammonia loss from intensive cattle feedlots: application of lignite. *Scientific Reports* 5(Article Number 16689): 1–5.

Chen, Y. and Schnitzer, M. (1976). Water adsorption on soil humic substances. *Canadian Journal of Soil Science* 56: 521–524.

Chirase, N.K., Greene, L.W., McCollum, F.T., et al. (2000). Effect of Bovipro™ on performance and serum metabolite concentrations of beef steers. *Journal of Animal Science* 78(Supplement 2): 105–106.

Church, D.C. (1991). *Livestock Feeds and Feeding*, 3rd ed., 15–36. Englewood Cliffs, NJ: Prentice-Hall.

Cieschi, M.T., Polyakov, A.Y., Lebedev, V.A., et al. (2019). Eco-friendly iron-humic nanofertilizers synthesis for the prevention of iron chlorosis in soybean (*Glycine max*) grown in calcareous soil. *Frontiers in Plant Science* 10: 1–17. doi: 10.3389/fpls.2019.00413.

Cimrin, K.M., Turkmen, O., Turan, M., and Tuncer, B. (2010). Phosphorous and humic acid application alleviate salinity stress of pepper seedling. *African Journal of Biotechnology* 9(36): 5845–5851.

Clapp, C.E., Hayes, M.H.B., and Swift, R.S. (1993). Isolation, fractionation, functionalities, and concepts of structure of soil organic macromolecules. In: *Organic Substances in Soil and Water: Natural Constituents and Their Influence on Contaminant Behaviour* (eds. A.J. Beck, K.C. Jones, M.H.B Hayes, and U. Mingelgrin), 31–69. Cambridge, UK: Royal Society of Chemistry.

Colorado Department of Agriculture (2005). Commercial fertilizers & soil conditioners law. *Title 35 Article 12* 1–4. Denver, CO.

Cornell Law School (2020). Electronic code of federal regulations (e-CFR). http://www.law.cornell.edu/cfr/text (accessed 19 February 2020).

Cozzi, R., Nicolai, M., Perticone, P., et al. (1993). Desmutagenic activity of natural humic acids: inhibition of mitomycin C and maleic hydrazide mutagenicity. *Mutation Research* 299(1): 37–44.

Cozzolino, A. and Piccolo, A. (2002). Polymerization of dissolved humic substances catalyzed by peroxidase. Effects of pH and humic composition. *Organic Geochemistry* 33: 281–294.

Cusack, P.M.V. (2008). Effects of a dietary complex of humic and fulvic acids (FeedMAX 15^{TM}) on the health and production of feedlot cattle destined for the Australian domestic market. *Australian Veterinary Journal* 86(1-2, January): 46–49.

Da Rocha, L.F., Cavalcante, L.F., Nunes, J.C., et al. (2019). Crop management advances for nutritional status and fruit yield of guava in semiarid. *Journal of Plant Nutrition* 42(8): 823–833.

Danilov, R.A. and Ekelund, N.G.A. (2001). Effects of solar radiation, humic substances and nutrients on phytoplankton biomass and distribution in Lake Solumsjo, Sweden. *Hydrobiologia* 444(1-3): 203–212.

Davison, J. (1988). Plant beneficial bacteria. *Nature Biotechnology* 6: 282–286.

Dawes, E.A. (1962). *Quantitative Problems in Biochemistry*, 2nd ed., 124–154. Edinburg, UK: Livingstone.

De Sanfilippo, E.C., Arguello, J.A., Abdala, G., and Orioli, G.A. (1990). Content of auxin-, inhibitor-and gibberellin-like substances in humic acids. *Biologia Plantarum (Praha)* 32(5): 346–351.

Dekker, J., Glen, F., and Medlen, C.E. (2003). Oxihumic acid and its use in the treatment of various conditions. US Patent 6,630,179 B1, filed 22 September 1999 and issued 29 June 2001, 1–39.

Developpement Durable, Environnement et Parcs. (2008). Guidelines for the beneficial use of fertilising residuals. *Reference Criteria and Regulatory Standards* 47–50. Quebec, QC.

Dogan, Y., Togay, Y., Togay, N., and Kulaz, H. (2014). Effect of humic acid and phosphorus applications on the yield and yield components in lentil (*Lens culinaris* Medic.). *Legume Research* 37(3): 316–320.

DTG (2019). *Futtergold* 1. Hauptstr, Germany.

Duiker, S.W. (2004). *Effects of Soil Compaction* 1–12. University Park, PA: College of Agricultural Sciences, Pennsylvania State University.

Duplessis, G.L. and MacKenzie, A.F. (1983). Effects of leonardite applications on phosphorous availability and corn growth. *Canadian Journal of Soil Science* 63: 749–751.

Ekin, Z. (2019). Integrated use of humic acid and plant growth promoting rhizobacteria to ensure higher potato productivity in sustainable agriculture. *Sustainability* 11: 1–13. doi: 10.3990/su11123417.

Eliasson, L., Bertell, G., and Bolander, E. (1989). Inhibitory action of auxin on root mediated elongation not mediated by ethylene. *Plant Physiology* 91: 310–314.

El-Zaiat, H.M., Morsy, A.S., El-Wakeel, E.A., et al. (2018). Impact of humic acid as an organic additive on ruminal fermentation constituents, blood parameters and milk production in goats and their kids growth rate. *Journal of Animal and Feed Sciences* 27: 105–113.

Eshghi, S. and Garazhian, M. (2015). Improving growth, yield and fruit quality of strawberry by foliar and soil drenched applications of humic acid. *Iran Agricultural Research* 34(1): 14–20.

Esringu, A., Kant, C., Yildirim, E., et al. (2011). Ameliorative effect of foliar nutrient supply on growth, inorganic ions, membrane permeability, and leaf relative water content of physalis plants under salinity stress. *Communications in Soil Science and Plant Analysis* 42(4): 408–423.

Esringu, A., Kaynar, D., Turan, M., and Ercisli, S. (2016). Ameliorative effect of humic acid and plant growth-promoting rhizobacteria (PGPR) on Hungarian vetch plants under salinity stress. *Communications in Soil Science and Plant Analysis* 47(5): 602–618.

Esringu, A., Sezen, I., Aytatli, B., and Ercisli, S. (2015). Effect of humic and fulvic acid application on growth parameters in *Impatients walleriana* L. *Akademik Ziraat Dergisi* 4(1): 37–42.

Fabian-Wheeler, E., Brandt, R.C., Hile, M.L., and Mikesell, R. (2015). Humic manure additive reduces odor from Pennsylvania swine finishing operation. *2015 Annual International Meeting Paper* 1–15. New Orleans, LA: American Society of Agricultural and Biological Engineers.

Fagbenro, J.A. and Agboola, A.A. (1993). Effect of different levels of humic acid on the growth and nutrient uptake of teak seedlings. *Journal of Plant Nutrition* 16(8): 1465–1483.

Feng, L., Xu, W., Sun, N., et al. (2020). Efficient improvement of soil salinization through phytoremediation induced by chemical remediation in extreme arid land northwest China. *International Journal of Phytoremediation* 22(3): 334–341.

Ferrara, G. and Brunetti, G. (2010). Effects of the times of application of a soil humic acid on berry quality of table grape (*Vitis vinifera* L.) cv Italia. *Spanish Journal of Agricultural Research* 8(3): 817–822.

Filip, Z. and Berthelin, J. (2001). Analytical determination of the microbial utilization and transformation of humic acids extracted from municipal refuse. *Fresenius Journal of Analytical Chemistry* 371(5): 675–681.

Filip, Z., Kanazawa, S., and Berthelin, J. (2000). Distribution of microorganisms, biomass ATP, and enzyme activities in organic and mineral particles of a long-term wastewater irrigated soil. *Journal of Plant Nutrition and Soil Science* 163(2): 143–150.

Florida Department of Agriculture and Consumer Services (2019). *How to Comply Fertilizer Manual* 1–14. Tallahassee, FL.

Follett, R.H., Murphy, L.S., and Donahue, R.L. (1981). *Fertilizers and Soil Amendments* 302–392, 424–457, and 458–506. Englewood Cliffs, NJ: Prentice-Hall.

Freter, R., Brickner, H., Botney, M., et al. (1983). Mechanisms that control bacterial populations in continuous-flow culture models of mouse large intestinal flora. *Infection and Immunity* 39(2): 676–685.

Fuchs, B., Orda, J., Pres, J., and Muchowicz, M. (1995). The effect of feeding piglets up to the 100th day of their life with peat preparation on their growth and physiological and biochemical indices. *Archivum Veterinarium Polonicum* 35(1-2): 97–107.

Gamble, D.S. and Khan, S.U. (1988). Atrazine hydrolysis in aqueous suspension of humic acid at 25.0 °C. *Canadian Journal of Chemistry* 66(10): 2605–2617.

Gao, Y., He, J., He, Z., et al. (2017). Effects of fulvic acid on growth performance and intestinal health of juvenile loach *Paramisgurnus dabryanus* (Sauvage). *Fish and Shellfish Immunology* 62: 47–56.

George, R.J., McFarlane, D.J., and Nulsen, B. (1997). Salinity threatens the viability of agriculture and ecosystems in Western Australia. *Hydrogeology Journal* 5(1): 6–21.

Gerrard, S. (2006). Phytoremediation of brine-contaminated soil with halophytic plants. *MSc Thesis* 1–95. University of Saskatchewan, Saskatoon, SK.

Govindasmy, R. and Chandrasekaran, S. (1992). Effect of humic acids on the growth, yield and nutrient content of sugarcane. *Science of the Total Environment* 117/118: 575–581.

Gramss, G., Ziegenhagen, D., and Sorge, S. (1999). Degradation of soil humic extract by wood-and soil-associated fungi, bacteria, and commercial enzymes. *Microbial Ecology* 37(2): 140–151.

Grattan, S.R. and Grieve, C.M. (1999). Salinity-mineral nutrient relations in horticultural crops. *Scientia Horticulturae* 78: 127–157.

Greene, G. and Steelink, C. (1962). Structure of soil humic acid. II. Some copper oxide oxidation products. *Journal of Organic Chemistry* 27(1): 170–174.

Grichko, V.P. and Glick, B.R. (2001). Amelioration of flooding stress by ACC deaminase-containing plant growth-promoting bacteria. *Plant Physiology and Biochemistry* 39(1): 11–17.

Hafez, M.M., Shafeek, M.R., Mahmoud, A.R., and Ali, A.H. (2015). Beneficial effects of nitrogen fertilizer and humic acid on growth, yield and nutritive values of spinach (*Spinacia olivera* L.). *Middle East Journal of Applied Sciences* 5(2): 597–603.

Hartz, T.K. and Bottoms, T.G. (2010). Humic substances generally ineffective in improving vegetable crop nutrient uptake or productivity. *HortScience* 45(6): 906–910.

Havlin, J.L., Beaton, J.D., Tisdale, S.L., and Nelson, W.L. (2005). *Soil Fertility and Fertilizers: an Introduction to Nutrient Management*, 7th ed., 362–416. Upper Saddle River, NJ: Pearson-Prentice Hall.

Haworth, R.D. (1970). The chemical nature of humic acid. *Journal of Soil Science* 111(1): 71–79.

Healy, W.B. (1968). Ingestion of soil by dairy cows. *New Zealand Journal of Agricultural Research* 11: 487–499.

Hering, J.G. and Morel, F.M. (1988). Humic acid complexation of calcium and copper. *Environmental Science and Technology* 22(10): 1234–1237.

Hile, M. (2015). Swine manure odor reduction using a humic amendment: on-farm demonstration. *Animal Manure Management March* 13: 1–3.

Hofrichter, M., Ziegenhagen, D., Sorge, S., et al. (1999). Degradation of lignite (low-rank coal) by Ligninolytic basidiomycetes and their manganese peroxidase system. *Applied Microbiology and Biotechnology* 52(1): 78–84.

Huang, D., Wang, L., and Luo T. (2016). Effects of applying humic fertilizers on tea's yield, nutritional quality and soil fertility in a tea garden. *Proceedings of the 3rd International Conference on Materials Engineering, Manufacturing Technology and Control* 1046–1051. Taiyuan, China: Atlantis Press.

Huck, T.A., Porter, N., and Bushell, M.E. (1991). Positive selection of antibiotic-producing soil isolates. *Journal of General Microbiology* 137(10): 2321–2329.

Hufstedler, G.D., Faulkenbery, J.M., Greene, L.W., and Byers, F.M. (1994). In vitro dry matter disappearance of coastal Bermudagrass hay incubated with humic acid. *Journal of Animal Science* 72(Supplement 1): 237.

Hufstedler, G.D., Greene, L.W., and Byers, F.M. (1993). In vitro digestibility of coastal Bermudagrass hay incubated with different levels and sources of substances containing humate. *Journal of Animal Science* 71(Supplement 1): 31.

HumaTech Inc. (2018). Agri-Hume. http://www.humatech.com/agrihume.php (accessed 7 June 2018).

Humic Products Trade Association (2017). California Department of Food and Agriculture (CDFA) considers switching to the HPTA approved test method. *News and Events, November 2017* 1. Spring Green, WI.

Humic Products Trade Association (2019). *Notice to HPTA Members* 1–4. Williston, ND.

Humic Products Trade Association (2020). CDFA approves use of HPTA test method seal. *News and Events, March 2020* 1. Williston, ND.

Ibrahim, A., Abdel-Razzak, H., Wahb-Allah, M., et al. (2019). Improvement in growth, yield, and fruit quality of three red sweet pepper cultivars by foliar application of humic and salicylic acids. *HortTechnology* 29(2): 170–178.

Ibrahim, S.M. and Goh, T.B. (2002). Report on aggregation, sand and copper distribution in amended mine tailing and mechanism of soil aggregation. *Technical Report* 1–15. Winnipeg, MB: Department of Soil Science, University of Manitoba.

Ipek, H., Avci, M., Iriadam, M., et al. (2008). Effects of humic acid on some hematological parameters, total antioxidant capacity and lying performance in Japanese quails. *Archiv fur Geflugelkunde* 72(2): 56–60.

Jarosova, M., Klejdus, B., Kovacik, J., et al. (2016). Humic acid protects barley against salinity. *Acta Physiologiae Plantarum* 38(6, June): 1–9.

Ji, F., McGlone, J.J., and Kim, S.W. (2006). Effects of dietary humic substances on pig growth performance, carcass characteristics, and ammonia emission. *Journal of Animal Science* 84: 2482–2490.

Justi, M., Morais, E.G., and Silva, C.A. (2019). Fulvic acid in foliar spray is more effective than humic acid via soil in improving coffee seedlings growth. *Archives of Agronomy and Soil Science* 65(14): 1969–1983.

Justia US Law (2012). *WI Stat #94.72 (2012 Through Act 45)* 1–11. Mountain View, CA.

Kandil, H. (2014). Response of pea plants (Pisum sativum L.) to phosphorus levels and humic acid levels. *International Conference of Agricultural Engineering Zurich* 1–8. Lausanne, Switzerland: Swiss Federal Institute of Technology Lausanne.

Khaled, H. and Fawy, H.A. (2011). Effect of different levels of humic acids on the nutrient content, plant growth, and soil properties under conditions of salinity. *Soil and Water Research* 6(1): 21–29.

Khan, M.A., Ungar, I.A., and Showalter, A.M. (2000). Effects of salinity on growth, water relations and ion accumulation of the subtropical perennial halophyte *Atriplex griffthii* var. *Stocksii*. *Annals of Botany* 85(2): 225–232.

Khan, M.Z., Ahmed, H., Ahmed, S., et al. (2019a). Formulation of humic substances coated fertilizer and its use to enhance K fertilizer use efficiency for tomato under greenhouse conditions. *Journal of Plant Nutrition* 42(6): 626–633.

Khan, S.A., Khan, S.U., Qayyum, A., et al. (2019b). Integration of humic acid with nitrogen wields an auxiliary impact on physiological traits, growth and yield of maize (*Zea mays* L.) varieties. *Applied Ecology and Environmental Research* 17(3): 6783–6799.

Killorn, R. and Voss, R.D. (1986). *Salt Index of Fertilizers* 1–2. Des Moines, IA: Iowa State University.

Kim, K., Bae, I., Cho, J., et al. (2019). Effects of humic acid and blueberry leaf powder supplementation in feeds on the productivity, blood and meat quality of finishing pigs. *Food Science of Animal Resources* 39(2): 276–285.

Klavins, M., Sire, J., Purmalis, O., and Melecis, V. (2008). Approaches to estimating humification indicators for peat. *Mires and Peat* 3(Article 7): 1–15.

Kocabagli, N., Alp, M., Acar, N., and Kahraman, R. (2002). The effects of dietary humate supplementation on broiler growth and carcass yield. *Poultry Science* 81: 227–230.

Krogmeier, M.J. and Bremner, J.M. (1989). Effects of phenolic acids on seed germination and seedling growth in soil. *Biology and Fertility of Soils* 8: 116–122.

Kucukersan, S., Kucukersan, K., Colpan, I., et al. (2005). The effects of humic acid on egg production and egg traits of laying hen. *Veterinarni Medicina* 50(9): 406–410.

Kulikova, N.A., Stepanova, E.V., and Koroleva, O.V. (2005). Mitigating activity of humic substances: direct influence on biota. In: *Use of Humic Substances to Remediate Polluted Environments: from Theory to Practice* (eds. I.V. Perminova, K. Hartfield, and N. Hertkorn), 285–309. Dordrecht, the Netherlands: Springer.

Ladd, J.N. and Butler, J.H.A. (1969). Inhibition and stimulation of proteolytic enzyme activities by soil humic acids. *Australian Journal of Soil Research* 7: 253–261.

Ladd, J.N. and Butler, J.H.A. (1971). Inhibition by soil humic acids of native and acetylated proteolytic enzymes. *Soil Biology and Biochemistry* 3: 157–160.

Lamar, R.T. and Talbolt, K.H. (2009). Critical comparison of humic acid test methods. *Communications in Soil Science and Plant Analysis* 40: 2309–2322.

Lazarovits, G. and Nowak, J. (1997). Rhizobacteria for improvement of plant growth and establishment. *HortScience* 32(2): 188–192.

Lee, Y.S. and Bartlett, R.J. (1976). Stimulation of plant growth by humic substances. *Soil Science Society of America Journal* 40: 876–879.

Liem, E. (2016). Humus enhances soil productivity: physical/chemical properties and increased crop yields. *Fortify Soil Health Conference* 1–54. Lethbridge, AB: R.A. West International.

Liem, L.E., Facey, R.M., Carrigan, I., et al. (2003a). Enhanced hydrocarbon bioremediation through the addition of humic substances. *Mining and the Environment III Conference* 1–8. Sudbury, ON: Canadian Land Reclamation Association/Centre for Environmental Monitoring.

Liem, L.E., Facey, R.M, and Fischer, M. (2003b). Enhanced hydrocarbon bioremediation through the addition of humic substances part II: improved land treatment and static-pile composting. *ESAA Remediation Technologies Symposium* 1–13. Banff, AB: Environmental Services Association of Alberta.

Lipski, M., Slawinski, J., and Zych, D. (1999). Changes in the luminescent properties of humic acids induced by UV radiation. *Journal of Fluorescence* 9(2): 133–138.

Lovley, D.R., Coates, J.D., Blunt-Harris, E.L., et al. (1996a). Humic substances as electron acceptors for microbial respiration. *Nature* 382(1 August): 445–448.

Lovley, D.R., Woodward, J.C., and Chapelle, F.H. (1996b). Rapid anaerobic benzene oxidation with a variety of chelated Fe(III) forms. *Applied and Environmental Microbiology* Jan(1): 288–291.

Lugtenberg, B.J.J., de Weger, L.A., and Bennett, J.W. (1991). Microbial stimulation on plant growth and protection from disease. *Current Opinion in Biotechnology* 2(3): 457–464.

Lynch, J.M. (1990). Beneficial interactions between micro-organisms and roots. *Biotechnology Advances* 8(2): 335–346.

Mackowiak, C.L., Grossl, P.R., and Bugbee, B.G. (2001). Beneficial effects of humic acid on micronutrient availability to wheat. *Soil Science Society of America Journal* 65: 1744–1750.

Mahoney, K.J., McCreary, C., Depuydt, D., and Gillard, C.L. (2016). Fulvic and humic acid fertilizers are ineffective in dry bean. *Canadian Journal of Plant Science* 97: 202–205.

Malcolm, R.E. and Vaughan, D. (1979). Humic substances and phosphatase activities in plant tissues. *Soil Biology and Biochemistry* 11: 253–259.

Masciandaro, G. and Ceccanti, B. (1999). Assessing soil quality in different agro-ecosystems through biochemical and chemico-structural properties of humic substances. *Soil and Tillage Research* 51(1): 129–137.

Mayland, H.F., Florence, A.R., Rosenau, R.C., et al. (1975). Soil ingestion by cattle on semiarid range as reflected by titanium analysis of feces. *Journal of Range Management* 28(6): 448–452.

McDonald, M., Ross, J., Orcutt, R., et al. (2010). Field trials for the treatment of potassium silicate based drilling waste using calcium humate. *AADE Conference and Exhibition* 1–6. Houston, TX: Association of American Drilling Engineers.

McKenzie, R.H. (2010). Agricultural soil compaction: causes and management. *Agdex 510-1* 1–10. Lethbridge, AB: Alberta Agriculture and Rural Development.

McMurphy, C.P., Duff, G.C., Sanders, S.R., et al. (2011). Effects of supplementing humates on rumen fermentation in Holstein steers. *South African Journal of Animal Science* 41(2): 134–140.

Meharg, A.A. and Killham, K. (1991). A novel method of quantifying root exudation in the presence of soil microflora. *Plant and Soil* 133: 111–116.

Meharg, A.A. and Killham, K. (1995). Loss of exudates from the roots of perennial ryegrass inoculated with a range of micro-organisms. *Plant and Soil* 170(2): 345–349.

Merwad, A.-R.M.A. (2017). Effect of humic and fulvic substances and Moringa leaf extract on Sudan grass plants grown under saline conditions. *Canadian Journal of Soil Science* 97(4): 703–716.

Method 942.05; MIN; ASH (2006). *Official methods of analysis proximate analysis and calculations ash determination (ash)-item 51*, 17th ed. Gaithersburg, MD: Association of Analytical Communities.

Method 954.02; LIPD; FA (2006). *Official methods of analysis proximate analysis and calculations crude fat (CF)-item 23*, 17th ed. Gaithersburg, MD: Association of Analytical Communities.

Method 978.10; CHO; FIB (2006). *Official methods of analysis proximate analysis and calculations crude fiber (CFiber)-item 4*, 17th ed. Gaithersburg, MD: Association of Analytical Communities.

Method 990.03; NITR; NT (2006). *Official methods of analysis proximate analysis and calculations total Nitrogen or crude protein (CP)-item 76*, 17th ed. Gaithersburg, MD: Association of Analytical Communities.

Minister of Justice (2019a). Feeds regulation. *SOR/83-593* 1–26 and 53–127. Ottawa, ON.

Minister of Justice (2019b). Feeds act. *R.S.C., 1985, c.F-9* 1–3. Ottawa, ON.

Minister of Justice (2015). Fertilizers regulations. *C.R.C., c.666* 1–6 and 24–31. Ottawa, ON.

Minister of Justice (2020). Fertilizers act. *R.S.C., 1985, c.F-10* 1–2. Ottawa, ON.

Morrison, R.I. (1963). Products of the alkaline nitrobenzene oxidation of soil organic matter. *Journal of Soil Science* 14(2): 201–216.

Moura, M.N., Martin, M.J., and Burguillo, F.J. (2007). A comparative study of the adsorption of humic acid, fulvic acid and phenol onto Bacillus subtilis and activated sludge. *Journal of Hazardous Materials* 149(1): 42–48.

Muday, G.K., Rahman, A., and Binder, B.M. (2012). Auxin and ethylene: collaborators or competitors? *Trends in Plant Science* 17(April, 4): 181–195.

Mudronova, D., Karaffova, V., Pesulova, T., et al. (2020). The effect of humic substances on gut microbiota and immune response of broilers. *Food and Agricultural Immunology* 31(1): 137–149.

Munns, R. and Tester, M. (2008). Mechanisms of salinity tolerance. *Annual Review of Plant Biology* 59: 651–681.

Murphy, T. and Cargill, C. (2004). The effects of indoor air pollutants on the health and production of growing pigs. *Pig News and Information* 25: 35N–44N.

Mylonas, V.A. and McCants, C.B. (1980). Effects of humic and fulvic acids on growth of tobacco. *Plant and Soil* 54: 485–490.

Nagaraju, R., Reddy, B.S.V., Gloridoss, R., et al. (2014). Effect of dietary supplementation of humic acids on performance of broilers. *Indian Journal of Animal Sciences* 84(4): 447–452.

Nagasawa, K., Wang, B., Nishiya, K., et al. (2016). Effects of humic acids derived from lignite and cattle manure on antioxidant enzymatic activities of barely root. *Journal of Environmental Science and Health, Part B* 51(2): 81–89.

Nahid, S. and Gomah, A.H.M. (1991). Response of wheat to dual inoculation with Va-Mycorrhiza and Azospirillum, fertilized with NPK and irrigated with sewage effluent. *Arid Soil Research and Rehabilitation* 5(2): 83–96.

Nardi, S., Pizzeghello, D., Gessa, C., et al. (2000). A low molecular weight humic fraction on nitrate uptake and protein synthesis in maize seedlings. *Soil Biology and Biochemistry* 32: 415–419.

Nardi, S., Pizzeghello, D., Musculo, A., and Vianello, A. (2002). Physiological effects of humic substances on higher plants. *Soil Biology and Biochemistry* 34: 1527–1536.

National Research Council of the National Academies (2003). *Air Emissions from Animal Feeding Operations: Current Knowledge, Future Needs* 14–25. Washington, DC.

Neri, D., Lodolini, E.M., Savini, G., et al. (2002). Foliar application of humic acids on strawberry (cv Onda). *Acta Horticulturae* 594: 297–302.

Neware, M.R., Deotale, R.D., and Manapure, P.R. (2017). Influence of humic acid and NAA on oil content and yield parameters of linseed. *International Journal of Recent Scientific Research* 8(9): 20433–20436.

Neyveli Lignite Corporation Ltd. (2008). *Humic Acid (Potassium Humate)* 1–4. Neyveli, India.

Niu, X., Bressan, R.A., Hasegawa, P.M., and Pardo, J.M. (1995). Ion homeostasis in NaCl stress environments. *Plant Physiology* 109(3): 735–742.

Noroozisharaf, A. and Kaviani, M. (2018). Effect of soil application of humic acid on nutrients uptake, essential oil and chemical compositions of garden thyme (*Thymus vulgaris* L.) under greenhouse conditions. *Physiology and Molecular Biology of Plants* 24(3): 423–431.

Norvell, W.A. (1991). Reactions of metal chelates in soils and nutrient solutions. In: *Micronutrients in Agriculture*, 2nd ed. (eds. J.J. Mortverdt, F.R. Cox, L.M. Shuman, and R.M. Welch), 187–228. Madison, WI: Soil Science Society of America.

NPK (1983). *Handbook on Environmental Aspects of Fertilizer Use* 58–65. The Hauge, Holland: Martinus Nijhoff.

O'Donnell, R.W. (1973). The auxin-like effects of humic preparations from leonardite. *Soil Science* 116(2): 106–112.

Olk, D.C., Dinnes, D.L., Scoresby, J.R., et al. (2018). Humic products in agriculture: potential benefits and research challenges-a review. *Journal of Soils and Sediments* 18(8): 2281–2891.

Orcutt, R., Logan, C., Ross, J., et al. (2009). Calcium humate reduces disposal costs by lowering electrical conductivity and sodium adsorption ratio in potassium silicate based drilling waste. *ESAA Remediation Technologies Symposium* 1–10. Banff, AB: Environmental Services Association of Alberta.

Orcutt, R., McDonald, M., Logan, C., and Liem, E. (2010). Calcium humate reduces disposal costs in advanced water-based mud systems waste. *Canadian Well Construction Journal* March(3): 8–10.

Oregon Department of Agriculture (2019). *Oregon Fertilizer Program Guide* 2–4. Salem, OR.

Oregon Department of Agriculture (2020). *Feed Licensing and Registration* 1–2. Salem, OR.

Organic Materials Review Institute (2017a). OMRI generic materials list. *OMRI Standards Manual for NOP Review* 4–80. Eugene, OR.

Organic Materials Review Institute (2017b). OMRI generic materials list. *Supplement to the 2017 OMRI Generic Materials List* 1–8. Eugene, OR.

Parida, A.K. and Das, A.B. (2005). Salt tolerance and salinity effects on plants: a review. *Journal of Ecotoxicology and Environmental Safety* 60(3): 324–349.

Parks, C., Ferket, P.R., Thomas, L.N., and Grimes, J.L. (1996). Growth performance and immunity of turkeys fed high and low crude protein diets supplemented with Menefee humate. *Poultry Science* 75(Supplement 1): 138–143.

Patti, A.F., Jackson, W.R., Norng, S., et al. (2013). Commercial humic substances stimulate tomato growth. In: *Functions of Natural Organic Matter in Changing Environment* 1079–1084. Dordrecht, Holland: Zheijang University Press and Springer Science + Business Media.

Philippe, F.-X., Cabaraux, J.-F., and Nicks, B. (2011). Ammonia emissions from pig houses: influencing factors and mitigation techniques. *Agriculture, Ecosystems and Environment* 141: 245–260.

Phillips, D.A., Fox, T.C., King, M.D., Bhuvaneswari, T.V., and Teuber, L.R. (2004). Microbial products trigger amino acid exudation from plant roots. *Plant Physiology* 136: 2887–2894.

Phuong, H.K. and Tichy, V. (1976). Activity of humus acids from peat as studied by means of some growth regulator bioassays. *Biologia Plantarum (Prague)* 18(3): 195–199.

Piccolo, A., Celano, G., and Pietramellara, G. (1993). Effects of fractions of coal-derived humic substances on seed germination and growth of seedlings (*Lactuga sativa* and *Lycopersicum esculentum*). *Biology and Fertility of Soils* 16: 11–15.

Pilanal, N. and Kaplan, M. (2003). Investigation of effects on nutrient uptake of humic acid applications of different forms to strawberry plant. *Journal of Plant Nutrition* 26(4): 835–843.

Prakash, P., Samundeeswari, R., Vivek, C., and Chitra Devi, A. (2011). The influence of potassium humate on *Pleurotus Florida*. *World Journal of Science and Technology* 1(7): 28–31.

Prikryl, Z. and Vancura, V. (1980). Root exudates of plants: VI. Wheat root exudation as dependent on growth, concentration gradient of exudates and the presence of bacteria. *Plant and Soil* 57(1): 69–83.

Rader Jr., L.F., White, L.M., and Whittaker, C.W. (1943). The salt index-a measure of the effect of fertilizers on the concentration of the soil solution. *Soil Science* 55(3): 201–218.

Rady, M.M., Abd El-Mageed, T.A., Abdurrahman, H.A., and Mahdi, A.H. (2016). Humic acid application improves field performance of cotton (Gossypium barbadense L.) under saline conditions. *Journal of Animal and Plant Sciences* 26(2): 487–493.

Rao, A.S. (1990). Root flavonoids. *The Botanical Review* 56: 1–84.

Rath, N.C., Huff, W.E., and Huff, G.R. (2006). Effects of humic acid on broiler chickens. *Poultry Science* 85(3): 410–414.

Rauthan, B.S. and Schnitzer, M. (1981). Effects of a soil fulvic acid on the growth and nutrient content of cucumber (*Cucumis sativus*) plants. *Plant and Soil* 63(3): 491–495.

Rea, E. and Pierandrei, F. (1994). Effects of fertilization with humic acids on soil and metabolism: a multidisplinary approach. In: *Humic Substances in the Global Environment and Implications to Human Health* (eds. N. Senesi, and T.M. Miano), 343–348. Amsterdam, Holland: Elsevier Science.

Reeza, A.A., Ahmed, O.H., Majid, N.M.N.A., and Jalloh, M.B. (2009). Reducing ammonia loss from urea by mixing with humic and fulvic acids isolated from coal. *American Journal of Environmental Sciences* 5(3): 420–426.

Reynolds, A.G., Wardle, D.A., Drought, B., and Cantwell, R. (1995). Gro-Mate soil amendment improves growth of greenhouse-grown 'Chardonnay' grapevines. *HortScience* 30(3): 539–542.

Robertson, A.M. and Galbraith, H. (1971). Effect of ventilation on the gas concentration in a part-slatted piggery. *Farm Building R&D Studies* May: 17–28.

Rodriguez, N.C., Urrutia, E.C., Gertrudis, B.H., et al. (2011). Antioxidant activity of fulvic acid: a living matter-derived bioactive compound. *Journal of Food, Agriculture and Environment* 9(3&4): 123–127.

Rosa, A.H., Vicente, A.A., Rocha, J.C., and Trevisan, H.C. (2000). A new application of humic substances: activation of supports for invertase immobilization. *Fresenius Journal of Analytical Chemistry* 368(7): 730–733.

Rosa, S.D., Silva, C.A., and Maluf, H.J.G.M. (2019). Phosphorous availability and soybean growth in contrasting Oxisols in response to humic acid concentrations combined with phosphate sources. *Archives of Agronomy and Soil Science* 66(2): 220–235. doi: 10.1080/03650340.2019.1608527.

Sadiq, S.A., Baloch, D.M., Ahmed, N., and Hidayatullah. (2014). Role of coal-derived humic acid in the availability of nutrients and growth of sunflower under calcareous soil. *Journal of Animal and Plant Sciences* 24(6): 1737–1742.

Salman, S.R., Abou-Hussein, S.D., Abdel-Mawgoud, A.M.R., and El-Nemr, M.A. (2005). Fruit yield and quality of watermelon as affected by hybrids and humic acid application. *Journal of Applied Sciences Research* 1(1): 51–58.

Samson, G. and Visser, S.A. (1989). Surface-active effects of humic acids on potato cell membrane properties. *Soil Biology and Biochemistry* 21(3): 343–347.

Sanli, A., Karadogan, T., and Tonguc, M. (2013). Effects of leonardite applications on yield and some quality parameters of potatoes (*Solanum tuberosum* L.). *Turkish Journal of Field Crops* 18(1): 20–26.

Sari, N.P. and Abdoellah, S. (2017). Effectiveness of humic acid application on growth of coffee seedlings. *Pelita Perkebunan* 33(3): 188–194.

Schmidt, C. (2015). Soil health demonstration using cocktail cover crop and humalite. *Technical Report* 1–2. Oyen, AB: Chinook Applied Research Association.

Selim, E.-M. and Mosa, A.A. (2012). Fertigation of humic substances improves yield and quality of broccoli and nutrient retention in a sandy soil. *Journal of Plant Nutrition and Soil Science* 175(2): 273–281.

Seyedbagheri, M.M. (2010). Influence of humic products on soil health and potato production. *Potato Research* 53: 341–349.

Seyedbagheri, M.M. (2015). 30 years of research documents the influence of humic substances on soil health, fertilizer and water-use efficiency. *48th Annual University of Idaho Potato Conference* 1–6. Pocatello, ID: University of Idaho.

Shah, S.B. and Kolar, P. (2012). Evaluation of additive for reducing gaseous emissions from swine waste. *Agriculture Engineering International: Commission Internationale du Genie Rural Journal* 14(2): 10–20.

Shahein, M.M., Afifi, M.M., and Algharib, A.M. (2015). Study the effects of humic substances on growth, chemical constituents, yield and quality of two lettuce cultivars (cvs. Dark Green and Big Bell). *Journal of Materials and Environmental Science* 6(2): 473–486.

Sharaf, M.M., Bakry, Kh.A., and El-Gioushy, S.F. (2011). The influence of some bio and organic nutritive addenda on growth, productivity, fruit quality and nutritional status of Washington navel orange trees. *Egyptian Journal of Applied Science* 26(9): 253–269.

Sharif, M., Khattak, R.A., and Sarir, M.S. (2002). Effect of different levels of lignitic coal derived humic acid on growth of maize plants. *Communications in Soil Science and Plant Analysis* 33(19-20): 3567–3580.

Shermer, C.L., Maciorowski, K.G., Bailey, C.A., et al. (1998). Caecal metabolites and microbial populations in chickens consuming diets containing a mined humate compound. *Journal of the Science of Food Agriculture* 77: 479–486.

Shi, Y., Parker, D.B., Cole, N.A., et al. (2001). Surface amendments to minimize ammonia emissions from beef cattle feedlots. *Transactions of the American Society of Agricultural Engineers* May/June 44(3): 677–682.

Silva-Matos, R.R.S., Cavalcante, I.H.L., Junior, G.B.S, et al. (2012). Foliar spray of humic substances on seedlings production of watermelon cv. Crimson Sweet. *Journal of Agronomy* 11(2): 60–64.

Sollins, P., Homann, P., and Caldwell, B.A. (1996). Stabilization and destabilization of soil organic matter: mechanisms and controls. *Geoderma* 74: 65–105.

South Dakota Department of Agriculture (2020). *Fertilizer Program* 1–3. Pierre, SD.

State of Hawaii Department of Agriculture (2020). Frequently asked questions: do fertilizers require licensing/registration in Hawaii? http://hdoa.hawaii.gov/faqs/ (accessed 15 January 2020).

Stevenson, F.J. (1994). *Humus Chemistry: Genesis, Composition, Reactions*, 2nd ed., 24–58. New York, NY: Wiley-Interscience.

Stewart, K. and Janin, A. (2014). Leonardite and biochar for mine impacted water and soils. *Technical Report Prepared for Wapaw Bay Resources* 1–25. Yukon, NWT: Yukon Research Centre, Yukon College.

Suman, S., Spehia, R.S., and Sharma, V. (2017). Humic acid improved efficiency of fertigation and productivity of tomato. *Journal of Plant Nutrition* 40(3): 439–446.

Sun, L., Ma, Y., Liu, Y., et al. (2019). The combined effects of nitrogen fertilizer, humic acid, and gypsum on yield-scaled greenhouse gas emissions from a coastal saline rice field. *Environmental Science and Pollution Research* 26(19): 19502–19511.

Suntari, R., Retnowati, R., Soemarno, and Munir, M. (2013). Study on the release of N-available (NH_4^+ and NO_3^-) of urea-humate. *International Journal of Agriculture and Forestry* 3(6): 209–219.

Suntari, R., Retnowati, R., Soemarno, and Munir, M. (2015). Determination of urea-humic acid dosage of Vertisols on the growth and production of rice. *Agrivita* 37(2): 185–192.

Syltie, P.W. (1985). Effects of very small amounts of highly active biological substances on plant growth. *Biological Agriculture and Horticulture* 2(3): 245–269.

Takahashi, Y., Minai, Y., Ambe, S., et al. (1997). Simultaneous determination of stability constants of humate complexes with various metal ions using multitracer technique. *Science of the Total Environment* 198(1): 61–71.

Tan, K.H. and Tantiwiramanond, D. (1983). Effect of humic acids on nodulation and dry matter production of soybean, peanut, and clover. *Soil Science Society of America Journal* 47: 1121–1124.

Taylor, R., Smedley, K., and Soni, M. (2014). Literature review of humic substances in animal feeds. *17th Meeting of the International Humic Substances Society* 270–271. Ioannina, Greece: International Humic Substances Society.

Terry, S.A., Ramos, A.F.O., Holman, D.B., et al. (2018a). Humic substances alter ammonia production and the microbial populations within a RUSITEC fed a mixed hay-concentrate diet. *Frontiers in Microbiology* 9: 1–9. doi: 10.3389/fmicb.2018.01410.

Terry, S.A., Ribeiro, G.deO., Gruninger, R.J., et al. (2018b). Effect of humic substances on rumen fermentation, nutrient digestibility, methane emissions, and rumen microbiota in beef heifers. *Journal of Animal Science* 96: 3863–3877.

Texas A&M University System (2005). Feed control act. *Texas Agriculture Code (1981), Ch. 141 (as Amended June 18, 2005)* 1–4. College Station, TX.

Texas A&M University System (2019). Commercial feed rules. *Texas Administrative Code, Title 4. Agriculture, Ch. 61 (Amended January 1, 2019)* 1–3. College Station, TX.

The European Agency for the Evaluation of Medicinal Products (1999). Humic acids and their sodium salts. *Summary Report* 1–2. London, UK.

Tomassen, B.P.H. and Faust, R.H. (2000). The use of processed humic acid product as a feed supplement in dairy production in the Netherlands. *Conference Paper IFOAM: IFOAM 2000, August 2000* 339. Basel, Switzerland: The World Grows Organic International Scientific Committee.

Trckova, M., Matlova, L., Hudcova, H., et al. (2005). Peat as a feed supplement for animals: a review. *Veterinarni Medicina* 50(8): 361–377.

Turhan, A. (2019). The role of humic acid application in reducing detrimental effects of salt in cauliflower (*Brassica oleraceae* L. var. Botrytis). *KSU Journal of Agriculture and Nature* 22(6): 837–842.

Urrutia, O., Erro, J., Guardado, I., et al. (2013a). Physico-chemical characterization of humic-metal-phosphate complexes and their potential application to the manufacture of new types of phosphate-based fertilizers. *Journal of Plant Nutrition and Soil Science* 177(2): 128–136.

Urrutia, O., Guardado, I., Erro, J., et al. (2013b). Theoretical chemical characterization of phosphate-metal-humic complexes and relationships with their effects on both phosphorus soil fixation and phosphorous availability for plants. *Journal of the Science of Food and Agriculture* 93(2): 293–303.

US Environmental Protection Agency (2013). Complex Polymeric Polyhydroxy Acids; Exemption from the Requirement of a Tolerance. *Federal Register Vol. 78 No. 147/ Wednesday, July 31, 2013/Rules and Regulations* 1–3. Washington, DC.

US Environmental Protection Agency (2019). Draft guidance for plant regulator label claims, including plant biostimulants. http://www.epa.gov/pesticides/epa-releases-public-comment-draft-guidance-plant-regulators-including-plant-biostimulants (accessed 15 August 2019).

Van de Venter, H.A., Furter, M., Dekker, J. and Cronje, I.J. (1991). Stimulation of seedling root growth by coal-derived sodium humate. *Plant and Soil* 138: 17–21.

Van Rensburg, C.J., Van Rensburg, C.E.J., Van Ryssen, J.B.J., et al. (2006). In vitro and in vivo assessment of humic acid as an aflatoxin binder in broiler chickens. *Poultry Science* 85: 1576–1583.

Varanini, Z., Pinton, R., De Biasi, M.G., et al. (1993). Low molecular weight humic substances stimulate H^+-ATPase activity of plasma membrane vesicles isolated from oat (*Avena sativa* L.) roots. *Plant Soil* 153: 61–69.

Verlinden, G., Pycke, B., Mertens, J., et al. (2009). Application of humic substances results in consistent increases in crop yield and nutrient uptake. *Journal of Plant Nutrition* 32(9): 1407–1426.

Vigneault, B., Percot, A., Lafleur, M., and Campbell, P.G.C. (2000). Permeability changes in model and phytoplankton membranes in the presence of aquatic humic substances. *Environmental Science and Technology* 34(18): 3907–3913.

Visscher, C., Hankel, J., Nies, A., et al. (2019). Performance, fermentation characteristics and composition of the microbiome in the digest of piglets kept on a feed with humic acid-rich peat. *Frontiers in Veterinary Science* 6: 1–10. doi: 10.3389/fvets.2019.00029.

Visser, S.A. (1973). Some biological effects of humic acids in the rat. *Acta Biologica et Medica Germanica* 31: 569–581.

Visser, S.A. (1982). Surface active phenomena by humic substances of aquatic origin. *Revue des Sciences de L'Eau* 1: 285–296.

Visser, S.A. (1985a). Effect of humic acids on numbers and activities of micro-organisms within physiological groups. *Organic Geochemistry* 8(1): 81–85.

Visser, S.A. (1985b). Physiological action of humic substances on microbial cells. *Soil Biology and Biochemistry* 17(4): 457–462.

Visser, S.A. (1987). Effect of humic substances on mitochondrial respiration and oxidative phosphorylation. *The Science of the Total Environment* 62: 347–354.

Vujinovic, T., Zanin, L., Venuti, S., et al. (2020). Biostimulant action of dissolved humic substances from a conventionally and an organically managed soil on nitrate acquisition in maize plants. *Frontiers in Plant Science* 10: 1–14. doi: 10.3389/fpls.2019.01652.

Walker, D.J. and Bernal, M.P. (2005). Plant mineral nutrition and growth in a saline Mediterranean soil amended with organic wastes. *Communications in Soil Science and Plant Analysis* 35(17-18): 2495–2514.

Walker, D.J. and Bernal, M.P. (2008). The effects of olive mill waste compost and poultry manure on the availability and plant uptake of nutrients in a highly saline soil. *Bioresource Technology* 99: 396–403.

Walker, T.S., Bais, H.P., Grotewold, E., and Vivanco, J.M. (2003). Root exudation and rhizosphere biology. *Plant Physiology* 132(May): 44–51.

Wang, H., Wu, Z., Zhou, Y., et al. (2012). Effects of salt stress on ion balance and nitrogen metabolism in rice. *Plant, Soil, and Environment* 58(2): 62–67.

Wang, Q., Chen, Y.J., Yoo, J.S., et al. (2008). Effects of supplemental humic substances on growth performance, blood characteristics and meat quality in finishing pigs. *Livestock Science* 117: 270–274.

Wang, W.-X., Vinocur, B., and Altman, A. (2003). Plant responses to drought, salinity and extreme temperatures: towards genetic engineering to stress tolerance. *Planta* 218(1): 1–14.

Wang, X.J., Wang, Z.Q., and Li, S.G. (1995). The effect of humic acids on the availability of phosphorous fertilizers in alkaline soils. *Soil Use and Management* 11: 99–102.

Washington State Department of Agriculture (2005). RCW 15.53.901. *Definitions* 1–2. Olympia, WA.

Washington State Department of Agriculture (2018). Ch. 16-250 WAC-commercial feed. http://apps.leg.wa.gov/WAC/default.aspx?cite=16-250&full=true#16-250-008 (accessed 7 June, 2018).

Washington State Department of Agriculture (2020). *Commercial Fertilizers* 1–2. Olympia, WA.

Westerhoff, P., Aiken, G., Amy, G., and Debroux, J. (1999). Relationships between the structure of natural organic matter and its reactivity towards molecular ozone and hydroxyl radicals. *Water Research* 33(10): 2265–2276.

Wheeler, L. (2015). Group petitions FDA to change animal feed rule. http://thehill.com/regulation/228583-group-petitions-fda-to-change-animal-feed-rule (accessed 10 March, 2020).

Xudan, X. (1986). The effect of foliar application of fulvic acid on water use, nutrient uptake and yield in wheat. *Australian Journal of Agriculture Research* 37: 343–350.

Yan, N., Marschner, P., Cao, W., et al. (2015). Influence of salinity and water content on soil microorganisms. *International Soil and Water Conservation Research* 3: 316–323.

Yang, H.L., Chiu, H.C., and Lu, F.J. (1996). Effects of humic acid on the viability and coagulant properties of human umbilical vein endothelial cells. *American Journal of Hematology* 51(3): 200–206.

Ye, S.-Q., Lu, X.-Z., and Zhou, A.-G. 2009. In vitro evaluation of the efficacy of sodium humate as an aflatoxin B1 adsorbent. *Australian Journal of Basic and Applied Sciences* 3(2): 1296–1300.

Yoruk, M.A., Gul, M., Hayirli, A., and Macit, M. (2004). The effects of supplementation of humate and probiotic on egg production and quality parameters during the late laying period in hens. *Poultry Science* 83: 84–88.

Young, C.C. and Chen, L.F. (1997). Polyamines in humic acid and their effect on radical growth of lettuce seedlings. *Plant and Soil* 195(1): 143–149.

Zancani, M., Bertolini, A., Petrussa, E., et al. (2011). Fulvic acid affects proliferation and maturation phases in Abies cephalonica embryogenic cells. *Journal of Plant Physiology* 168(11): 1226–1233.

Zancani, M., Petrussa, E., Krajnakova, J., et al. (2009). Effect of humic acids on phosphate level and energetic metabolism of tobacco BY-2 suspension cell cultures. *Environmental and Experimental Botany* 65(2-3): 287–295.

Zanin, L., Tomasi, N., Cesco, S., et al. (2019). Humic substances contribute to plant iron nutrition acting as chelators and biostimulants. *Frontiers in Plant Science* 10: 1–10. doi: 10.3389/fpls.2019.00675.

Zhang, Y.A., Tanaka, J.A., Dosman, A., et al. (1998). Acute respiratory responses of human subjects to air quality in a swine building. *Journal of Agricultural Engineering Research* 70: 367–373.

5

Commercial Humic Products

5.1 Overview

While low-rank coal is rich in humic substances (HS, or "humic") and beneficial for agricultural applications, it has to be manufactured into commercial products before being accepted into the industry. Material sampling and selection in the mine site is critical to ensure that the commercial humic products are manufactured from a high quality of raw material. Understanding the characteristics of the commercial products is important, particularly how they differ from one another with respect to their physical and chemical properties. This is essential for proper product handling (including transfer, reprocessing, repackaging, storage, and field application) at the client or sub-contractor's site.

This chapter is dedicated to Alberta low-rank coal (humalite) as the raw material and commercial products manufactured from it, including dry HS, liquid humic acids (HA), liquid fulvic acids (FA), enhanced liquid products, pelletized HS, and dry soluble HS products. For other low-rank coal materials and their associated products, the information here may still be applicable due to the similarity of the raw material. This chapter is not relevant for other organic matter (OM)-containing materials and their derived humic-containing products, including carbonaceous shale, natural soil or water, peat or peat moss, manure, compost, and municipal or industrial waste.

5.2 Raw Material

Humalite is the weathered product of subbituminous coal and carbonaceous shale materials originated from Alberta (Hoffman et al. 1993). Due to the practice of collecting only low-rank coal material from the mine site, humalite is identical

Low-Rank Coal Applications in Agriculture: Humic Analyses, Products, and Performance, First Edition. L. Edwin Liem.

to Alberta low-rank coal. Throughout this chapter, the term "humalite" is used to identify the raw material collected from the mine site and manufactured into commercial humic products.

Table 5.1 shows historical data for chemical and biological parameters of humalite. Considering its acidity (pH), this material can be considered a weak acid. It is very high in HS (HA and FA) and low in ash, aluminum (Al), and sodium (Na). This material is very low in arsenic (As), cadmium (Cd), chromium (Cr), mercury (Hg), nickel (Ni), and lead (Pb). Toxic organics (dioxins/furans and

Table 5.1 Chemical and biological parameters.

Parameter	Value	Comment
pH	3.7 (0.10)	Weak acid
HA + FA (= HS)	91.8 (6.20)[a] %	Very high
	83.2 (5.05)[b] %	High
Ash	22.8 (2.33) %	Low
Al	2.1 (1.59) %	Low
Na	0.3 (0.29) %	Low
As	7 (3.0) ppm	Very low
Cd	< 1 ppm	Very low
Cr	4 (3.1) ppm	Very low
Hg	60 (3) ppb	Very low
Ni	9 (6.2) ppm	Very low
Pb	5 (2.5) ppm	Very low
Dioxins/furans	0.4 (0.05) ppt	Negligible
Aflatoxin B1	BDL[c]	Negligible
Aflatoxin B2	BDL[c]	Negligible
Aflatoxin G1	BDL[c]	Negligible
Aflatoxin G2	BDL[c]	Negligible
E. coli	BDL[c]	Negligible
Fecal coliform	BDL[c]	Negligible
Salmonella	Negative	Negligible

Source: Canadian Humalite International Inc. (2020a);
note:
[a] colorimetric method (dry matter analysis),
[b] ISO 19822 (dry matter analysis),
[c] below detection limit, in brackets are standard deviations for mean values, completed on different samples.

aflatoxins) and pathogens (*E. coli*, fecal coliform, and *Salmonella*) have been found to be negligible (Canadian Humalite International Inc. 2020a).

Table 5.2 shows historical data for physical properties of humalite including its Hardgrove Grindability Index (HGI), density, moisture, and water solubility (Canadian Humalite International Inc. 2020a). As indicated in this table, humalite is very much manageable with respect to its handling and further processing.

Table 5.3 shows historical data for safety parameters of humalite (Canadian Humalite International Inc. 2020a). This material is low in crystalline silica (Si), high in minimum ignition temperature (MIT), and weak with respect to its explosion severity (Kst). Humalite is considered "non-hazardous" according to the Globally Harmonized System of Classification and Labeling of Chemicals (GHS)/ Workplace Hazardous Materials Information System (WHMIS) and is considered to be "non-dangerous goods" according to the Transportation of Dangerous Goods (TDG) (Canadian Humalite International Inc. 2020b).

Due to its high quality, material safety, and manageable properties, humalite has a major advantage over other humic containing materials, including carbonaceous shale, natural soil or water, peat or peat moss, manure, compost, and

Table 5.2 Humalite physical parameters.

Parameter	Value	Comment
HGI	111	Very soft, manageable
Density	880 (32) kg/m^3	Manageable
Moisture	37 (4.2) %	Medium, manageable
Water solubility	4.2 (1.87) %	Very low

Source: Canadian Humalite International Inc. (2020a); *note:* in brackets are standard deviations for mean values.

Table 5.3 Humalite safety parameters.

Parameter	Value	Comment
Crystalline Si	1.7 (0.99) %	Low
MIT	270 °C	High
Kst	135	Weak
GHS	Non-hazardous	"Warning" label element
TDG	Non-dangerous	No assigned UN numbers

Sources: Canadian Humalite International Inc. (2020a); *note:* in bracket is standard deviation for mean value.

municipal or industrial waste. Still, a careful material selection has to be completed prior to its collection from the mine site. Humalite is typically found adjacent to other materials with an inferior quality that must be excluded at all cost, including overburden soil, coal, clay, sand, and rock. These materials typically contained < 1 to 39% HS. Numerous samples of raw material are to be taken from the site and analyzed for parameters as listed in Tables 5.1 and 5.2. These tables present historical data for humalite, which can be used as a guideline when selecting raw material from the site. Most importantly, each sample must meet its limits, as listed in Table 5.4.

Table 5.4 shows the practical limits for agricultural applications of humalite. Materials with moisture contents of less than 30% can be considered dry, those between 30 and 40% are wet, and those over 40% are very wet. High moisture contents create challenges during raw material handling, especially during winter. OM can be assumed to be the sum of C, H, O, N, and S, and ash = 100% – OM. Higher ash values result in low OM and HS. For most applications, OM higher 75% or ash lower 25% are desired. HS values of less than 85% (based on the colorimetric method, dry matter analysis) are not acceptable.

Table 5.4 also shows agricultural limits set by Canadian Council of Ministries of the Environment (2002a and 2002b), Canadian Food Inspection Agency (2017a, 2017b, and 2017c), and National Research Council (2005). The lowest applicable values among them are listed in the table. pH values between 6 and 8 are the target. The pH of humalite and its commercial products will not have any effects on soil pH due to their relatively low application rates to soil (see Chapters 4 and 6). Selected metals (Al, Na, As, Cd, Cr, Hg, Ni, and Pb) may naturally be found at the mine site, and therefore have to be monitored. High levels of these metals are to be avoided at all cost. Pathogen levels that exceed limits (*E. coli*, fecal coliform, and *Salmonella*) as well as toxic organics (dioxins or furans and aflatoxins) must also be avoided. These parameters indicate the possible presence of animals and/or human activities on the site.

Upon receiving acceptable results, the site excavation may commence. Any values which do not conform to the limits detailed above result in the abandonment of the site. Comparing Table 5.1 (historical data) and Table 5.4 (agricultural limits) on selected metals, pathogens, and toxic organics, it is clear that the raw material obtained from the mine site has been meeting all the requirements consistently throughout the years.

5.3 Dry HS

This product is the most basic commercial product manufactured from humalite. Dry HS of large and medium particulate sizing are used to provide a slow-release source of HS to soil. This improves its quality and is beneficial for crops. Dry HS

Table 5.4 Limits for agricultural applications.

Parameter	Agricultural soil	Livestock	Unit
Moisture	None	None	%
OM	> 75[a]	> 75[a]	%
Ash	< 25[a]	< 25[a]	%
HS	> 85[a,b]	> 85[a,b]	%
pH	6–8[c]	6–8[a]	–
Al	N/A[c]	0.1[f]	%
Na	N/A[c]	1.7[f,g]	%
As	< 12[c]	8[h]	ppm
Cd	< 1.4[c]	0.2[h]	ppm
Cr	< 64[c]	100[f]	ppm
Hg	< 1[d]	0.2[f]	ppm
Ni	< 36[d]	100[f]	ppm
Pb	< 70[c]	8[h]	ppm
Dioxins/furans	4[e]	0.75[h]	ppt
Aflatoxin B1	N/A	20[i]	ppb
Aflatoxin B2	N/A	20[i]	ppb
Aflatoxin G1	N/A	20[i]	ppb
Aflatoxin G2	N/A	20[i]	ppb
E. coli	N/A[d]	N/A[h]	#
Fecal coliform	1000[d]	N/A[h]	#
Salmonella	N/D[d]	N/D[h]	#

Note:
[a] practical limits for most applications (Canadian Humalite International Inc. 2020a),
[b] colorimetric method (dry matter analysis),
[c] Canadian Council of Ministries of the Environment (2002a),
[d] Canadian Food Inspection Agency (2017a),
[e] Canadian Council of Ministries of the Environment (2002b),
[f] National Research Council (2005),
[g] as NaCl,
[h] Canadian Food Inspection Agency (2017b),
[i] Canadian Food Inspection Agency (2017c), N/A = not available, N/D = not detected.

of small and very small particulate sizing are used as the input material for manufacturing other humic products.

Table 5.5 shows dry HS of large, medium, small, and very small particulate sizing and their important parameters, including moisture, particulate sizing, HS, density, particulate strength, water solubility, and pH. Based on 395 samples, each product contains at least 85.8% HS (mean = 93.7%, based on the colorimetric method, dry matter analysis) and a similar density (mean = 890 kg/m^3, analyzed according to ASTM D7263-09(2018)e2 2018) and pH (mean = 3.9, analyzed according to ASTM D4972-18 2018). Based on 30 samples taken from each product, mean moisture = 18 to 31% (analyzed according to ASTM D3302/D3302M-17 2017) and mean particulate sizing = 0.1 to 2.0 mm (analyzed according to ASTM D6913/D6913M-17 2017). Based on 30 samples, dry HS of large sizing has 30 to 41 (mean = 35) kPa particulate strength. This analysis has been completed using a force gauge measurement unit in accordance with ITM Instruments Inc. (2020). Due to their small sizing in particulates, no strength analyses have been completed for other products. During manufacturing, 362 samples have been taken from dry HS of large, medium, small, and very small sizing and analyzed for water solubility according to EEC A.6 (1992). Each product has been found to have a similar solubility in water (mean = 5.1%). Table 5.5 shows that all products are similar to the raw material.

Table 5.5 Dry HS.

Product	Particulate sizing	Mean moisture %	Mean particulate sizing mm	Mean HS[a] %
1	Large	31 (2.3)	2.0	All products: 93.7 (4.27)
2	Medium	25 (3.1)	0.9	
3	Small	25 (2.9)	0.3	
4	Very small	18 (2.6)	0.1	

Product	Mean density kg/m^3	Mean particulate strength kPa	Mean water solubility %	Mean pH
1	All products: 890 (28)	35 (3.0)	All products: 5.1 (2.02)	All products: 3.9 (0.17)
2		N/A		
3		N/A		
4		N/A		

Source: Canadian Humalite International Inc. (2020a);
note:
[a] colorimetric method (dry matter analysis), in brackets are standard deviations, N/A = not available.

Table 5.6 shows selected metals, based on one sample taken from each product. The sample has been analyzed according to ASTM D2013/D2013M-18 (2013), ASTM D6357-11 (2011), and ASTM D6722 (2011). This table shows that each product is of good quality, which is similar to the raw material (see Table 5.1). Under typical application rates, it will likely meet all limits as listed in Table 5.4. Figure 5.1 shows dry HS of very small (top left), small (top right), medium (bottom left), and large sizing (bottom right).

Dry HS can be delivered to the client or sub-contractor in bulk (38 t), polypropylene supersacks (500 to 1000 kg), or bags of multilayer paper materials plus a thin layer of polypropylene (10 to 25 kg). Arriving at the site, the product in bulk is unloaded onto a storage silo, while the product in supersacks or bags is stored within a storage facility. The product in bulk may also be repackaged prior to field application. During field work, dry HS of large or medium particulate sizing is

Figure 5.1 Dry HS. *(See color plate section for color representation of this figure).*

Table 5.6 Selected metals.

Parameter	Dry HS				Unit
	Large	Medium	Small	Very small	
Al	0.4	0.6	0.7	0.7	%
Na	0.1	0.2	0.2	0.2	%
As	< 1.0	1.3	3.8	7.0	ppm
Cd	< 0.2	< 0.2	< 0.2	< 0.2	ppm
Cr	< 5.0	< 5.0	5.8	7.3	ppm
Hg	< 50	< 50	< 50	71	ppb
Ni	6.8	9.9	10.6	10.1	ppm
Pb	< 5.0	5.7	8.0	7.3	ppm

Source: Canadian Humalite International Inc. (2020a).

loaded into a field applicator and subsequently applied. Dry HS of small or very small particulate sizing is used to manufacture other humic products.

While dry HS is considered safe, precautions still have to be taken, especially when handling dry HS of small and very small particulate sizing. Wearing the recommended personal protective equipment (PPE) is necessary for workers and the required preparations must be made in case incidents do occur, as required in the safety data sheet (SDS). When handling the product, the SDS and product application label should be used as the main reference. Product characteristics and important details on how to handle the product are provided, which the client or sub-contractor should be made aware of.

5.4 Liquid HA

Liquid HA is manufactured from dry HS of very small particulate sizing. The main objective with liquid HA is to provide a fast-release source of HS (containing mainly HA and some FA) to soil and crops. This product comes in neutral (7) and high (10) pH values, with HS strengths of 11 and 12%, respectively. Liquid HA 24% (pH = 10) offers the most concentrated HS contained in a liquid product. Liquid HA is also the input material used to manufacture other humic products.

Table 5.7 shows liquid HA 11, 12, and 24% and their important parameters, including HS, total solids, specific gravity, pH, kinematic viscosity, particulate sizing, and water solubility.

Table 5.7 Liquid HA.

Parameter	Liquid HA 11%	Liquid HA 12%
HA + FA (= HS)[a]	11.1 (0.16) %	12.0 (0.17) %
Total solids	8.8 (0.14) %	9.1 (0.08) %
Specific gravity	1.03 (0.007)	1.05 (0.007)
pH	6.9 (0.09)	10.2 (0.11)
Kinematic viscosity	1.3 mm^2/s	1.4 mm^2/s
Particulate sizing,	4.6 (3.67) μm	5.5 (3.69) μm
mean, maximum	21.6 μm	22.5 μm
Water solubility	98.8 (0.45) %	99.0 (0.15)

Parameter	Liquid HA 24%
HA + FA (= HS)[a]	24.3 (0.16) %
Total solids	21.7 (0.19) %
Specific gravity	1.09 (0.009)
pH	9.8 (0.25)
Kinematic viscosity	6.3 mm^2/s
Particulate sizing,	5.5 (4.51) μm
mean, maximum	33.0 μm
Water solubility	97.4 (0.20) %

Source: Canadian Humalite International Inc. (2020a);
note:
[a] colorimetric method (as-received analysis), in brackets are standard deviations for mean values.

Based on 22 samples taken from each product, liquid HA has been found to contain 11.1 to 24.3% HS (based on the colorimetric method, including moisture), 8.8 to 21.7% total solids (analyzed according to EPA Method 1684 2001), 1.03 to 1.09 specific gravity (analyzed according to the weight of a 10 mL sample), and 6.9 to 10.2 pH (analyzed according to APHA 4500-H^+ (1992). It is important to note that the analysis for HS has been completed based on the colorimetric method and that the results may exceed their total solids (see Section 3.3).

Based on one sample taken from each product, liquid HA contains 1.3 to 6.3 mm^2/s kinematic viscosity (analyzed at 40 °C according to ASTM D445-19 (2019). Each sample has also been analyzed using a particle size analyzer (in accordance with Beckman Coulter 2011), showing mean values of 4.6 to 5.5 μm, with a maximum particulate sizing of 33.0 μm.

Three samples taken from each product have been analyzed for water solubility according to EEC A.6 (1992). Liquid HA has been found to have almost 100% water solubility (ranging from 97.4 to 99.0%), which is a major improvement over dry HS. However, some insoluble components are present within the product, including small amounts of humin and inorganics. It is important to note that critical steps of this analysis include mixing the product with water and then using a centrifuge to separate any precipitates from the solution. The product solubility is then calculated gravimetrically as the soluble components in the mixture divided by the original product weight. More insoluble components within the product result in a lower water solubility. Here, insoluble components are more likely the larger molecular sizing of humic components. As expected, they are found mainly in liquid HA 24%. In addition to the fact that the product is not a true solution (it is more like a colloidal and/or suspension, instead), this analysis represents the worst-case scenario. This laboratory analysis uses a centrifuge to create a forced separation within the product, which never actually happens in real life, especially during product storage. These results, however, show the importance of the recommendation to "shake the product before use," which has to be included in the application label.

Table 5.8 shows selected metals, based on one sample taken from each product. The sample has been analyzed according to EPA Method 6020A (SW-846) (1998) and EPA Method 245.1 (1994). This table indicates that each product is of good quality, containing very small amounts of metals. Under typical application rates, it will likely meet all limits as listed in Table 5.4. Figure 5.2 (left) shows liquid HA 12%, including its diluted form prior to the colorimetric analysis.

Table 5.8 Selected metals.

	Liquid HA			
Parameter	**11% HA**	**12% HA**	**24% HA**	**Unit**
Al	323	458	797	ppm
Na	267	252	454	ppm
As	< 0.1	0.2	0.4	ppm
Cd	< 0.1	< 0.1	< 0.1	ppm
Cr	< 0.1	0.4	0.7	ppm
Hg	7	< 5	< 5	ppb
Ni	< 0.5	0.9	1.6	ppm
Pb	0.1	0.5	0.9	ppm

Source: Canadian Humalite International Inc. (2020a).

Figure 5.2 Liquid HA and FA. *(See color plate section for color representation of this figure).*

Liquid HA can be delivered to the client or sub-contractor in bulk (38 000 l) or high-density polyethylene totes with metal cages (1000 l). Arriving at the site, the product in bulk is unloaded onto a storage tank, while the product in totes is stored within a storage facility. The product in bulk may also be repackaged prior to field application. During field work, liquid HA is loaded into a field applicator and subsequently applied.

When handling the product, the SDS and product application label should be used as the main reference. Product characteristics and important details on how to handle the product are provided, which the client or sub-contractor should be made aware of.

5.5 Liquid FA

Liquid FA is manufactured from dry HS product of very small particulate sizing containing mainly FA and some HA. The main objective of liquid FA is to provide a fast-release source of FA to crops. Most applications are intended for foliar. Liquid FA is also the input material used to manufacture other humic products.

Table 5.9 shows liquid FA and its important parameters, including FA, HS, total solids, specific gravity, pH, E_4/E_6, kinematic viscosity, particulate sizing, and water solubility. Note: E_4 = absorbance at 465 nm and E_6 = absorbance at 665 nm; E_4/E_6 is also called the index of molecular condensation.

Based on nine samples taken from the product, liquid FA has been found to contain 0.7% FA (based on ISO 19822, including moisture), 1.0% HS (based on the colorimetric method, including moisture), 1.1% total solids (analyzed according to

Table 5.9 Liquid FA.

Parameter	Value	Unit
FA[a]	0.7 (0.11)	%
HA + FA (= HS)[b]	1.0 (0.11)	%
Total solids	1.1 (0.08)	%
Specific gravity	1.01 (0.005)	–
pH	9.9 (0.09)	–
E_4/E_6	> 10	–
Kinematic viscosity	0.7	mm^2/s
Particulate sizing,	1.1 (0.64)	μm
mean, maximum	2.4	μm
Water solubility	99.7 (0.11)	%

Source: Canadian Humalite International Inc. (2020a);
note:
[a] ISO 19822 (including moisture),
[b] colorimetric method (including moisture), in brackets are standard deviations for mean values.

EPA Method 1684 2001), 1.01 specific gravity (analyzed according to the weight of a 10 mL sample), and 9.9 pH (analyzed according to APHA 4500-H^+ (1992). The product has been found to have high E_4/E_6 values (over 10) analyzed using a UV-visible spectrophotometer. This indicates that the product contains small molecular weight and sizing of humic components, like in the case of FA (see Section 3.2.4).

One sample of the product has been analyzed to contain 0.7 mm^2/s kinematic viscosity (analyzed at 40 °C according to ASTM D445-19 (2019). The sample has also been analyzed using a particle size analyzer (in accordance with Beckman Coulter 2011), showing a mean value of 1.1 μm, with a maximum particulate sizing of 2.4 μm.

Three samples taken from liquid FA have been analyzed for water solubility according to EEC A.6 (1992). This product has been found to have 99.7% water solubility. As expected, liquid FA has a higher water solubility than that of liquid HA due to the smaller molecular weight and size of humic components within the product.

Table 5.10 shows selected metals, based on one sample taken from the product. The sample has been analyzed according to EPA Method 6020A (SW-846) (1998) and EPA Method 245.1 (1994). This table indicates that the liquid FA is of good quality, containing extremely small amounts of metals. Under typical application

Table 5.10 Selected metals.

Parameter	Value	Unit
Al	32	ppm
Na	139	ppm
As	0.17	ppm
Cd	< 0.1	ppm
Cr	< 0.1	ppm
Hg	< 0.5	ppb
Ni	< 0.5	ppm
Pb	< 0.1	ppm

Source: Canadian Humalite International Inc. (2020a).

rates, it will likely meet all limits as listed in Table 5.4. Figure 5.2 (right) shows liquid FA, including its diluted form, prior to the colorimetric analysis.

Liquid FA is delivered to the client or sub-contractor in high-density polyethylene totes with metal cages (1000 l) or heavy-duty polyethylene jugs (20 l). Arriving at the site, the product is stored within a storage facility. The product in totes may also be repackaged prior to field application. During field work, liquid FA is loaded into a field applicator and subsequently applied.

When handling the product, the SDS and product application label should be used as the main reference. Product characteristics and important details on how to handle the product are provided, which the client or sub-contractor should be made aware of.

5.6 Enhanced Liquid Products

Liquid HA or FA can be enriched with nutrients to produce enhanced liquid products. The main objective with enhanced liquid HA or FA products is to combine HA or FA with selected nutrients, so that they can be applied on the field as fully nutrient packaged humic products. Enhanced liquid HA products are also the input material used to manufacture other humic products.

The characteristics of each product depend on the types and amounts of nutrients incorporated within liquid HA or liquid FA. With respect to the crop application, these include six macronutrients (i.e. N, P, K, S, Ca, and Mg) and seven micronutrients (i.e. B, Cl, Cu, Fe, Mn, Mo, and Zn) (Havlin et al. 2005). For the livestock application, nutrients include energy (carbohydrates and fats), protein, minerals, and vitamins. "Minerals" in this case are similar to "nutrients" for the crop application, which include major minerals (i.e. Ca, P, Na, Cl, K, Mg, and S) and trace minerals (i.e. Fe, Zn, Co, Mn, Co, I, Mo, and Se) (Church 1991 and Alberta Agriculture and Forestry 2016). Na, Co, I, and Se are not for crops, while protein is a form of N. B is not for livestock.

Table 5.11 lists important parameters that determine characteristics of each product, including HS, FA, total solids (moisture = 100% – total solids), nutrients, specific gravity, pH, and kinematic viscosity. These values are different from one to another, and therefore have to be determined case by case. Subsequently, a sample can be taken from the product and analyzed for selected metals. The results are then compared with their limits as listed in Table 5.4.

Enhanced liquid HA and FA products can be delivered to the client or sub-contractor in bulk (38 000 l), high-density polyethylene totes with metal cages (1000 l), or heavy-duty polyethylene jugs (20 l). Arriving at the site, the product in bulk is unloaded onto a storage tank, while the product in totes or jugs is stored within a storage facility. The product in bulk and totes may also be repackaged prior to field application. During field work, the enhanced liquid HA or FA product is loaded into a field applicator and subsequently applied.

When handling the product, the SDS and product application label should be used as the main reference. It is important to note that the SDS for enhanced liquid HA or FA product may differ greatly from that of liquid HA or FA. Product characteristics and important details on how to handle the product are provided, which the client or sub-contractor should be made aware of.

Table 5.11 Enhanced liquid HA and FA.

Parameter	Value
HA + FA (= HS) or FA	TBD
Total solids or moisture	TBD
Nutrients	TBD
Specific gravity	TBD
pH	TBD
Kinematic viscosity	TBD
Al, Na, As, Cd, Cr, Hg, Ni, Pb	TBD

Note: TBD = to be determined.

5.7 Pelletized HS

Pelletized HS is manufactured from dry HS of small sizing. The main objective of pelletized HS is to offer a slow-release product that can better resist abrasion during product handling. This product may also contain selected nutrients. In this way, it can be applied to the field as a fully nutrient packaged humic product.

Table 5.12 shows pelletized HS (containing only HS without nutrients) and its important parameters, including HS, density, pH, moisture, particulate strength, particulate sizing, and water solubility. The following results have been obtained based on 33 samples taken from the product (all mean values): 78.0% HS (based on the colorimetric method, dry matter analysis), 870 kg/m^3 density (analyzed according to ASTM D7263-09(2018)e2 2018), 3.9 pH (analyzed according to ASTM D4972-18 2018), 19% moisture (analyzed according to ASTM D3302/D3302M-17 2017), 47 kPa particulate strength (according to ITM Instruments Inc. 2020), 4.5 mm particulate sizing (analyzed according to ASTM D6913/D6913M-17 2017), and 19.3% water solubility (according to EEC A.6 1992). This product is an improvement over dry HS. It is larger, stronger, and more water-soluble. Larger and stronger products better resist abrasion during product handling. Products with a higher water solubility are more readily available to crops.

Table 5.13 shows selected metals, based on one sample taken from the product. The sample has been analyzed according to ASTM D2013/D2013M-18 (2013), ASTM D6357-11 (2011), and ASTM D6722 (2011). This table shows that this product is of good quality, containing very small amounts of metals. Under typical

Table 5.12 Pelletized HS.

Parameter	Mean value
HA + FA (= HS)[a]	78.0 (2.63) %
Density	870 (17) kg/m^2
pH	3.9 (0.19)
Moisture	19 (1.9) %^{2}
Particulate strength	47 (2.6) kPa
Particulate sizing	4.5 (1.10) mm^2
Water solubility	19.3 (1.74) %

Source: Canadian Humalite International Inc. 2020a);
note:
[a] colorimetric method (dry matter analysis), in brackets are standard deviations.

Table 5.13 Selected metals.

Parameter	Value	Unit
Al	0.4	%
Na	0.3	%
As	3.8	ppm
Cd	< 0.2	ppm
Cr	7.2	ppm
Hg	< 50	ppb
Ni	10.3	ppm
Pb	6.2	ppm

Source: Canadian Humalite International Inc. (2020a).

application rates, it will likely meet all limits as listed in Table 5.4. Figure 5.3 (left) shows pelletized HS.

Pelletized HS can be enriched with nutrients to produce enhanced pelletized HS products. The main objective with these products is to combine HS and selected nutrients, so that they can be applied on the field as fully nutrient packaged humic products.

Characteristics of each product depend on the types and amounts of nutrients incorporated within the pellets. For the crop application, six macronutrients (i.e. N, P, K, S, Ca, and Mg) and seven micronutrients (i.e. B, Cl, Cu, Fe, Mn, Mo, and

Figure 5.3 Pelletized and dry soluble HS. *(See color plate section for color representation of this figure).*

Zn) can be added to the product (Havlin et al. 2005). For the livestock application, carbohydrates, fats, protein, major minerals (i.e. Ca, P, Na, Cl, K, Mg, and S), trace minerals (i.e. Fe, Zn, Co, Mn, Co, I, Mo, and Se), and vitamins can be added to the product (Church 1991 and Alberta Agriculture and Forestry 2016). Characteristics of the enhanced pelletized HS product, including the contents of HS, nutrients, and selected metals, can be generated and compared with their limits as listed in Table 5.4.

Pelletized and enhanced pelletized HS products can be delivered to the client or sub-contractor in bulk (38 t), polypropylene supersacks (500 to 1000 kg), or bags of multilayer paper materials plus a thin layer of polypropylene (10 to 25 kg). Arriving at the site, the product in bulk is unloaded onto a storage silo, while the product in supersacks or bags is stored within a storage facility. The product in bulk may also be repackaged prior to field application. During field work, the product is loaded into a field applicator and subsequently applied.

When handling the product, the SDS and product application label should be used as the main reference. It is important to note that the SDS for enhanced pelletized HS may differ greatly from that of pelletized HS. Product characteristics and important details on how to handle the product are provided, which the client or sub-contractor should be made aware of.

5.8 Dry Soluble HS

Dry soluble HS is manufactured from liquid HA. The main objective is to offer a dry product, which can be turned back into a liquid product prior to field application. It is ideal when transportation becomes a concern, such as shipping liquid HA for an overseas market.

Table 5.14 shows dry soluble HS and its important parameters, including HS, density, pH, moisture, particulate sizing, and water solubility. Based on 33 samples taken from the product, the following results have been obtained (all mean values): 85.6% HS (based on the colorimetric method, dry matter analysis), 750 kg/m^3 density (analyzed according to ASTM D7263-09(2018)e2 2018), 9.7 pH (analyzed according to ASTM D4972-18 2018), 12% moisture (analyzed according to ASTM D3302/D3302M-17 2017), 50 μm particulate sizing (analyzed according to ASTM D6913/D6913M-17 2017), and 91.9% water solubility (according to EEC A.6 1992). The high water solubility indicates that it can practically be turned back into solution prior to field application.

Table 5.15 shows selected metals, based on one sample taken from this product. The sample has been analyzed according to ASTM D2013/D2013M-18 (2013), ASTM D6357-11 (2011), and ASTM D6722 (2011). This table shows that this

Table 5.14 Characteristics of dry soluble HS.

Parameter	Mean value
HA + FA (= HS)[a]	85.6 (1.2) %
Density	750 (16) kg/m^3
pH	9.7 (0.35)
Moisture	12 (2.9) %
Particulate sizing	50 (range: 20–106) μm
Water solubility	91.9 (0.96) %

Source: Canadian Humalite International Inc. (2020a);
note:
[a]colorimetric method (dry matter analysis), in brackets are standard deviations.

Table 5.15 Selected metals.

Parameter	Value	Unit
Al	0.5	%
Na	0.2	%
As	2.0	ppm
Cd	0.4	ppm
Cr	6.9	ppm
Hg	< 50	ppb
Ni	20.7	ppm
Pb	< 5.0	ppm

Source: Canadian Humalite International Inc. (2020a).

product is of good quality, containing very small amounts of metals. Under typical application rates, it will likely meet all limits as listed in Table 5.4. Figure 5.3 (right) shows dry soluble HS.

Enhanced dry soluble HS products are manufactured from enhanced liquid HA products. The main objective with these products is to combine HS and selected nutrients, so that they can be applied on the field as fully nutrient packaged humic products. Characteristics of the enhanced dry soluble HS product, including the contents of HS, nutrients, and selected metals, can be generated and compared with their limits as listed in Table 5.4.

Dry soluble HS products can be delivered to the client or sub-contractor in bags of multilayer paper materials plus a thin layer of polypropylene (10 to 25 kg). Arriving at the site, the product is stored within a storage facility.

When handling the product, the SDS and product application label should be used as the main reference. Product characteristics and important details on how to handle the product are to be included on the product application label, which the client or sub-contractor should be made aware of.

Another important consideration when handling this product is how to turn it back into solution practically and safely before field application. Dry soluble HS contains a very small particulate sizing and a low moisture content. Therefore, a direct mixing with water will result in the creation of excessive dust and a non-uniform solution. Section 6.5 discusses the applicability of dry soluble HS in more detail, including its mixing instruction.

5.9 Product Handling

It is critical to note that after its manufacturing, the product receives a lot of handling by the client or sub-contractor even before being applied on the field. Product handling has to be evaluated and planned carefully. An improper handling can lead to negative effects on the integrity of the product, the functioning of both plant and field equipment, as well as the overall practicality and safety of the work. The SDS and product application label are two of the most important documents that the client or sub-contractor must have in hand prior to handling the product.

For each product, an SDS must be generated. A proper sampling of the product must be completed, in which critical parameters are quantified, ideally by third-party laboratories. These parameters, in addition to other information, are needed to provide complete answers for the 16 items requested in the document:

1) product identification;
2) hazard identification;
3) composition/information on ingredients;
4) first-aid measures;
5) fire-fighting measures;
6) accidental release measures;
7) handling and storage;
8) exposure controls/personal protection;
9) physical and chemical properties;
10) stability and reactivity;
11) toxicological information;
12) ecological information;
13) disposal consideration;

14) transport information;
15) regulatory information; and
16) other information.

The main reference of this document is the GHS of Classification and Labeling of Chemicals or the WHMIS for Canada (Canadian Centre for Occupational Health and Safety 2019a). Another act and set of regulations to be considered is the TDG by Transport Canada, which is adopted by all Canadian provinces. TDG promotes public safety when dangerous goods are being handled, ordered for transport, or transported by road, rail, or water. It also establishes safety requirements. Dangerous goods are the ones listed in Schedules 1 and 3 (and have particular UN numbers according to the UN Dangerous Goods List) or meeting the classification criteria outlined in Part 2 of the regulations (Canadian Centre for Occupational Health and Safety 2019b).

Since low-rank coal is not classed as dangerous goods, its derivative humic products are most likely non-hazardous and therefore are not classed as dangerous goods as well. Still, a unique SDS has to be generated for each product. When other ingredients are included in the case of enhanced humic products, a unique SDS is definitely needed for each of them. Each SDS must be revisited once every three years, in which an update or modification is to be completed to represent the most current information of the product.

Among other things, the SDS shows important characteristics of the product, including pH, density or specific gravity, hardness (abrasiveness) or viscosity, stability and reactivity, and others that the client or sub-contractor must be aware of. This document also informs how to handle and store the product in general. A proper selection on types and materials of equipment as well as the optimum method to handle the product can therefore be made accordingly.

For each product, an application label must be generated. Other than providing information on its registration, compatibility with nutrients and other soil amendments, and recommended application rates for the intended application, this document also informs the client and sub-contractor how to handle the product, especially during field application. Critical parameters include pH, density or specific gravity, hardness (abrasiveness) or viscosity, and stability and reactivity. This information determines the suitable types and materials of equipment as well as the optimum method to handle the product.

Common practices when handling dry and liquid fertilizers, soil amendments, and other products/chemicals commonly sold in market can be used as a guideline (L. E. Liem Consulting 2020). When requested or deemed necessary, the information can be included in the product SDS and/or application label presented to the client or sub-contractor to ensure the overall practicality and safety of the work.

Auger conveyors have to be avoided when transferring dry (and fragile) products of large sizing. Belt conveyors and bucket elevators are the better alternatives, as interactions between the moving and static components of the equipment are minimal. For dry products of small sizing, bucket elevators and auger conveyors, or tubular drag conveyors, are the choices to minimize the release of dust to the surrounding area. For dry products of very small sizing, tubular drag conveyors and pneumatic tubes of the pressure-dense phase system are the best, as no release of dust to the surrounding area occurs. All transfer equipment can be made from carbon steel or hard steel when necessary.

When dealing with neutral to relatively high pH liquid products, carbon steel or polyvinylchloride is the choice piping material. For an extremely low pH, polyvinylchloride can be used; and for an extremely high pH, stainless steel or polyvinylchloride can be used. For nominal pipe sizing of 76 to 152 mm, recommended velocities are 90 to 220 m/min, respectively. Its maximum pressure loss is 70 kPa/100 m. Minimum velocities to allow a proper cleaning for the piping from possible solids buildup are 45 to 75 m/min. For pump suction and discharge lines, the maximum pressure drops are 5 to 20 kPa/100 m, respectively. Due to the presence of solids in the piping, flat on bottom reducers/expanders and butterfly valves are recommended. For liquid products with minimum amounts of solids, any reducers/expanders are applicable, and ball valves are the best as the isolation mechanism to provide a complete liquid isolation.

A centrifugal type of pump made from carbon steel with a stainless steel semi-closed impeller, double mechanical seals, coated bushings, and nitrile rubber components is the best for viscous liquid products with neutral to relatively high pH containing significant amounts of solids. For small flowrates, an open impeller type of pump is acceptable. For an extremely low pH, use a centrifugal magnetic drive type of pump without seals, made from polypropylene with a ceramic shaft/polypropylene impeller, ceramic bushings, and fluoropolymer rubber components. For an extremely high pH, use a centrifugal magnetic drive type of pump without seals, made from polypropylene with a stainless steel shaft/polypropylene impeller, coated bushings, and fluoropolymer rubber components.

Dry storage silos of carbon steel material (plus internal coating with slip additives and a mechanical vibrator attached to the bottom cone to prevent product bridging) are typically located outdoors. Each has to be foam insulated and best traced with electric heat tapes to avoid product freezing during winter. Liquid storage tanks of high-density polyethylene tanks are ideally placed inside the building. If not, they have to be foam insulated and traced with electric heat tapes.

All dry and liquid products (in supersacks, bags, totes with metal cages, and jugs) are to be stored inside the building. An adequate heating, ventilation, and air conditioning (HVAC) system is a must to capture dust particulates. A dust collector containing bag filters is best to control the dust within the facility. The building is to have the radiant and floor water/glycol-heating system and is maintained under an optimum temperature and humidity. If the building does not have a heated floor, a wooden pallet is needed to protect the product from possible freezing and moisture on the floor. They are also required to provide a stable support and a greater mobility for the forklift.

All electrical devices (motors, most importantly) located within the building have to be enclosed according to their ratings/usages. When dealing with a very dusty environment, all of them have to be rated as totally enclosed explosion proof Class 2. For a less severe dust presence within the building, a step-down requirement can be implemented. Each motor can be rated as totally enclosed fan cooled. Both mechanical and electrical control rooms have to properly be ventilated.

5.10 Closure

Historical data of humalite collected from the mine site indicates that the raw material has been of good quality and consistent throughout the years. This material is safe and manageable and can be manufactured into various commercial humic products.

Table 5.16 summarizes commercial humic products manufactured from humalite, starting from dry HS. Dry HS is the input material for pelletized HS, liquid HA, and liquid FA. With the addition of nutrients, liquid HA and liquid FA can be manufactured into enhanced liquid HA and FA products. Enhanced pelletized HS products are manufactured by combining dry HS and nutrients. Dry

Table 5.16 Commercial humic products.

Product	Water solubility	Nutrient content
Dry HS	Low	No
Liquid HA	High	No
Liquid FA	High	No
Pelletized HS	Low	No
Enhanced pelletized HS	Low	Yes
Enhanced liquid HA or liquid FA	High	Yes
Dry soluble HS	High	No
Enhanced dry soluble HS	High	Yes

Source: Canadian Humalite International Inc. (2020a).

soluble HS is manufactured from liquid HA and enhanced dry soluble HS products are manufactured from enhanced liquid HA products. For products with nutrients, their formulations are determined according to the client's requests. Dry HS is the least water-soluble, followed by pelletized HS, dry soluble HS, liquid HA, and liquid FA. In general, the presence of nutrients within a product does not affect its water solubility.

Understanding the basic properties of each product is critical to determining its optimum rate and method of application during field work. Each product is unique, therefore none is better than the other. When applied properly on the field, each product gives the best result for its intended purpose. Dry and pelletized HS products are designed to benefit the soil first, then improve the growth of crops. Liquid HA is applied to both soil and crops, while liquid FA is intended for crops only. Dry soluble HS is to be made into liquid HA and then applied accordingly. Dry HS is the only humic product that has been used in the livestock application.

Product handling on the client and sub-contractor's side has to be evaluated and planned carefully. An improper handling can lead to negative effects on the integrity of the product, the functioning of both plant and field equipment, as well as the overall practicality and safety of the work. For each product, the SDS and application label must be reviewed in advance. With respect to product handling, the provided information determines the following:

1) the identification or classification of the product;
2) the equipment types and materials to handle the product; and
3) the method used to handle the product.

Considerations here are informed by the worst-case scenario to ensure maximal practicality and safety during product handling. Work may commence upon acknowledging and accommodating all necessary conditions and requirements as specified.

Reference

APHA 4500-H^+ (1992). pH value. In: *Standard Methods for the Examination of Water and Wastewater*, 18th ed. (eds. A.E. Greenberg, L.S., Clesceri, and A.D. Eaton), 4-65 to 4-69. Washington, DC: American Public Health Association.

Alberta Agriculture and Forestry (2016). Nutrients for cattle. *Agdex 400/50-3* 1–6. Edmonton, AB.

ASTM D2013/D2013M-18 (2013). *Standard practice for preparing coal samples for analysis*. West Conshohocken, PA: ASTM International.

ASTM D3302/D3302M-17 (2017). *Standard test method for total moisture in coal*. West Conshohocken, PA: ASTM International.

ASTM D445-19 (2019). *Standard test method for kinematic viscosity of transparent and opaque liquids (and calculation of dynamic viscosity)*. West Conshohocken, PA: ASTM International.

ASTM D4972-18 (2018). *Standard test methods for pH of soils*. West Conshohocken, PA: ASTM International.

ASTM D6357-11 (2011). *Test methods for determination of trace elements in coal, coke, & combustion residues from coal utilization processes by Inductively Coupled Plasma atomic emission, Inductively Coupled Plasma mass, & graphite furnace Atomic Absorption spectrometry*. West Conshohocken, PA: ASTM International.

ASTM D6722 (2011). *Standard test method for total mercury in coal and coal combustion residues by direct combustion analysis*. West Conshohocken, PA: ASTM International.

ASTM D6913/D6913M-17 (2017). *Standard test methods for particle-size distribution (gradation) of soils using sieve analysis*. West Conshohocken, PA: ASTM International.

ASTM D7263-09(2018)e2 (2018). *Standard test methods for laboratory determination of density (unit weight) of soil specimens*. West Conshohocken, PA: ASTM International.

Beckman Coulter (2011). Instruction for use. *LS 13 320 Laser Diffraction Particle Size Analyzer* 1–246. Brea, CA.

Canadian Centre for Occupational Health and Safety (2019a). Global harmonized system (GHS). http://www.ccohs.ca/oshanswers/ chemicals/ghs.html (accessed 31 July 2019).

Canadian Centre for Occupational Health and Safety (2019b). Transportation of dangerous goods (TDG) – review. http://www.ccohs.ca/oshanswers/legisl/tdg/tdg_overview.html (accessed 31 July 2019).

Canadian Council of Ministries of the Environment (2002a). Agricultural. *Soil Quality Guidelines for the Protection of Environmental and Human* 1–9. Hull, QC.

Canadian Council of Ministries of the Environment (2002b). Polychlorinated Dibenzo-p-Dioxins and Polychlorinated Dibenzofurans (PCDD/Fs). *Soil Quality Guidelines for the Protection of Environmental and Human Health* 1–13. Hull, QC.

Canadian Food Inspection Agency (2017a). Safety guidelines for fertilizers and supplements. *T-4-93* 1–11. Ottawa, ON.

Canadian Food Inspection Agency (2017b). *Proposal-Maximum Chemical Contaminant Levels in Livestock Feeds* 1–30. Ottawa, ON.

Canadian Food Inspection Agency (2017c). *Proposal-Contaminant Standards for Aflatoxins, Deoxynivalenol, Fumonisins, Ergot Alkaloids and Salmonella in Livestock Feeds* 1–18. Ottawa, ON.

Canadian Humalite International Inc. (2020a). Raw material and products. *Laboratory Analyses and Technical Data* 1–656. Edmonton, AB.

Canadian Humalite International Inc. (2020b). Raw material and products. *Safety Data Sheets* 1–55. Edmonton, AB.

Church, D.C. (1991). *Livestock Feeds and Feeding*, 3rd ed., 15–36. Englewood Cliffs, NJ: Prentice-Hall.

EEC A.6 (1992). Water solubility. Dir 92/69/EEC (O.J. L383 A). In: *Official Journal of the European Communities No. L 383/113* (member K. Van Miert), 1–9. Aberdeen, UK: Commission Directive 92/69/EEC.

EPA Method 1684 (2001). *Total, fixed, and volatile solids in water, solids, and biosolids*. Washington, DC: Office of Water, Office of Science and Technology, Engineering and Analysis Division (4303), US Environmental Protection Agency.

EPA Method 245.1 (1994). *Determination of mercury in water by cold vapor atomic absorption spectrometry*. Cincinnati, OH: Environmental Monitoring Systems Laboratory Office of Research and Development, US Environmental Protection Agency.

EPA Method 6020A (SW-846) (1998). *Inductively Coupled Plasma-mass spectrometry*. Washington, DC: US Environmental Protection Agency.

Havlin, J.L., Beaton, J.D., Tisdale, S.L., and Nelson, W.L. (2005). *Soil Fertility and Fertilizers: an Introduction to Nutrient Management*, 7th ed., 362–416. Upper Saddle River, NJ: Pearson-Prentice Hall.

Hoffman G.L., Nikols, D.J., Stuhec, S., and Wilson, R.A. (1993). Evaluation of leonardite (humalite) resources of Alberta. *Open File Report 1993-18* 1–18. Edmonton, AB: Alberta Research Council.

ITM Instruments Inc. (2020). Mark-10 M3-500. *Digital Force Gauge Series 3* 1–2. Calgary, AB.

L. E. Liem Consulting (2020). *Design and Engineering Work* 1–54. Edmonton, AB.

National Research Council (2005). *Mineral Tolerance of Animals*, 2nd rev., 10–14. Washington, DC.

6

Product Performance

6.1 Overview

This chapter presents the most important information on low-rank coal applications in agriculture: whether commercial products manufactured from this material can be applied on the field and provide the intended benefits. In short, whether they can fulfill their advertised function. The evaluation is focused on humic substances (HS, or "humic"), as well as their components – humic acids (HA) and fulvic acids (FA), the active components within low-rank coal – on the productivity of soil, crops, and livestock. All information presented here has been obtained from original work completed using commercial humic products manufactured from Alberta low-rank coal (humalite). The first part of the work concerns the applicability of dry, liquid, and dry soluble humic products, including their compatibility with common nutrients. The second part of the work concerns the efficacy of the products on soil, crops, and livestock, with all data having been generated from actual projects. All projects but one have been completed on soil and crops, with one project completed on livestock.

For soil and crop applications, dry and liquid humic products can be directly applied or blended with nutrients prior to application. Dry soluble humic products are typically firstly made into liquid products before being applied. Important factors to be considered when investigating product performance include the following:

1) Product types, application methods, and product compatibility with common fertilizers. Dry humic products, due to their low water solubility, require much larger effective rates than those of liquid and dry soluble humic products that have close to 100% water solubility. Applications over the field require larger amounts of product per area than do those applications close to the target

Low-Rank Coal Applications in Agriculture: Humic Analyses, Products, and Performance, First Edition. L. Edwin Liem.

crops (this holds for both dry and liquid products). It is important to remember that dry humic products are soft and porous, and tend to absorb water from their surroundings. Liquid humic products may separate in water at low acidity (pH), and form insoluble matter with certain nutrients.

2) Types of soil and crops, available nutrients in soil, and product application timing and rates. Products with low water solubility are applied much earlier than water-soluble products to allow them to break down in soil. Low water solubility products are sometimes identified as "slow release" and water-soluble products as "fast release" (instantaneous) products. The most dramatic results have been observed when crops are grown under stressed conditions, i.e. in soil with low organic matter (OM), high salinity, low nutrient contents, and reduced nutrient inputs. Under simulated growing conditions, it is possible to evaluate how much nutrients can be reduced while maintaining (or even enhancing) crop yields with the incorporation of humic products in the fertilizing program. It is also critical to identify the optimum rates of humic product use before responses peak or become negative at higher rates.

In the case of the livestock application, one product efficacy project has been completed using a dry humic product blended together with feed. This area of work has not been well explored, but its market sector has been growing significantly in recent years. The main focus here is on the improvement of animal health and feed efficiency (the feed to yield ratio), and on the reduction of byproducts (mainly in gaseous forms) from the whole operation.

6.2 Applicability of Dry and Pelletized HS

Alberta low-rank coal (humalite) is processed to manufacture dry HS of large, medium, small, and very small particulate sizing products (see Section 5.3). Dry HS (and fertilizers) can be broadcasted on the surface of soil before planting (this is known as the surface broadcast or basal application) or it can be broadcasted over grown crops (this is known as the top-dress application for small grains, pastures, and turf). The typical equipment used for this application includes a storage tank connected by an auger or belt conveyor to one or two high-capacity spinner(s) positioned above but close to the ground. All equipment is mounted on a mobile unit pulled by a tractor (Cornell University 2010 and Green and Eliason 1999). Typical application rates are between approximately 180 (at the center of the swath) and 440 kg/ha (Follett et al. 1981). A belt conveyor is a better choice than an auger conveyor since this minimizes the possible grinding of dry HS. The same product can also be broadcasted using a boom spraying system. This consists of a storage tank, an air pumping/metering unit, connecting hoses, a main horizontal pipe, and spraying orifices. All equipment is typically mounted on a truck. This method of application offers a more uniform product distribution across the field (Garvey 2017).

For both methods of application, dry HS of large particulate sizing is best. It should be applied to the ground at a reasonable distance, with minimum amounts of dust created. Dry HS of medium particulate sizing may still be applicable using the boom spraying system. Both humic products of small and very small particulate sizing are not applicable. Even in the case of products with large and medium particulate sizing, caution must be taken. These products contain significant amounts of moisture and combining one of them with certain nutrients, including urea, may result in a mushy blend within the storage tank.

Dry HS (and fertilizers) can also be applied during planting. They should be applied so as to be near to developing roots but not so close that they may cause injury to the germinating seeds. This is called the band application. The seeds can be applied in one row above and to the side of the product (and fertilizers), or in two rows above and on both sides of the product (and fertilizers). Both seeds and the product (and a small amount of fertilizer, so as not to harm the seeds) can also be applied in the same furrow, which is called a seed placement or pop-up application. Dry HS (and fertilizers) can also be applied in between rows after young crops emerge, which is called a side-dress application (Cornell University 2010, Green 1999a, and Follett et al. 1981).

The typical equipment used for these applications is air seeders or air drills (consisting of seed and fertilizer tanks), a central pneumatic for seeds and nutrients delivery, and a ground opener for seed and fertilizer placement (with or without supporting wheels). All equipment is mounted on a mobile unit. In a single shoot system, there is one air line for both seeds and nutrients delivered from the storage tank to the ground opener. This system is not ideal for dry HS since the high moisture content may react with certain nutrients forming a mushy blend. The double shoot system is the only option for humic products. It involves two air lines connected from two different storage tanks to the ground opener. Typically, this opens two separate furrows, one for the seeds and one for the fertilizers. There is a variation consisting of two storage tanks with one air line for a dry product and/or seeds and one liquid line for a liquid fertilizer. There may even be three lines from three different storage tanks. The principle remains the same: the first tank stores seeds that are blended with the humic product. This blend will not become mushy and can be delivered in a pressurized air line to the first furrow. The fertilizer or a blend of different fertilizers is stored in the other storage tank, to be delivered in a separate line to the second furrow. Sometimes the seed and humic product line branches into two rows on both sides of the fertilizer band, but this does not make any difference with respect to the humic product application (Green 1999a and Green 1999b).

Pelletized HS is an improvement over dry HS of large and medium particulate sizing. It is manufactured from dry HS of small particulate sizing (see Section 5.7). The particles are harder and drier, and are able to resist abrasion during product handling and a moist environment better. Most importantly, this pelletized product can be custom formulated to contain the desired amounts of HS and nutrients. The central aim here is to create a single product, thus eliminating the need to blend different materials.

Table 6.1 summarizes the characteristics of dry HS of large and medium particulate sizing, and pelletized HS. This pelletized product contains only HS, no nutrients are included within the formulation. This table shows that the pelletized product has the largest particulate sizing (mean = 4.5 mm), the least moisture content (mean = 19%), and the strongest particulate hardness (mean = 47 kPa). In addition, this product shows a higher water solubility (19.3%). This suggests that the pelletized humic product has some major advantages over other products and is therefore recommended for most field application.

Tables 6.2 to 6.4 summarize product compatibility results, which are based on laboratory-scale studies blending each product with common nutrients at ratios from 1:1 to 10:1 and vice versa by weight. The nutrients to be considered include six macronutrients, i.e. nitrogen (N), phosphorus (P), potassium (K), sulfur (S), calcium (Ca), and magnesium (Mg); as well as two micronutrients, i.e. boron (B) and chloride (Cl). P, K, and S may also be expressed as diphosphorus pentoxide (P_2O_5), dipotassium oxide (K_2O), and sulfate (SO_4), respectively. The responses include dry (D, which is the only acceptable response), wet (W), mushy (M), rather wet (RW), and rather mushy (RM); recorded two hours after blending (Canadian Humalite International Inc. 2020b). Other responses indicate

Table 6.1 Characteristics of dry HS.

Product	Large particulate sizing	Medium particulate sizing	Pelletized humic product
Mean particulate sizing (mm)	2.0	0.9	4.5
Mean moisture (%)	31 (2.3)	25 (3.1)	19 (1.9)
Mean particulate hardness (kPa)	35 (3.0)	N/A	47 (2.6)
Mean solubility in water (%)	5.1 (2.02)		19.3 (1.74)
Mean pH	3.9 (0.17)		3.9 (0.19)
Mean density (kg/m^3)	890 (28)		870 (17)
Mean HS (%)	93.7 (4.27)		78.0 (2.63)

Source: Canadian Humalite International Inc. (2020a); *note:* in brackets are standard deviations, N/A = not available.

Table 6.2 Compatibility of dry HS of large particulate sizing.

By weight		Urea				
Humic	Nutrient	46N	43N	$Ca(NO_3)_2$ 15N 18Ca	NH_4NO_3 34N	MAP 11N $52P_2O_5$
1	10	W	D	D	D	D
1	4	W	D	D	D	D
1	2	M	D	M	M	D
1	1	M	RW	M	M	RW
2	1	M	D	M	M	D
4	1	M	D	M	M	D
10	1	RM	D	RM	RM	D

By weight				
Humic	Nutrient	Ulexite B[a] 9B 15Ca 3Mg	$MgCl_2$[a] 22Mg 66Cl	Epsom salt[a] 10Mg 13S
1	10	D	W	D
1	4	D	M	D
1	2	D	M	D
1	1	D	M	D
2	1	D	M	D
4	1	D	M	D
10	1	D	M	D

By weight					
Humic	Nutrient	KCl[a] $60K_2O$ 45Cl	$CaCl_2$[a] 30Ca 53Cl	$CaCO_3$[a] 36Ca	Phosrock[a] $12P_2O_5$
1	10	D	W	D	D
1	4	D	M	D	D
1	2	D	M	D	D
1	1	D	M	D	D
2	1	D	M	D	D
4	1	D	RM	D	D
10	1	D	RM	D	D

Note: D = dry, W = wet, M = mushy, RW = rather wet, RM = rather mushy,
[a] organic material according to USDA-NOP (as long as non-synthetic, some with restrictions).

Table 6.3 Compatibility of dry HS of medium particulate sizing.

By weight		Urea				
Humic	**Nutrient**	**46N**	**43N**	**$Ca(NO_3)_2$ 15N 18Ca**	**NH_4NO_3 34N**	**MAP 11N $52P_2O_5$**
1	10	D	D	D	D	D
1	4	W	D	M	M	D
1	2	M	D	M	M	D
1	1	M	RW	M	M	RW
2	1	M	D	M	M	D
4	1	M	D	M	M	D
10	1	RM	D	RM	RM	D

By weight				
Humic	**Nutrient**	**Ulexite B[a] 9B 15Ca 3Mg**	**$MgCl_2$[a] 22Mg 66Cl**	**Epsom salt[a] 10Mg 13S**
1	10	D	W	D
1	4	D	M	D
1	2	D	M	D
1	1	D	M	D
2	1	D	M	D
4	1	D	M	D
10	1	D	M	D

By weight					
Humic	**Nutrient**	**KCl[a] $60K_2O$ 45Cl**	**$CaCl_2$[a] 30Ca 53Cl**	**$CaCO_3$[a] 36Ca**	**Phosrock[a] $12P_2O_5$**
1	10	D	D	D	D
1	4	D	RM	D	D
1	2	D	RM	D	D
1	1	D	M	D	D
2	1	D	M	D	D
4	1	D	M	D	D
10	1	D	RM	D	D

Note: D = dry, W = wet, M = mushy, RW = rather wet, RM = rather mushy,
[a] organic material according to USDA-NOP (as long as non-synthetic, some with restrictions).

Table 6.4 Compatibility of pelletized HS.

By weight		Urea				
Humic	Nutrient	46N	43N	$Ca(NO_3)_2$ 15N 18Ca	NH_4NO_3 34N	MAP 11N $52P_2O_5$
1	10	D	D	D	D	D
1	4	D	D	D	D	D
1	2	RW	D	D	RW	D
1	1	RW	D	D	W	D
2	1	RW	D	D	RM	D
4	1	RW	D	D	RM	D
10	1	D	D	D	W	D

By weight				
Humic	Nutrient	Ulexite B[a] 9B 15Ca 3Mg	$MgCl_2$[a] 22Mg 66Cl	Epsom[a] 10Mg 13S
1	10	D	D	D
1	4	D	W	D
1	2	D	M	D
1	1	D	M	D
2	1	D	M	D
4	1	D	M	D
10	1	D	M	D

By weight					
Humic	Nutrient	KCl[a] $60K_2O$ 45Cl	$CaCl_2$[a] 30Ca 53Cl	$CaCO_3$[a] 36Ca	Phosrock[a] $12P_2O_5$
1	10	D	D	D	D
1	4	D	D	D	D
1	2	D	D	D	D
1	1	D	RM	D	D
2	1	D	RM	D	D
4	1	D	RM	D	D
10	1	D	W	D	D

Note: D = dry, W = wet, M = mushy, RW = rather wet, RM = rather mushy,
[a] organic material according to USDA-NOP (as long as non-synthetic, some with restrictions).

outcomes which are to be avoided, especially M. In general, urea (46-0-0), calcium nitrate ($Ca(NO_3)_2$), ammonium nitrate (NH_4NO_3), magnesium chloride ($MgCl_2$), and calcium chloride ($CaCl_2$) are not compatible with dry HS. Turning dry HS into pelletized humic products improves overall compatibility. Polymer coated urea (43-0-0) is a different version of urea intended for a slow release application. This product, as well as granular mono ammonium phosphate (MAP), is more compatible with all humic products. In general, ulexite B, Epsom salt ($MgSO_4.7H_2O$), potassium chloride (KCl), limestone ($CaCO_3$), and phosrock (containing mainly P, Ca, and Mg) are compatible with all humic products.

Also mentioned under the table are materials that are considered to be organic according to US Department of Agriculture-National Organic Program (USDA-NOP) (Canadian Humalite International Inc. 2020c).

6.3 Applicability of Liquid HA

This product is manufactured from dry HS of very small particulate sizing to contain mainly HA and some FA (see Section 5.4). Liquid HA (and fertilizers) can be broadcasted on the surface of soil and seedlings, or the leaves of grown crops (known as "foliar application") using a sprinkling system of a center-pivot or lateral-move type. In addition to a storage tank and a pumping/metering unit, this system consists of horizontal piping mounted on top of motorized towers (with wheels for movement). A center-pivot system moves around a designated point, while a lateral-move system moves along a straight path. It is intended for use with traditional field crops. The liquid is applied to the soil surface via overhanging pipes (typically of 100 to more than 300 mm in diameter) connected to screens and then to sprinkler nozzles. The nozzle is either of an impact or spray type, and it can either be fixed or rotating. The nozzles range from size 9 (1.9 mm hole size) to 50 (4.3 mm hole size). Its operating pressure can be as high as 500 kPa but is typically 250 kPa or less (Evans 2001 and Valley 2018).

Liquid HA (and fertilizers) can be broadcasted using a boom spraying system. This system consists of a storage tank, a liquid pumping/metering unit, connecting hoses, a main horizontal pipe, screens, and spraying nozzles. When in operation, the whole unit is dragged across the field using a tractor. The liquid is sprayed over the soil surface at a closer distance than that of a sprinkling system. The nozzle's flowrate and the unit's speed are interconnected. The nozzles can either be a flat-fan, flood, or cone type. Its operating pressure ranges between 70 and 690 kPa (typically 410 to 480 kPa). The nozzle is classified based on the size of the droplets formed, from very fine ($< 145\ \mu m$) to extremely coarse ($> 500\ \mu m$). Flowrates on the nozzle tip depend on the operating pressure and on the type and size of the nozzle, ranging from 0.4 to 6.0 l/minute (Grisso et al. 2013 and Wolf 2018).

Liquid HA (and fertilizers) can be applied to smaller areas using a trickle irrigation system. It consists of a storage tank, a liquid pumping/metering unit, a filtration unit (i.e. involving screens and a media filter) and a network of pipes, tubing, and in-line or on-line emitters. The emitters can be drippers, micro-spray heads, or irrigation mats. The whole unit is very much fixed on the soil's surface or subsurface. The liquid is sprayed at a desired flowrate over the soil surface or root zone. This unit is intended for tress and vine crops. Greenhouses and row crops such as vegetables and some fruit orchards use in-line emitters at operating pressures of 15 to 210 kPa and flowrates of 3 to 4 l/minute. Land applications use on-line emitters at operating pressures of 35 to 410 kPa and flowrates of 120 to 450 l/minute (Goyal 2015).

Lastly, liquid HA (and fertilizers) can be applied in a hydroponic system, in which crops are grown in water. Some variations include growing crops in growing media other than soil, where a nutrient solution is fed to make them grow (Agriculture and Agri-Food Canada 2017). The aquaponic system is a slightly expanded version of this. It combines a soilless environment and aquaculture (fish farming) with beneficial microorganisms contributing to crop growth (Lethbridge College 2020). In general, a hydroponic system consists of water containers, plumbing, solution tanks, lighting, air/heat/moisture supply, pumps/flow meters, and control equipment (Brechner and Both 2013). In this application, liquid HA is blended with water and nutrients in the solution tank and then applied to the crops accordingly.

With respect to the compatibility of liquid HA and nutrients, a small-scale laboratory study is recommended. This includes the following components: a storage tank, a pumping/metering unit, a screen, and a nozzle. The capacities of the tank and pump are selected based on the desired flowrate and pressure, as well as the type and size of the nozzle used. These have to be discussed with the client in view of the intended operation on the field. The typical flowrates, operating pressures, and nozzle types and sizes as previously discussed can be used as guidelines. A 50 mesh (280 μm opening) or 100 mesh (140 μm opening) screen is a typically used for this kind of operation in agriculture.

The study can be completed by blending a humic product with a nutrient. Liquid nutrients to be considered include urea ammonium nitrate (UAN, 28-0-0) and ammonium polyphosphate (APP, 10-34-0). Dry nutrients include all sources of macronutrients, including potassium nitrate (KNO_3), powder MAP, mono potassium phosphate (MKP), and potassium sulfate (K_2SO_4). It is also important to remember to use the same type of water for dilution, i.e. tap water from the city, well water from the site, or river water close by. Each type of water has different characteristics that may affect the blend. City water may have some trace chlorine, while well water may contain high pH, alkalinity, hardness, and electrical conductivity (EC). Sodium (Na), iron (Fe), and manganese (Mn) are commonly found in well water. River water may contain high suspended solids, organic carbon (C), alkalinity, hardness, and EC. In addition to cations like Na, Fe, and Mn, it may contain SO_4 and Cl, coming from runoff fertilizers upstream or leached from its base rocks. It is important to avoid precipitates and/or gels formed during or after mixing. If the screen and/or the selected nozzle are plugged, this is an indication that the mixture is not compatible and therefore has to be avoided.

Liquid HA of 12% HS strength (based on the colorimetric method, including moisture) and 10 pH has shown incompatibilities with some liquid and dry nutrients at certain blending ratios. For blends with liquid fertilizers, the ratios evaluated range from 0 to 100%, and vice versa (see Table 6.5). For blends with dry fertilizers, the ratios evaluated range from 1:1 to 40:1 (see Table 6.6). Outcomes of the evaluation are as follows: fluid (F, which is ideal), thick (T, which is acceptable with advance notice given to the client), solid (S), and precipitate or separate (P, which has to be avoided at all costs); recorded two hours after blending (Canadian Humalite International Inc. 2020b). Liquid HA is compatible with some N (urea 46-0-0) but is not always compatible with others.

Tables 6.5 and 6.6 are general guidelines representing the worst-case scenario. In most cases, the mixture of this product and nutrients is diluted with water inside the storage tank before being applied. This specific mixing ratio has to firstly be evaluated in the laboratory. Also mentioned under the table are materials that are considered to be organic according to USDA-NOP (Canadian Humalite International Inc. 2020c).

Note: samples of liquid HA have been completely frozen, then allowed to thaw at room temperature. Particulates contained within the products have been found to settle out. A rigorous agitation has been required to put them back into

Table 6.5 Liquid HA and liquid fertilizers.

Humic	Liquid UAN (28-0-0)	Response
0–30%	70–100%	F
30–90%	10–70%	T
90–100%	0–10%	F

Humic	Liquid APP (10-34-0)	Response
0–10%	90–100%	F
10–30%	70–90%	S
30–90%	10–70%	T
90–100%	0–10%	F

Note: F = fluid, T = thick, S = solid, P = precipitate/separate.

Table 6.6 Liquid HA and dry fertilizers.

By weight					
Humic	**Nutrient**	**$Ca(NO_3)_2$ 15N 18Ca**	**Urea 46N**	**KNO_3 13N $46K_2O$**	**NH_4NO_3 34N**
1	1	S	F	S	P
2	1	T	F	T	P
4	1	T	F	T	F
10	1	T	F	T	F
20	1	T	F	F	F
40	1	F	F	F	F

By weight					
Humic	**Nutrient**	**MAP 12N $60P_2O_5$**	**MKP $52P_2O_5$ $34K_2O$**	**$MgCl_2$[a] 22Mg 66Cl**	**Epsom salt[a] 10Mg 13S**
1	1	S	S	S	T
2	1	S	S	S	T
4	1	T	T	T	T
10	1	T	S	S	T
20	1	S	S	T	T
40	1	F	S	F	F

By weight				
Humic	**Nutrient**	**KCl[a] $60K_2O$ 45Cl**	**K_2SO_4[a] $52K_2O$ 18S**	**$CaCl_2$[a] 30Ca 53Cl**
1	1	PT	S	PS
2	1	PT	S	PS
4	1	T	S	S
10	1	TF	S	S
20	1	F	S	T
40	1	F	S	F

Note: F = fluid, T = thick, S = solid, P = precipitate/separate,
[a] organic material according to USDA-NOP (as long as non-synthetic, some with restrictions).

solution. This indicates that the integrity of the product is compromised at a low temperature, and therefore this has to be avoided at all costs.

6.4 Applicability of Liquid FA

This product is manufactured from dry HS of very small sizing to contain mainly FA and some HA (see Section 5.5). Liquid FA (and fertilizers) can be applied foliar to seedlings or grown crops. Applying them to soil will result in a major loss of humic components due to their relatively small particulate sizing. The same equipment as is used for liquid HA can be used for liquid FA. Another acceptable practice is to apply liquid FA (and fertilizers) in a hydroponic or aquaponic system foliar to leaves or mixed with the nutrient solution and absorbed by crop roots.

With respect to any compatibility issues of liquid FA and nutrients, a small-scale laboratory test is recommended (involving the same equipment and set-up used for liquid HA). Liquid nutrients to be considered include UAN and APP. Dry nutrients include all sources of macro- and some micronutrients (copper sulfate, $CuSO_4$; iron sulfate, $FeSO_4$; manganese sulfate, $MnSO_4$; zinc chloride, $ZnCl_2$; disodium octoborate, $Na_2B_8O_{13}.4H_2O$; and disodium molybdate, $Na_2MoO_4.2H_2O$), since liquid FA is mainly for foliar application. As with liquid HA, it is important to avoid precipitates, foams, and/or gels forming during or after mixing. If the screen and/or the selected nozzle is plugged, this is an indication that the mixture is not compatible and is therefore to be avoided.

Liquid FA of 0.7% FA and 0.3% HA (total HS = 1.0%, based on ISO 19822, including moisture) with 10 pH has shown incompatibilities with some liquid and dry nutrients at certain blending ratios. However, in general, this product performs better than liquid HA of 12% HS strength. For blends with dry fertilizers, the ratios evaluated range from 1:1 to 40:1 (see Table 6.6). The outcomes here can be fluid (F), foam (f), thick (T), solid (S), and precipitate/separate (P); recorded two hours after blending (Canadian Humalite International Inc. 2020b). F is ideal but T and f may be acceptable with advance notice given to the client. It is essential to avoid S and P. Tables 6.7 and 6.8 are general guidelines representing the worst-case scenario. In most cases, the mixture of this product and nutrients is diluted with water inside the storage tank before being applied. Also mentioned under the table are materials that are considered to be organic according to USDA-NOP (Canadian Humalite International Inc. 2020c).

Note: samples of liquid FA have been completely frozen and then allowed to thaw at room temperature. Particulates contained within the product have been found to settle out. Different from liquid HA, however, they have been found to turn back into solution with only a slight agitation. While the integrity of the product is not greatly affected by temperature, liquid FA product is still best stored at room temperature.

Table 6.7 Liquid FA and liquid fertilizers.

Humic	Liquid UAN (28-0-0)	Response
0–10%	90–100%	F
10–30%	70–90%	F
30–90%	10–70%	F
90–100%	0–10%	F

Humic	Liquid APP (10-34-0)	Response
0–10%	90–100%	F
10–30%	70–90%	F
30–90%	10–70%	F
90–100%	0–10%	F

Note: F = fluid, T = thick, S = solid, P = precipitate/separate, f = foam.

Table 6.8 Liquid FA and dry fertilizers.

By weight					
Humic	Nutrient	Urea 46N	$Ca(NO_3)_2$ 15N 18Ca	KNO_3 13N $46K_2O$	$MgCl_2$[a] 22Mg 66Cl
1	1	F	Ff	P	TP
2	1	F	Ff	F	TP
4	1	F	Ff	F	TP
10	1	F	P	F	TP

By weight				
Humic	Nutrient	$Na_2B_8O_{13}.4H_2O$[a] 20B	$CuSO_4$[a] 25Cu 12S	$FeSO_4$[a] 29Fe 16S
1	1	TP	Tf	T
2	1	TP	T	T
4	1	T	T	T
10	1	F	T	T

(Continued)

Table 6.8 (Continued)

By weight		$MnSO4^{a}$	$Na_2MoO_4.2H_2O^{a}$	$ZnCl_2$
Humic	Nutrient	32Mn 18S	46Mo	46Zn 51Cl
1	1	F	Tf	TPf
2	1	F	F	TP
4	1	F	F	TP
10	1	F	F	TP

Note: F = fluid, T = thick, S = solid, P = precipitate/separate, f = foam,
[a] organic material according to USDA-NOP (as long as non-synthetic, some with restrictions).

6.5 Applicability of Dry Soluble HS

This product is manufactured from liquid HA to contain a minimum amount of moisture (see Section 5.8). Dry soluble HS is to be made into liquid HA prior to their application on the field. This product has a very small particulate sizing and low moisture content. Pouring dry soluble HS into the mix tank first then adding water is not recommended, as excessive dust may be formed, the product may cake at the bottom, and it will not blend with the water. The use of a surface impeller is not ideal, as excessive foam may form on the surface. The impeller is best positioned close to the bottom. It is better to provide a circulation line by having a suction inlet at the bottom and a discharge outlet in the middle of the tank. Having a discharge outlet at the top is not recommended as this may lead to the formation of foam.

The best method for blending dry soluble HS with water is to use an inline blending apparatus (Canadian Humalite International Inc. 2020b). This consists of an inlet water line, a product suction line, a product container, an inline pump, control/isolation valves, an outlet water line, and a liquid tank. The diameter of the inlet water line is reduced and then enlarged to its original size, creating a high negative pressure point. A product suction line of a smaller diameter is then connected to that point. This line is connected to a product container, ideally positioned above it. The humic product, due to very small particulate sizing and low moisture content, is sucked into the product line and instantaneously blended with water within the water line. This mixture is then sent to the outlet water line and liquid tank using the pump. Valves are needed to control both the product suction and water flows, determining the percentage of HS in the liquid. This product can be made into liquid HA with strengths from 7 to 24% HS (see Figure 6.1).

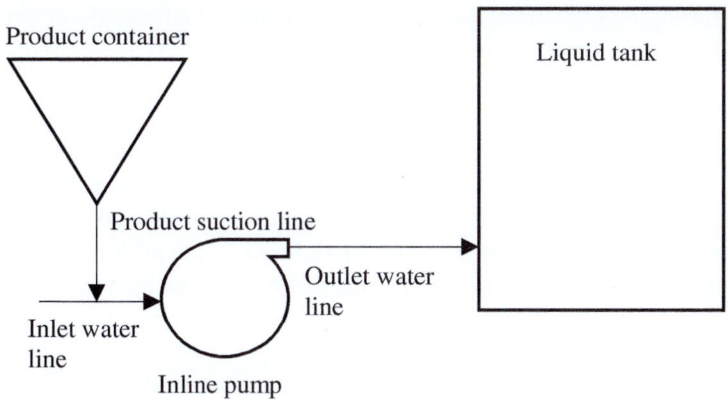

Figure 6.1 Inline mixing apparatus.

Table 6.9 Liquid HA from dry soluble HS and liquid fertilizers.

Humic	Liquid fertilizers	Response
0–100%	0–100%	F, T, S, or P

Table 6.10 Liquid HA from dry soluble HS and dry fertilizers.

By weight		
Humic	**Dry fertilizers**	**Response**
1–40	1	F, T, S, or P

After this product has been made into liquid HA (typically at 12% HS strength), its compatibility with nutrients can be evaluated in a small laboratory test (as previously discussed). Tables 6.9 and 6.10 show the compatibility studies with both liquid and dry nutrients, in which responses are to be recorded as: fluid (F), thick (T), solid (S), and precipitate/separate (P); recorded two hours after blending.

6.6 Product Efficacy Projects

In total, there were 25 efficacy projects completed from 2011 to 2020 using commercial humic products manufactured from Alberta low-rank coal (humalite), including dry HS, liquid HA, liquid FA, and pelletized HS. Each product might vary slightly during manufacturing and was therefore individually analyzed for each project. Twenty-four projects were completed to evaluate the efficacy of the products on soil and crops. Twenty-one projects were completed on outdoor fields to improve saline and compacted soils, and to improve yields of rice, wheat, canola, lentil, barley, lettuce, chrysanthemum, tea, tomato, coffee, watermelon, corn, soybean, and shrub. Three laboratory projects were completed on the growth of lettuce, the emergence of lentil and pea seeds, and the growth of a beneficial type of fungus (see Tables 6.11 and 6.24). In 2017, one project on livestock was completed in the laboratory and a test pen to evaluate feed digestion efficiency and emission of gaseous by-products (Canadian Humalite International Inc. 2020d).

Projects with significant statistical results or showing critical information are discussed in sub-sections and, when possible, the economy of humic products is also discussed. Not every project has complete data or shows significant statistical

Table 6.11 Soil and crop product efficacy projects.

No	Year	Target	Product	Location	Parameter
1	2011	Rice	Dry HS, Liquid HA	Field, Vietnam	Yield
2	2013	Wheat	Liquid FA	Field, Alberta	Yield
3	2013	Wheat	Liquid HA	Field, Alberta	Yield
4	2014	Soil	Dry HS, Liquid HA	Field, Alberta	Salinity
5	2014	Lettuce	Liquid FA	Laboratory, Alberta	Yield
6	2014	Canola	Liquid HA	Field, Alberta	Yield
7	2016	Lentil	Liquid HA	Field, Alberta	Yield
8	2015 to 2017	Soil	Dry HS, Liquid HA	Field, Alberta, Chile	Compaction
9	2018	Barley	Liquid FA	Field, Alberta	Yield
10	2019	Wheat	Pelletized HS	Field, Alberta	Yield
11 to 24	2011 to 2020			See Table 6.24	

results, as they have been completed within practical and operational constraints during field work. They are included as they show strong trends and interesting information. This chapter ends with a cost–benefit study on the overall feasibility of humic products for four crop applications (Canadian Humalite International Inc. 2020d).

6.6.1 Rice

Two humic products with low and high water solubility were evaluated in this project. The dry HS had a mean particulate sizing of 2.1 mm and 4.1 pH. While it contained a concentrated amount of HS (60.2% HS, based on the colorimetric method, including moisture), it had a low water solubility at 5.1%. It contained low amounts of nutrients and very low amounts of toxic metals, as well as negligible amounts of toxic organics and pathogens. Liquid HA (pH = 10.2) was practically 100% water-soluble at 7.0 pH and above. It contained 11.9% HS (based on the colorimetric method, including moisture). Other than K_2O at 1.5%, it contained negligible amounts of other nutrients, toxic metals, toxic organics, and pathogens.

The project was completed on yields of rice in the Mekong Delta, Vietnam, during the dry season of August to November 2011. Soil in the area was alluvial and had the following properties: pH = 5.5, OM = 2.7%, total N = 0.2%, available Olsen's P = 2.8 ppm, exchangeable K = 4.9 cmol/kg, and cation exchange capacity (CEC) = 20 cmol/kg. The crop being evaluated was OM 5451, a common variety of rice with a 90-day harvest.

The project was completed in plots of 5.5 m × 6.0 m (or 33.0 m^2) each. Each was separated by 1.0 m distance by furrow drains (0.2 × 0.3 m^2) to avoid flooding between plots. Experiments were designed as shown in Table 6.12. For each of nine treatments, nutrients as N-P_2O_5-K_2O and in the amounts of 60-24-18, 80-32-24, and 100-40-30 were applied before sowing, representing 60, 80, and 100% nutrient rates, respectively. Control 1, Control 2, and Control 3 corresponded to 60, 80, and 100% nutrient inputs, respectively. Dry HS at 100 kg/ha was applied to soil for each of Treated 1-1 to 3-2 before sowing. Liquid HA at 3.9 l/ha in total was applied foliar to plants during growing season as follows: 2.6 l/ha on day 40 after sowing, plus 1.3 l/ha on day 55 after sowing for each of Treated 1-1, 2-1, and 3-1, respectively; and 1.3 l/ha on days 15, 40, and 55 after sowing for each of Treated 1-2, 2-2, and 3-2, respectively. The same amount of liquid product was applied for each of Treated 1-1, 2-1, 3-1, 1-2, 2-2, and 3-2, but with different timing, i.e. late vs. evenly distributed throughout the growing season. Each of nine treatments was replicated three times. For each treatment, yields were recorded in t/ha and averaged (see Figure 6.2). No further statistical analyses were completed. However, the general trends were obvious, and the results provided important information on the product optimum rate.

Table 6.12 Design of experiment.

Treatment	NPK	Dry HS	Liquid HA	#
Control 1	60%	0	0	
Control 2	80%			
Control 3	100%			
Treated 1-1	60%	100 kg/ha before sowing	2.6 l/ha on day 40 after sowing + 1.3 l/ha on day 55 after sowing (= 3.9 l/ha in total)	3
Treated 2-1	80%			
Treated 3-1	100%			
Treated 1-2	60%		1.3 l/ha each on days 15, 40, and 55 after sowing (= 3.9 l/ha in total)	
Treated 2-2	80%			
Treated 3-2	100%			

Note: 60% nutrients = 60N-24P_2O_5-18K_2O kg/ha, 80% nutrients = 80N-32P_2O_5-24K_2O, 100% nutrients = 100N-40P_2O_5-30K_2O kg/ha, # = number of replications.

As expected, increasing nutrients from 60N-24P_2O_5-18K_2O (60% nutrients, Control 1) up to 80N-32P_2O_5-24K_2O kg/ha (80% nutrients, Control 2) and then to 100N-40P_2O_5-30K_2O kg/ha (100% nutrients, Control 3) resulted in higher crop yields from 5.55 to 6.15 then 6.63 t/ha (10.8 and 7.8% higher), respectively. No significant differences were observed between late vs. evenly distributed applications of the liquid product during growing season (Treated 1-1, 2-1, and 3-1 compared to Treated 1-2, 2-2, and 3-2). Treated 1-1 and 1-2, Treated 2-1 and 2-2, and Treated 3-1 and 3-2 could therefore be assumed as duplicates.

Adding both the dry and liquid products to 60% nutrients (Treated 1-1 and 1-2) enhanced yields up to 6.17 t/ha (11.2% higher than Control 1). Adding both products to 80% nutrients (Treated 2-1 and 2-2) enhanced crop yields, with the best result found to be 6.77 t/ha (10.1% higher than Control 2). However, adding both products to 100% nutrients (Treated 3-1 and 3-2) resulted in reduced crop yields, worst at 5.83 t/ha (12.1% lower than Control 3), suggesting that the combined nutrients and HS at these levels were too much for the plants.

Most importantly, using 80% nutrients with both products (Treated 2-1 and 2-2) maintained or even slightly increased crop yields up to 6.77 t/ha compared to using 100% nutrients that produced 6.63 t/ha (Control 3). Using 60% nutrients with both humic products (Treated 1-1 and 1-2) maintained or even slightly increased crop yields (up to 6.17 t/ha) compared to using 80% nutrients that produced 6.15 t/ha (Control 2). These results found that the same yields could be maintained by reducing nutrient inputs by 20% with the incorporation of both dry and liquid products. At the highest nutrients, however, the addition of HS was not

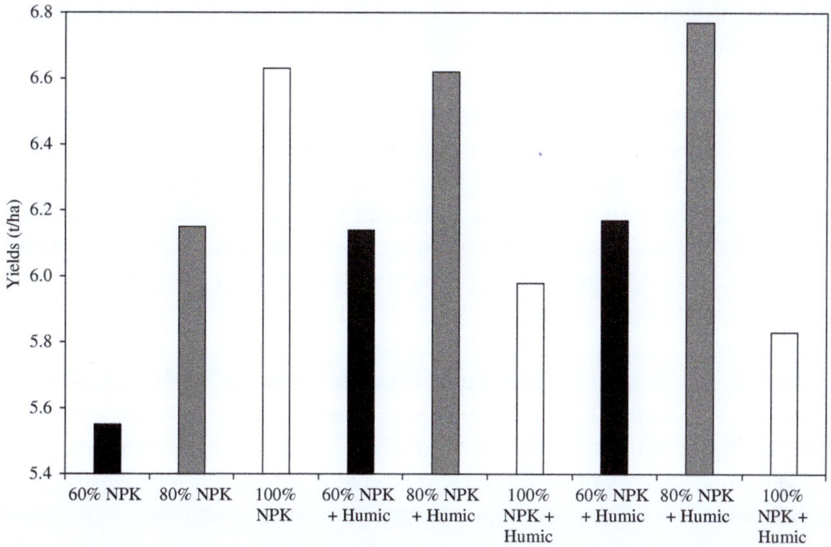

Figure 6.2 Treatment vs. yields.

beneficial. This indicated that applying more HS and nutrients would not always produce better results, and an optimum rate must be carefully determined.

In summary, 100 kg/ha dry HS and 3.9 l/ha liquid HA enhanced yields of rice by up to 11.2% over the control, or allowed yields to be maintained while nutrients were reduced by 20%.

6.6.2 Wheat (1)

This project evaluated liquid FA with 9.8 pH and close to 100% water solubility at any pH. It contained 0.3% HA and 0.7% FA (total HS = 1.0%, based on ISO 19822, including moisture). It had 0.8% K and negligible amounts of other nutrients, toxic metals, toxic organics, and pathogens.

The project was completed on 27 outdoor test plots measuring 1.4 m × 6.7 m (or 9.4 m^2) each and located in Alberta. Wheat of the Harvest variety was planted in each plot in May 2013. The soil in the area was loam (with almost equal proportions of sand, silt, and clay) with Solonetzic clay at the base, pH = 6.1, EC = 0.3 mS/cm, and OM = 6.0%. The available macronutrients in the soil were 57 kg N/ha, 48 kg P_2O_5/ha, 707 kg K_2O/ha, and 84 kg SO_4/ha. The available micronutrients in the soil included 0.8 ppm Cu, 0.9 ppm B, 4 ppm Zn, 21 ppm Mn, and 160 ppm Fe. Soil analyses showed that it was deficient in N, marginal in P and Cu, adequate in B and Zn, and optimum in other nutrients.

Nutrients of 67N and $22P_2O_5$ kg/ha were applied on each plot during seeding. In order to meet these requirements, urea (46-0-0) and MAP (11-52-0) were combined at a ratio of 76:24. This blend was applied at a rate of 178 kg/ha. Micronutrients were sprayed two weeks later on the seedlings. On each of Control 1, 2, and 3, $CuSO_4$ was sprayed at a rate of 0.48 kg /ha (providing 0.12 kg Cu/ha), $FeSO_4$ at 2.14 kg/ha (providing 0.62 kg Fe/ha), and $ZnCl_2$ at 0.61 kg/ha (providing 0.28 kg Zn/ha), respectively. On each of Treated 1-1 to 3-2, liquid FA was sprayed at a rate of 400 ml/ha in combination with each of the micronutrients. The micronutrient rates were 0.12 and 0.06 kg Cu/ha (Treated 1-1 and 1-2), 0.62 and 0.31 kg Fe/ha (Treated 2-1 and 2-2), and 0.28 and 0.14 kg Zn/ha (Treated 3-1 and 3-2), respectively.

Each of the nine treatments was replicated three times (see Table 6.13). Harvest was made in September 2013 and yields from each treatment were recorded in t/ha and averaged. All data were analyzed using Minitab™ for means, standard deviations, and analysis of variance (ANOVA) at $p = 0.05$. Fisher's least significant difference ($\alpha = 0.05$) was completed for all treatments, and they were found to be statistically insignificant. However, the trends were obvious, and the results contained important information on the benefit of HS for the effort to reduce nutrient inputs while enhancing or maintaining yields.

As shown in Figure 6.3, crop yields increased from 4.34 to 4.39 t/ha (Control 1, 2, and 3) to 4.60 to 4.73 t/ha when liquid FA was incorporated (Treated 1-1, 2-1, and 3-1). In comparison to the control, Zn experienced the lowest increase from

Table 6.13 Design of experiment.

			Micronutrients		Liquid FA	
Treatment	**Plot size m^2**	**NP kg/ha**	**Rate**	**kg/ha**	**ml/ha**	**#**
Control 1			100% Cu	0.12		
Control 2			100% Fe	0.62	0	
Control 3			100% Zn	0.28		
Treated 1-1	1.4 m × 6.7 m (= 9.4 m^2)	67-22	100% Cu	0.12		
Treated 2-1			100% Fe	0.62		3
Treated 3-1			100% Zn	0.28	400	
Treated 1-2			50% Cu	0.06		
Treated 2-2			50% Fe	0.31		
Treated 3-2			50% Zn	0.14		

Note: # = number of replications.

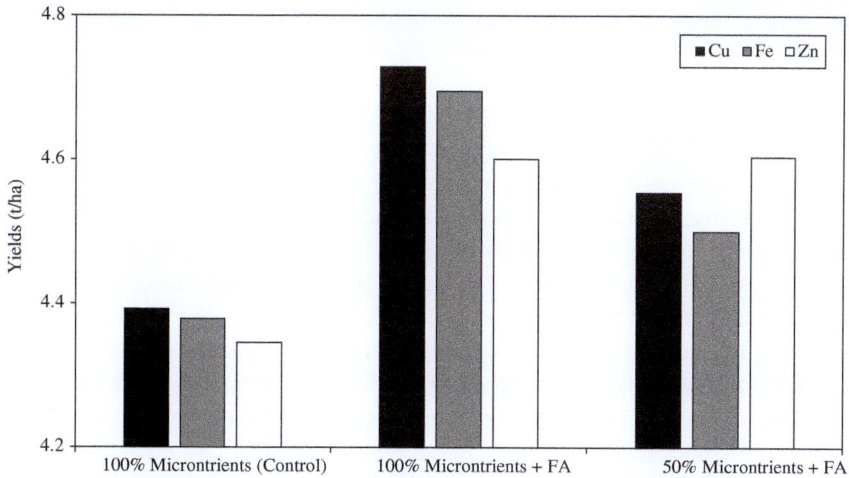

Figure 6.3 Treatment vs. yields.

4.34 to 4.60 t/ha (6.0% higher), while Cu increased the most from 4.39 to 4.73 t/ha (7.7% higher). Even when the micronutrient applications were reduced to approximately 50% of the original rates, yield increases were still observed at 4.50 to 4.60 t/ha when liquid FA was incorporated (Treated 1-2, 2-2, and 3-2, see Figure 6.3). In comparison to the control, Fe had the lowest increase from 4.38 to 4.50 t/ha (2.7% higher), while Zn the highest increase at 4.60 t/ha (6.0% higher).

Although detailed mechanisms were not investigated in this project, the results clearly suggested that HS in liquefied form facilitated the plant's utilizing the applied micronutrients more efficiently, resulting in higher crop yields or maintaining them at reduced nutrient inputs.

The economy of this application was evaluated based on the addition of liquid FA and the reduction of micronutrients, while maintaining crop yields. Prices for liquid FA, $CuSO_4$, $FeSO_4$, and $ZnCl_2$ were \$18.03/l, \$3.78/kg, \$1.21/kg, and \$4.85/kg, respectively (delivered to site, Canadian Humalite International Inc. 2020d). The addition of liquid FA at a rate of 400 ml/ha increased production costs by \$7.21/ha. Liquid FA and micronutrients were compatible and could be applied together in one pass. The reduction of micronutrients by 50% (0.24, 1.07, and 0.31 kg/ha of $CuSO_4$, $FeSO_4$, and $ZnCl_2$, respectively) lowered production costs by \$3.71/ha. This indicated that the use of liquid FA at this high application rate was not economically worthwhile. As shown later in Section 6.6.9, an application rate as low as 70 ml/ha was found to be effective and economically worthwhile.

In summary, 400 ml/ha liquid FA, applied foliar, slightly enhanced or maintained yields of wheat when micronutrients were reduced by 50%. Due to a very high product application rate, the use of liquid FA was not found to be economically worthwhile.

6.6.3 High Salinity Soil (1)

High salt contents resulted in low soil productivity (due to reduced nutrients and water intake by plants and inferior soil structures and organisms), as well as reduced growth and health in most plants (Khan et al. 2000, Grattan and Grieve 1999, Yan et al. 2015, and Wang et al. 2012). The presence of salts in soil would commonly be measured by its EC and sodium adsorption ratio (SAR, defined as the ratio between soluble Na ions and root square of half the sum of soluble Ca and Mg ions). Based upon its EC, SAR, and pH, soil could be classed as optimum, high or low pH, sodic or saline, and saline-sodic. Amendments, usually in large amounts, were added with the aim of adjusting a particular parameter (increasing either water-soluble Ca or Mg) and thus lowering the SAR. Soil with acceptable SAR, EC, and pH would facilitate a better soil condition for plants to grow (McKenzie 2010a).

The addition of large amounts of amendments would result in a high product cost and create additional challenges in terms of practicality during field application and impact over a longer term. Gypsum ($CaSO_4.2H_2O$), for example, indeed lowered SAR, but at the same time added water-soluble Ca and SO_4 to soil, increasing its EC. Epsom salt added water-soluble Mg and SO_4 to soil. Instead of applying these amendments to the soil and then observing for any changes in EC, SAR, and pH, a different solution was tried: apply OM (including HS) to soil with the aim of directly improving the productivity of soil. The mechanism was straight forward. Due to its unique chemical and physical properties, OM (and HS) was able to bind Na, making it rather insoluble in water so that it would not adversely affect osmotic pressure in the root zones. At the same time, it lowered SAR while maintaining EC. The addition of OM products was found to overcome the detrimental effect of salinity (Walker and Bernal 2005 and 2008).

A product efficacy project was completed in 2013 on a 64-ha commercial field located in Alberta. The soil was 40% dark brown Solodized Solonetz, 40% dark brown Solod, and 20% Orlic dark brown Chernozem. Soil OM was recorded at 3.6%. The field had three zones (7 ha in total), shown in Figure 6.4 as saline (S), very saline (S+), and extremely saline (S++) zones. The zones had 12.4, 29.6, and 62.9 of SAR; 7.1, 7.5, and 7.8 of pH; and 1.5, 7.0, and 38.3 mS/cm of EC, respectively. The rest of the field had 9.7, 7.0, and 0.8 mS/cm of SAR, pH, and EC, respectively; shown in the same figure as the normal (N) zone. Liquid HA was the evaluated product and had 10.2 pH, 12.0% HS (based on the colorimetric method, including moisture), almost 100% water solubility at 7.0 pH and above, 1.5% K_2O, as well as negligible amounts of other nutrients, toxic metals, toxic organics, and pathogens.

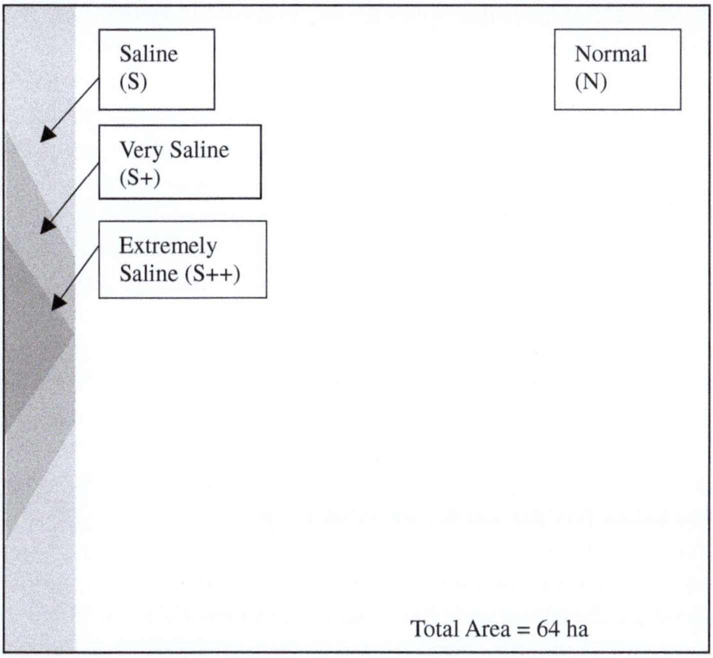

Figure 6.4 Test site.

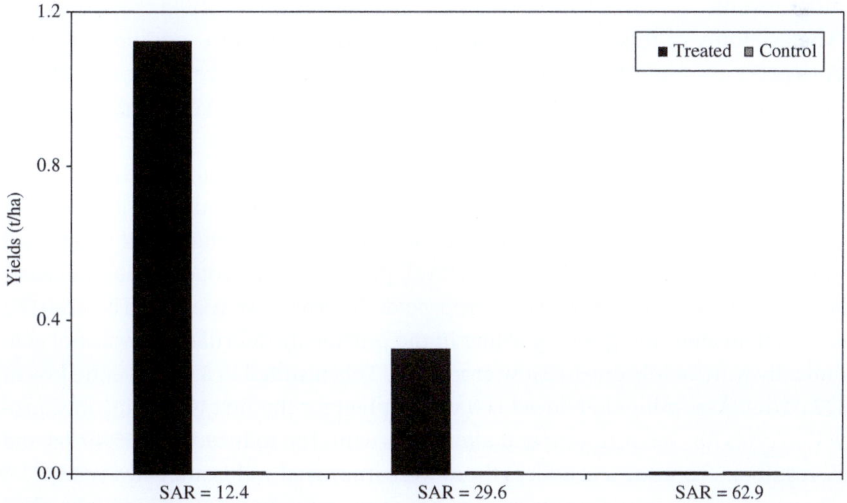

Figure 6.5 Treatment vs. yields.

From 2009 to 2012, the three saline zones (S, S+, and S++) together with the rest of the field (N) were alternately seeded with canola and wheat of the CDC Go variety. Every year, dry urea (46-0-0) at 123 kg/ha was applied during seeding. For past four years, no yields were produced during harvest from the saline zones (identified as Control in Figure 6.5). In 2013, these three zones and the rest of the field were again seeded with CDC Go wheat. However, this time during seeding, liquid HA was broadcasted over the saline zones at a rate of 95 l/ha, together with dry urea (46-0-0) at 123 kg/ha. During harvest (identified as Treated in Figure 6.5), 0.32 and 1.12 t/ha yields were observed from two zones with SAR of 29.6 (S+) and 12.4 (S), respectively. This was an achievement that had never been observed from 2009 to 2012.

Figure 6.6 shows early plant emergences on the saline soil with SAR of 12.4 (S). No yield was observed from the worst zone with SAR of 62.9 (S++). The application of HS under this extreme condition did not help. The rest of the field (N) showed a harvest of 1.86 t/ha. Dry urea (46-0-0) was applied at 123 kg/ha, however the liquefied humic product was not included.

Although detailed mechanisms were not investigated, this project clearly demonstrated the efficacy of HS in this liquid form: it made soil with high salinity productive. To produce significant yield increases under normal soil conditions, a typical application rate of the same product would be as low as 9 l/ha. For soil with high salt contents, a high product application rate might be needed for the first two years only. After that, the quality of soil would have significantly improved. Also, should amendments such as gypsum or Epsom salt be applied, much higher rates of these products could be required to achieve the same or similar results.

The economy of this application was evaluated based on the addition of liquid HA to previously infertile land to make it productive. The price for liquid HA was $1.03/l, and at a product rate of 95 l/ha, it increased costs by $97.85/ha. The price for urea (46-0-0) was $0.62/kg, and at a product rate of 123 kg/ha, it increased costs by $76.26/ha (delivered to site, Canadian Humalite International Inc. 2020d). Other production costs (excluding fertilizer and land-related costs) to grow spring wheat were estimated at $579.23/ha. The target price for spring wheat was $250/t (Manitoba Agriculture and Resource Development 2020a). Total production costs became $753.34/ha, and at 1.12 t/ha crop yield, the gross revenue would be $280.00/ha. This indicated that growing wheat in the previously infertile land was not economically worthwhile due to a low crop yield. This resulted in a net revenue loss of $473.34/ha. Assuming that liquid HA was applied for the first two years only, production costs for the third year and afterwards would be reduced to $655.49/ha and still result in a revenue loss unless there were improved yields after two years.

However, it was important to notice that even in the non-saline zone, the yield of 1.86 t/ha was not great. Manitoba Agriculture and Resource Development

Figure 6.6 Early plant emergences. *(See color plate section for color representation of this figure).*

(2020a), for example, had a target yield for spring wheat of 3.93 t/ha. The following factors were likely the main cause of these low yields:

1) the quality of soil in this farmland (including in the normal zone) was very poor; and
2) nutrient requirements for the farm were not properly determined and therefore the applied nutrients were not optimized.

In summary, 95 l/ha liquid HA applied to soil with high salinity was able to significantly enhance its quality, and a yield of wheat of up to 1.12 t/ha was produced from previously infertile soil. However, to make the effort economically worthwhile, a complete, detailed evaluation on nutrient requirements for the whole farm would have to firstly be completed in order to optimize crop yields. Another option would be growing a different type of crop with a higher target price.

6.6.4 High Salinity Soil (2)

High soluble salts in soil reduced water and nutrients uptake by plants, deteriorated soil structures, and inhibited the growth of most crops. The main determining parameters to consider were SAR, EC, and pH. SAR > 13 and/or EC > 2.0 mS/cm, together with pH > 8.5, would commonly be considered detrimental to soil (McKenzie 2010a). Remediation of this type of soil could be achieved in different ways. One strategy was to allow soluble salts to accumulate in areas away from the primary root zone, not interfering with any plant activities. This "managed accumulation" approach would commonly be achieved through crop bedding and surface irrigation systems. While this practice could be effective, it might not be practical, since periodic leaching of salts down from the soil surface and below the root zone would be required from time to time (Cardon et al. 2007).

The presence of high soluble salts in soil could be managed by adding OM (including HS) containing large molecular compounds with strong affinities to most ions (including Na salts), making them rather insoluble in water. While they would not be utilized by most crops, these HS-bound salts would not interfere with any plant activity either. This idea came from the more common practice of applying OM (including HS) to bind nutrients (such as NO_3) in soil to prevent them from leaching to the groundwater table. In this application, the nutrients-bound HS would still be available, and in fact be in the best form to be taken up by crops (Stevenson 1994).

In order to evaluate the efficacy of HS for remediating soil with high soluble salts, a project was completed on a 64-ha farmland in Alberta from October 2013 to July 2014. The soil was 40% dark brown Solod, 40% dark brown Solodized Solonetz, and 20% Orlic dark brown Chernozem. The project was completed on a 1.6-ha area located on the lowest topography of the farmland. It had not been productive for years and was not seeded in 2013. Three test plots (6 m^2 each) were randomly selected from this area and identified as the control, treated 1, and treated 2 plots.

In October 2013, one 0.25 kg soil sample was randomly taken from each plot at 15 cm below surface. The samples were then aggregated and analyzed for soluble Ca, soluble Mg, soluble Na, EC, pH, and OM. The results were 4.7 meq/l, 1.0 meq/l, 23.0 meq/l, 2.8 mS/cm, 7.3, and 3.6%, respectively, representing the control for October 2013. SAR was calculated at 15.3 (the ratio between soluble Na ions and root square of half the sum of soluble Ca and Mg ions). While the pH was acceptable, the soil had undesirable levels of SAR and EC.

Dry HS had a mean particulate sizing of 2.1 mm and a low water solubility at 5.0%. It had 4.1 pH and contained 59.8% HS (including moisture), 1.1% Ca, 0.3% Mg, and 0.3% Na. This product contained negligible amounts of toxic metals, toxic organics, and pathogens. In October 2013, this product was broadcasted on the treated 1 plot and tilled to 15 cm below surface. The rate of application was 2.24 t/ha, corresponding to 1.34 t HS/ha.

Liquid HA had 10.2 pH and contained 11.9% HS (including moisture), 1.7% K_2O, 0.2% Ca, and 0.3% combined Mg, Na, and other elements. This product contained negligible amounts of toxic metals, toxic organics, and pathogens. In October 2013, it was applied over the treated 2 plot and tilled to 15 cm below

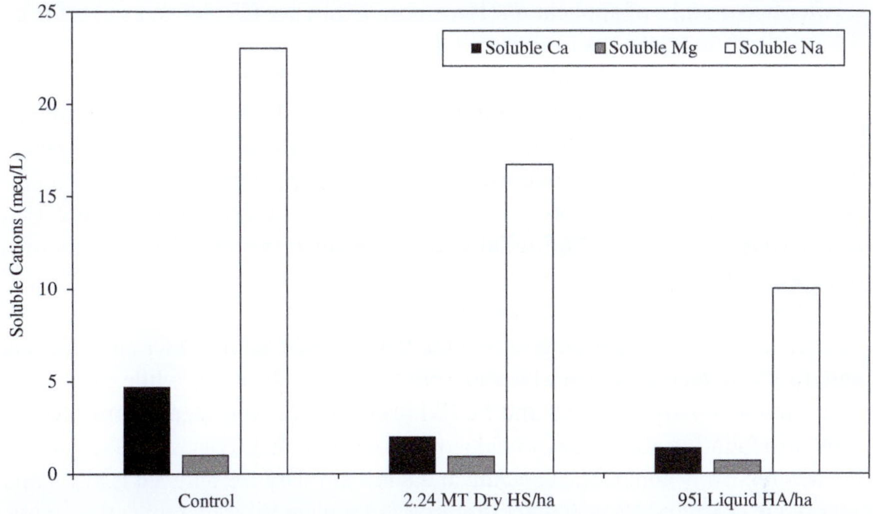

Figure 6.7 Treatment vs. soluble cations (after 2 months).

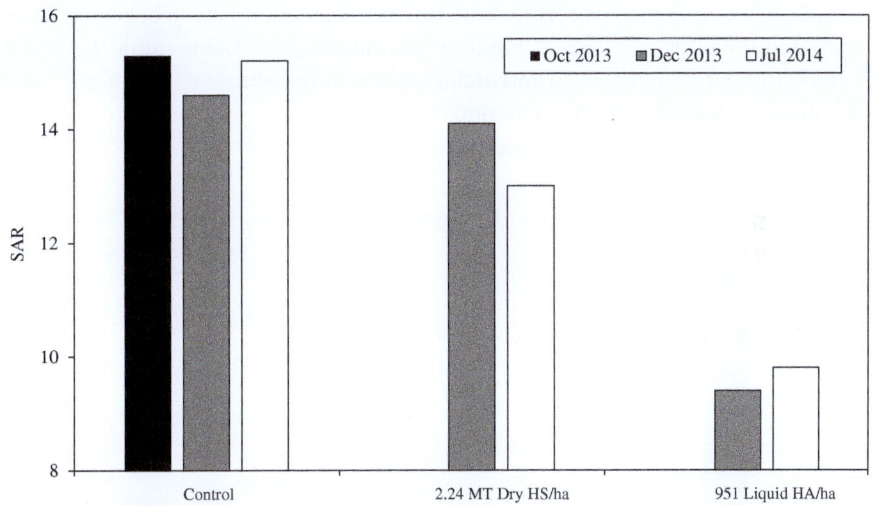

Figure 6.8 Treatment vs. SAR.

surface. The application rate was 95 l/ha, corresponding to 11.9 kg HS/ha (note: its specific gravity was 1.05).

Three samples weighing 0.25 kg each were taken from each plot at 15 cm below surface in December 2013 and July 2014. The samples for each plot were then aggregated and analyzed for soluble Ca, soluble Mg, soluble Na, EC, and pH.

After two months of application (December 2013), dry HS reduced soil pH from 7.3 to 7.1, while liquid HA increased it to 7.5, and the soil remained around these levels after nine months (July 2014). This corresponded with the nature of the products: dry HS was acidic and liquid HA was basic. Soil soluble Ca was reduced from 4.7 to 2.0 and 1.4 meq/l (57.4 and 70.2% lower), soluble Mg from 1.0 down to 0.9 and 0.7 meq/l (10.0 and 30.0% lower), and soluble Na from 23.0 down to 16.7 and 10.0 meq/l (27.4 and 56.5%) with the addition of dry HS and liquid HA, respectively (see Figure 6.7). Similar trends were also observed after nine months of application (July 2014).

SAR was then calculated from soluble Ca, Mg, and Na ions. The control SAR was relatively constant, averaging at 15.0. Dry HS reduced it to 14.1 and 13.0 (6.0 and 13.3% lower) after two and nine months of application, while liquid HA reduced it to 9.4 and 9.8 (37.3 and 34.7% lower), respectively (see Figure 6.8). HS were also found to reduce soil soluble ions (measured as EC) as well. Background EC was relatively constant, averaging at 2.8 mS/cm. Dry HS reduced it to 1.9 and 1.2 mS/cm (32.1 and 57.1% lower) after two and nine months of application, while liquid HA reduced it to 1.2 and 1.1 mS/cm (57.1 and 60.7% lower), respectively (see Figure 6.9).

Although detailed mechanisms were not investigated, this project clearly demonstrated the efficacy of HS in binding soil soluble salts. Using liquid HA, SAR and EC of approximately 15.0 and of 2.8 mS/cm were reduced to less than 13 and 2.0 mS/cm, respectively, in a period of two months, and therefore met the

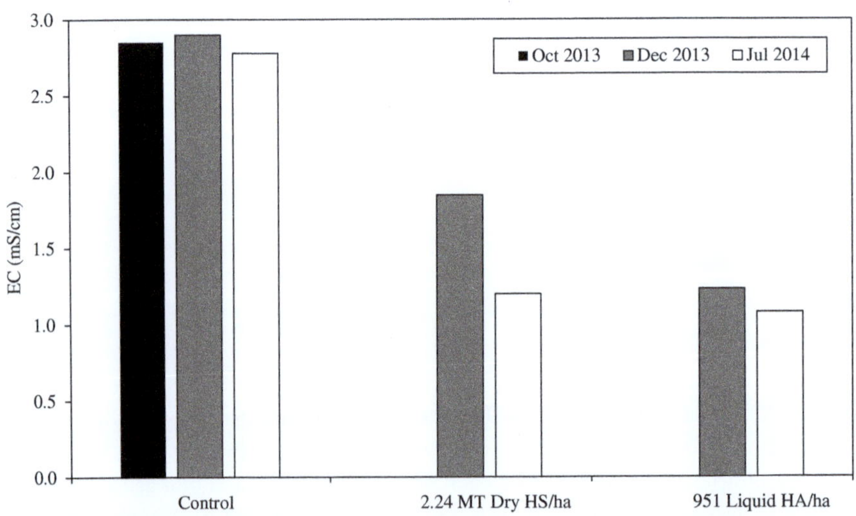

Figure 6.9 Treatment vs. EC.

acceptable targets. While the salts were still within the soil matrix, their accumulation in soil was managed and would therefore not interfere with any plant activity.

Most importantly, these findings emphasized the importance of product availability (i.e. water solubility) over quantity. A large amount of dry HS (1.34 t HS/ha), a raw material with small particulates, did not perform that well even after nine months of application. When the material was solubilized in water (i.e. liquid HA), a much smaller amount of product (11.9 kg HS/ha) resulted in much better reductions of SAR and EC after only two months of application. The application of HS to remediate soil with high soluble salts would therefore best be achieved using a water-soluble product.

In summary, the "managed accumulation" approach to remediate soil with high soluble salts could practically and efficiently be achieved with the addition of HS. After only two months of humic application in a water-soluble form (liquid HA at a rate of 95 l/ha), soil SAR and EC were reduced from approximately 15.0 and 2.8 mS/cm to less than 13 and 2.0 mS/cm, respectively.

6.6.5 Lettuce[1]

The benefits of HS (including HA and FA) on the growth of plants and seedlings were well documented. Liquid humic products at a rate as low as 500 mg/l promoted root initiation of geranium cuttings grown in solution (O'Donnell 1973). During a laboratory experiment, the growth of cucumber plants grown in solution was enhanced with the incorporation of a dry purified humic product with optimum rates between 100 to 300 mg/kg (Rauthan and Schnitzer 1981). It was reported that 4000 mg/kg rate of a dry humic product benefitted the growth of vineyards planted on sand (Reynolds et al. 1995). Geranium seedlings were found to have a significant root growth when substrate-drenched with a liquid humic product at a rate as low as 2500 mg/l (Hartwigsen and Evans 2000). A liquid humic product evaluated in a laboratory experiment at an optimum rate of 2500 mg/kg promoted the growth of different ornamental plant seedlings grown in germination papers (Evans and Li 2003). The growth of zinnia and marigold seedlings and transplants grown in sand and peat media was improved with the use of a dry humic product at a rate of 31 250 mg/kg (Dudley et al. 2004). Greenhouse experiments found that wheat planted on sandy clayey soil grew well with the addition of either a dry humic product to soil at rate of 1000 mg/kg, or a liquid humic product to leaves at a rate of 1000 mg/l (Asik et al. 2009). Field experiments showed a growth increase of broad beans grown in sandy

1 The revised and shorter version of this work was published in Verdi et al. (2015), in which L.E. Liem was the main and corresponding author.

soil when a liquid humic product was foliar applied at a rate of 4000 mg/l (Shafeek et al. 2013).

These findings showed that optimum rates of humic products varied widely from one study to another and mainly depended on their water solubility. When short-term results were desired, liquid (processed) products, due to their almost complete solubility in water, would generally be more effective than dry (unprocessed) products, which were practically insoluble in water. Optimum rates for liquid products would generally be much lower than those of dry products. Still, other factors had to be taken into consideration, such as types and conditions of the growing media, application methods, and types and varieties of plants, as well as their growing phases (Chen and Aviad 1990).

Since most product efficacy projects were completed in agricultural fields, a study would be needed on the use of commercial humic products in a hydroponic system. In this system, nutrients would be delivered to the crops more efficiently. Lower nutrient applications would therefore be expected to achieve the same yield. The growing condition would also be easier to control and a better performance could be observed (Agriculture and Agri-Food Canada 2017 and Brechner and Both 2013).

Lettuce was selected as the target crop due to its robustness and simplicity for analysis, which involved only roots and leaves. The crop was grown with limited nutrients so that low rates of HS could be expected to have a significant effect on its productivity. This hypothesis was based on findings presented earlier that the benefits of HS would be more obvious when the growing condition was not optimum. The main objective of this project was to determine liquid FA's optimum rate and to observe any negative responses at higher rates.

6.6.5.1 Materials and Methods

Liquid FA was selected as the humic product to be evaluated as it contained water-soluble humic components. This meant that the results could be almost instantaneously observed. This product had 10.2 pH and was practically 100% water-soluble at any pH. It contained 0.6% HA and 0.8% FA (total HS = 1.4%, including moisture). The analyses were completed using the Humic Products Trade Association (HPTA) method (Lamar et al. 2014) which was based on the Methods of Soil Analysis (Swift 1996). This method was later adopted and known as ISO 19822 (ISO 19822:2018(E) 2018).

Total C corresponded to OM originating from the raw low-rank coal material. HA were defined as components of HS that were soluble in aqueous alkali solution and insoluble in aqueous acid solution. FA were defined as components of HS that were soluble in both aqueous alkali and acid solutions (Stevenson 1965). The hydrophobic FA were isolated by means of the DAX-8 resin (Stevenson 1994). This product was manufactured to contain small HS particulates within the solution in order to maximize their absorption by plant roots (see Table 6.14). The mean diameter of these particulates was 1.1 ± 0.64 (standard deviation) μm.

All work was completed in 2014 in a laboratory with controlled temperature (24 ± 0.5 °C) and humidity (76 ± 1%). The plants were grown in solution, containing limited amount of nutrients, as well as negligible amounts of OM and microbial activities, thus creating a stressed growing condition. The experiments were designed so that optimum conditions due to the application of HS and responses at higher rates could be observed. Much higher rates were also included so that negative responses (if any) due to excessive amounts of HS could be observed.

Two hundred eighty-eight seeds of lettuce of the Grand Rapids variety were washed and cleaned using deionized water and planted on growing media of 50% perlite and 50% sphagnum. The media contained 1000 mg/kg N, 500 mg/kg P, and 1000 mg/kg K. Moisture contents of the media were between 34 to 37%. Survival rates of the seedlings after three weeks were recorded at 85%. For the 43 seedlings that perished, it was most likely this happened randomly. Eighty-four surviving seedlings were randomly selected from the media and washed using deionized water to remove as much as soil matter as possible. They were then transferred to seven deep-water culture-growing units.

Each culture-growing unit had dimensions of 50 cm (long) × 40 cm (wide) × 13 cm (deep) for a total volume of 26 l. Each had a cover on top containing 12 holes, in which one seedling could be suspended. The seedlings were randomly placed in the units. Twelve plants were grown in one unit.

One unit was randomly selected from the seven available units and designated as the control unit. This unit was filled up to a 10 cm depth with a control solution (for an effective volume of 20 l). Plant roots were all submerged within the solution. This treatment, identified as the control, was evaluated on four plants and replicated three times (for a total of 12 plants in this unit). As shown in Table 6.14, the control solution contained limited amounts of macro- and micronutrients as well as negligible amounts of OM (i.e. total C) and microbial activities (i.e. heterotrophic plate count, HPC). This would ensure that the plants grew under a stressed condition.

Six units were randomly selected from the seven available units and designated as the treated units. Each unit was filled with the control solution mixed with different rates of liquid FA up to a 10 cm depth. Plant roots were all submerged in the solution. Rates of liquid FA were 70, 500, 4000, 6000, 14 000, and 42 000 mg/l, providing 1, 7, 55, 83, 193, and 580 mg/l of HS (calculated as liquid FA rate × 1.38%), respectively. One rate corresponded to one treatment, identified as treatments 1 to 6. Each treatment was evaluated on four plants, replicated three times, or for 12 plants in total per unit.

Four Sun Blaster® growing lamps of 5022 lm/m each were suspended over the growing units. A timer controlled their operations as follows: 10 hours "on"

Table 6.14 Characteristics of control solution and liquid FA.

Parameter	Control solution	Liquid FA
N	224 mg/l	110 mg/l
P	448 mg/l	49 mg/l
K	288 mg/l	0.76%
S	16 mg/l	41 mg/l
Ca	44 mg/l	107 mg/l
Mg	12 mg/l	7 mg/l
B	< 1 mg/l	3 mg/l
Cu	< 1 mg/l	< 1 mg/l
Fe	< 1 mg/l	< 1 mg/l
Mn	< 1 mg/l	< 1 mg/l
Mo	< 1 mg/l	< 1 mg/l
Zn	< 1 mg/l	< 1 mg/l
Cl	6 mg/l	36 mg/l
HPC bacteria	0/100 ml	0/100 ml
EC	421.1 mS/cm	23.1 mS/cm
Total C	2 mg/l	0.41%
HA	< 0.01%	0.55%
FA	< 0.01%	0.83%
HA + FA (= HS)	< 0.01%	1.38%
Particulate diameter, maximum	N/A	2.4 μm
Particulate diameter, mean	N/A	1.1 μm
Particulate diameter, minimum	N/A	0.4 μm
Particulate diameter, standard deviation	N/A	0.64 μm
Water solubility	~ 100%	~ 100%
pH	6.4	10.2

Source: Canadian Humalite International Inc. (2020a); *note:* N was analyzed according to British Standards (Water Quality 2003), other nutrients according to APHA 3120B (1992), Cl using APHA 4110B (1992), HPC bacteria according to APHA 9215D (1992), EC according to APHA 2510 (1992), total C according to APHA 5310B (1992), HA and FA according to the HPTA method (Lamar et al. 2014) later known as ISO 19822 (ISO 19822:2018(E) 2018), particulate diameter according to Beckman Coulter (2011), water solubility according to EEC A.6 (1992), pH according to APHA 4500-H^+ (1992), N/A = not available.

during the day and 14 hours "off" in the evening. One air pump with a 5.0 l/minute capacity was connected to flexible air tubing with a 0.3 cm diameter, which branched out and was connected to each unit. Using control valves, fresh ambient air was equally distributed among the units and continuously bubbled within the solution to ensure the solution's freshness, providing oxygen to the roots.

After two months, each plant was harvested. Roots and leaves were measured for lengths (with a tolerance of 0.1 cm), and leaves were counted. Roots and leaves were carefully dried using paper towels and put in a desiccator at room temperature for 24 hours to remove all surface moisture. They were aggregated and weighed (with a tolerance of 0.01 g) to provide one measurement of plant weight. The plants were oven dried for 72 hours at 75 ± 0.5 °C and weighed with a tolerance of 0.01 g.

The survival rates for all treatments were recorded. Root, leaf, and plant (root + leaf) length measurements were analyzed using Minitab® for means, standard deviations, and ANOVA at $p = .05$. The analyses were completed both at the replication and treatment levels. Only the results at the treatment level were presented. Fisher's least significant difference ($\alpha = 0.05$) analysis was completed for all treatments. Leaf counts and plant weight measurements were recorded.

6.6.5.2 Results and Discussion

As shown in Table 6.15, only one plant perished (Treatment 5), which was most likely random. This indicated that the culture-growing units, in which limited amount of nutrients and HS were directly fed to the roots, worked effectively, providing adequate nutrients to the plants.

Table 6.15 Plant survival rates and study design.

Treatment	Liquid FA mg/l	HS mg/l	Plants per treatment	Survival rate %
Control	0	0	12	100
T1	70	1	12	100
T2	500	7	12	100
T3	4000	55	12	100
T4	6000	83	12	100
T5	14 000	193	12	92[a]
T6	42 000	580	12	100

Note:
[a] only 1 plant from the total of 12 plants died.

As shown in Table 6.16 and Figure 6.10, the control (0 mg/l HS) had the shortest root at 8.8d cm by a significant margin. Treatment 6 (580 mg/l HS) had the longest root at 21.7a cm, followed by Treatment 2 (7 mg/l HS) at 16.8b cm, Treatment 1 and 5 (1 and 193 mg/l HS) at 15.2bc and 13.8bc cm, respectively, and Treatment 3 and 4 (55 and 83 mg/l HS) at 12.7c and 12.8c cm, respectively. Note: different alphabetical groupings (a to d) indicated that the values were statistically different.

Concentrations of HS between 1 and 7 mg/l (or 70 and 500 mg/l of liquid FA) were found to be optimum, resulting in an increase of up to 90.9% in root lengths over the control. These optimum low concentrations were likely caused by the following factors: 1) all plant roots were submerged in the growing solution,

Table 6.16 Length measurements.

	Root cm			Leaf cm		
Treatment	**Mean**	**SD**	**ANOVA**	**Mean**	**SD**	**ANOVA**
Control	8.8d	3.53	6.46 to 11.03	9.6C	1.75	8.33 to 10.93
T1	15.2bc	4.55	12.91 to 17.56	10.7C	1.79	9.41to 12.01
T2	16.8b	4.28	14.50 to 19.15	10.8C	1.65	9.28 to 11.88
T3	12.7c	4.81	10.42 to 15.06	9.5C	2.89	8.18 to 10.78
T4	12.8c	3.96	10.53 to 15.17	10.6C	2.53	9.31 to 11.90
T5	13.8bc	3.78	11.44 to 16.08	12.9B	2.89	11.56 to 14.17
T6	21.7a	4.52	19.33 to 23.98	17.0A	1.89	15.74 to 18.34

	Plant cm		
Treatment	**Mean**	**SD**	**ANOVA**
Control	18.4s	4.99	15.09 to 21.74
T1	25.9qr	6.08	22.62 to 29.28
T2	27.6q	5.67	24.08 to 30.74
T3	22.2rs	7.07	18.89 to 25.55
T4	23.4qr	6.01	20.12 to 26.78
T5	26.7qr	5.85	23.29 to 29.95
T6	38.7p	4.53	35.36 to 21.74

Note: SD = standard deviation, ANOVA = analysis of variance.

therefore all nutrients and HS were efficiently absorbed by the plants; 2) the control solution contained a limited amount of nutrients and the benefits of HS would mainly be observed when plants were grown in a condition of stress; and 3) liquid FA contained small particulate sizing of HS, maximizing their absorption by the plant roots.

While many efforts were made to compare the efficacy of HA with that of FA, the use of liquid FA in these experiments provided a different view on how commercial humic products could be labeled. FA were commonly defined as components of HS soluble at any pH, typically at a pH less than 2. Most commercial fulvic products on the market therefore had low pH. Liquid FA had 10.2 pH. However, it was manufactured using a technology that captured the smaller molecular weight of the humic components (Canadian Humalite International Inc. 2020a). As noted by Wang et al. (1995) and Thurman et al. (1982), FA, in most cases, happened to be smaller in sizing (i.e. had a smaller molecular weight) than HA. Therefore, while liquid FA had high pH, it indeed had components of HS with a small molecular weight. These would be FA and some HA of smaller molecular weight; both were small enough in sizing to be absorbed well by root hairs, which contributed to their efficacy.

The results showed that the roots were shorter at higher rates of liquid FA. These agreed with the findings of Whitehead (1964), Morrison (1963), and Krogmeier and Bremner (1989) that certain types of phenolic acids found in soil, as well as from the degradation of HA, could have a negative effect on plant growth when

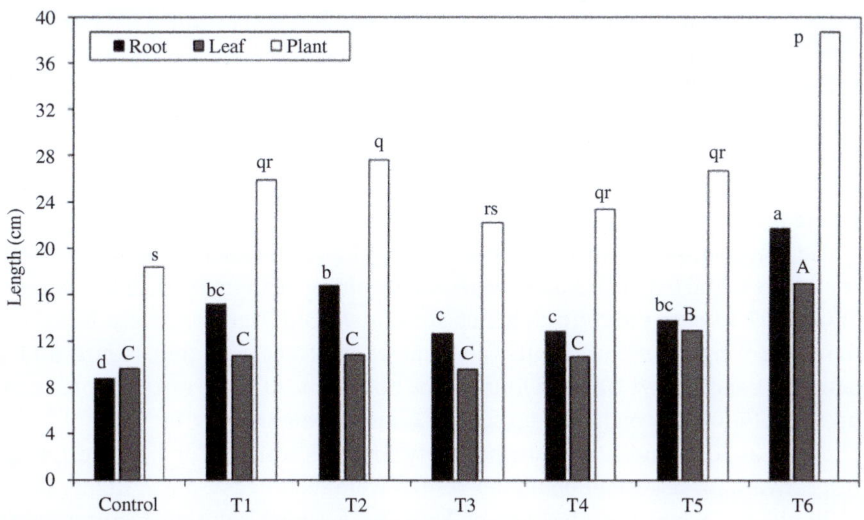

Figure 6.10 Root, leaf, and plant lengths.

Table 6.17 Estimated addition of nutrients in control solution.

Treatment	Estimated addition of K mg/l	Estimated increase of NPK %
Control	0	0
1	< 1	< 1
2	4	< 1
3	30	3
4	46	5
5	106	11
6	319	33

present in excessive amounts. Laboratory and greenhouse experiments completed by others confirmed this hypothesis, showing reduced plant growth at relatively high strengths of HA (Reynolds et al. 1995, Asik et al. 2009, and Shafeek et al. 2013).

Interestingly, when the rates were furtherly increased, positive results were again observed. Although the detailed mechanisms behind this were not investigated, the results suggested that this anomaly might have been caused by the nutrients (i.e. K) present in the product. At lower rates, its effects were negligible, and therefore the increased (or decreased) root lengths were most likely attributable to HS. At higher rates, the effect of K became more and more influential (see Table 6.17). So, the high presence of K in the solution was assumed to be responsible for the growth of the plant.

The same pattern was also observed for the leaves (see A to C alphabetical groupings) and the plants (see p to s alphabetical groupings). Product rates between 1 and 7 mg/l HS produced the best results, followed by reduced responses at higher rates, and positive responses at much higher rates (see Table 6.16 and Figure 6.10).

The leaf counts shown in Table 6.18 also showed the same trend. The best result was found between 1 and 7 mg/l HS, resulting in an increase in leaf counts of up to 27.3% over the control, which stood at 98 and 77, respectively. Based on the plant weight measurements (also shown in Table 6.18 and Figure 6.11), between 1 and 7 mg/l HS were found to be optimum. At these rates, increases of up to 62.1% fresh weight (or up to 26.8% dried weight) over the control were observed at 36.21 and 22.34 g, respectively, for fresh weights (or 1.89 and 1.49 g, respectively, for dry weights). Better results were recorded at much higher rates of liquid FA, when the presence of K in the growing solution became more influential.

Table 6.18 Leaf counts and plant weight measurements.

Treatment	Leaf counts	Plant fresh weight g	Plant dried weight g
Control	77	22.34	1.49
1	74	25.19	1.87
2	98	36.21	1.89
3	74	25.08	1.71
4	87	32.20	2.00
5	103	60.54	3.69
6	108	85.49	5.44

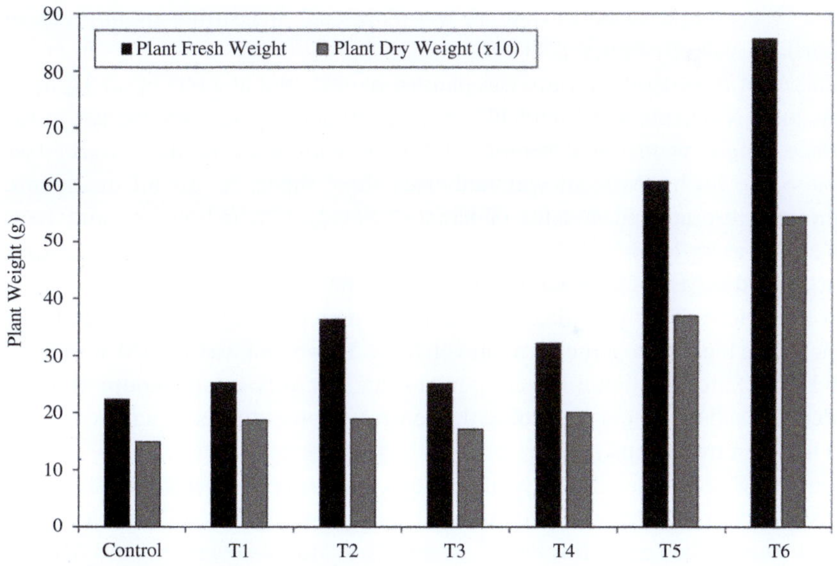

Figure 6.11 Plant fresh and dry weights.

6.6.5.3 Conclusion

These experiments demonstrated the efficacy of HS at low concentrations on the growth of lettuce. The plants were grown in deep- water culture-growing units containing limited amounts of nutrients, representing a stressed growing condition. Their growth was significantly enhanced with the addition of liquid FA at rates between 70 and 500 mg/l (representing 1 and 7 mg/l HS). Higher rates did not necessarily result in better responses. This finding was critical in the context of sustainable horticulture, where maximum yields from minimum inputs would be desirable.

6.6.6 Canola

A product efficacy project was completed in Alberta in 2014. Liquid HA was selected as the humic product to be evaluated. It contained 10.1 pH, 12.1% HS (based on the colorimetric method, including moisture), and 1.8% K_2O, as well as negligible amounts of other nutrients, toxic organics, and pathogens.

Soil in the area was clay loam, containing 5.8 pH, 0.3 mS/cm EC, and 8.3% OM. It had available macronutrients at 19 ppm N, 18 ppm P_2O_5, 565 ppm K_2O, 23 ppm SO_4, 1730 ppm Ca, and 327 ppm Mg. The available micronutrients were 0.8 ppm Cu, 1.0 ppm B, 5.0 ppm Cl, 6.8 ppm Zn, 27.2 ppm Mn, and 222 ppm Fe. The project was completed on 27 outdoor test plots, each measuring 1.4 m × 6.6 (or 9.2 m^2).

Canola of the 6060RR variety was planted in each plot at a rate of 6.7 kg/ha in May 2014. Nutrients and liquid HA at different rates were then applied to the surface. There were ten different treatments completed in this project (see Table 6.19). Each treatment was replicated three times. Across all treatments, nutrients were applied at rates of 85% (57-19-0.2), 92% (62-20-0.3), and 100% (67-22-0.3) as kg/ha N-P_2O_5-K_2O. Liquid HA was added at 0.0 (labelled as "HA0"), 4.5 (labelled as "HA1"), 9 (labelled as "HA2"), and 19 l/ha (labelled as "HA3"). N and P nutrients were provided in the form of urea (46-0-0) and MAP (11-52-0), which were blended together at a ratio of 76:24. This blend was applied at a rate of 178 kg/ha. Since the liquid HA contained some K, a very small amount of it was added for each treatment (including the control) in order to ensure a fair comparison between treatments.

Harvest was completed in September 2014, in which yields from three replications of each treatment were recorded in t/ha and averaged (see Figure 6.12). All data were analyzed using Minitab™ for means, standard deviations, and ANOVA ($P = 0.05$). Fisher's least significant difference ($\alpha = 0.05$) was completed for all treatments, and they were found to be statistically insignificant. However, the general trends were obvious and the results contained important information with respect to the determination of the optimum rate of liquid HA (see Section 6.6.1 for a similar case).

At 100% nutrients, it was found that the control produced the lowest yield at 2.17 t/ha. The application of liquid HA improved crop yields best for Treated 3-2 at 2.47 t/ha. At its highest rate (Treated 3-3), the yield was reduced, but still better than that of the control. At 85% nutrients and the lowest liquid HA rate (Treated 1-1), the yield was 1.8% lower than that of the control (at 2.13 t/ha). Crop yields

Table 6.19 Experimental protocol.

Treatment	Plot size m^2	NPK Rate	NPK kg/ha	Liquid HA Rate	Liquid HA l/ha	#
Control	1.4 m × 6.6 m (= 9.2 m^2)	100%	67-22-0.3	HA0	0.0	3
Treated 1-1		85%	57-19-0.2	HA1	4.5	
Treated 2-1		92%	62-20-0.3			
Treated 3-1		100%	67-22-0.3			
Treated 1-2		85%	57-19-0.2	HA2	9	
Treated 2-2		92%	62-20-0.3			
Treated 3-2		100%	67-22-0.3			
Treated 1-3		85%	57-19-0.2	HA3	19	
Treated 2-3		92%	62-20-0.3			
Treated 3-3		100%	67-22-0.3			

Note: # = number of replications.

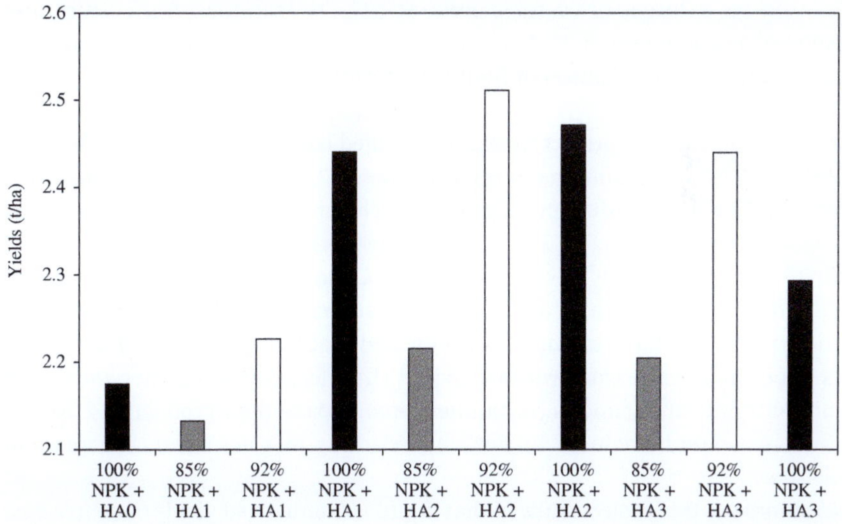

Figure 6.12 Treatment vs. yields.

were increased by up to 3.8% over Treated 1-1 with higher application rates of liquid HA. The best result (at 2.21 t/ha) was achieved at the medium rate of liquid HA (Treated 1-2). At 92% nutrients, crop yields were improved by up to 15.7% over the control with the application of liquid HA. The best result (at 2.51 t/ha) was achieved at the medium rate of liquid HA (Treated 2-2).

As hypothesized, HS indeed helped to improve the productivity of soil in the form of enhanced crop yields. Similar to the results obtained in Section 6.6.1, higher amounts of HS and nutrients did not always produce better responses. Very high amounts of nutrients resulted in more drastic yield reductions at very high HS rates. This project showed the importance of conducting a product efficacy project for determining the right amounts of HS and nutrients to produce the best response.

The economy of this project was evaluated based on the application of liquid HA (at 9 l/ha) to increase crop yields at the recommended nutrients from 2.17 (Control) to 2.47 (Treated 3-2) t/ha. The price for liquid HA was $1.03/l (delivered to site, Canadian Humalite International 2020d), and at that rate of application, it increased production costs by $9.27/ha. Applying this product on the field using a high clearance sprayer increased production costs by $12.29/ha (Manitoba Agriculture and Resource Development 2020b). The total increase in production costs therefore became $21.56/ha. An increase in crop yields of 0.30 t/ha (at a target price for canola at $477/t, Manitoba Agriculture and Resource Development 2020a) would result in $143.10/ha gross revenue. This indicated that the application of liquid HA resulted in a net revenue increase of $121.54/ha.

The economy of this project was also evaluated based on the application of liquid HA (at 9 l/ha) to maintain crop yields between 2.17 and 2.21 t/ha (Control and Treated 1-2) at a 15% reduction of nutrients. The application of liquid HA would increase production costs by $21.56/ha. The prices for urea and MAP were $0.62 and 0.72/kg, respectively (delivered to site, Canadian Humalite International Inc. 2020d). At 178 kg/ha for these nutrients at a ratio of 76:24, a 15% reduction would give $17.19/ha in savings. This indicated that the application of liquid HA was not economically worthwhile. If liquid HA was applied together with other liquid nutrients or chemicals (if required) in one pass, then production costs for liquid HA would be reduced to $9.27/ha, resulting in a net revenue increase of $7.92/ha.

In summary, this project showed that liquid HA enhanced yields of canola best at 92% nutrients (62-20-0.3 as N-P_2O_5-K_2O kg/ha) and 9 l/ha liquid HA. Increasing nutrients to 100% (67-22-0.3) or reducing them to 85% (57-19-0.2) resulted in lower yields. The incorporation of HS at its optimum rate at reduced nutrients would maintain crop yields. The economy of this project was worthwhile when the humic product was applied to increase crop yields at the same nutrient input, or when applied together with reduced nutrients in one pass to maintain crop yields.

6.6.7 Lentil

Liquid HA was evaluated in this project. The product contained small amounts of macronutrients (at 0N-0P_2O_5-1.8K_2O), very small amounts of micronutrient and toxic metals, as well as negligible amounts of pathogens and toxic organics. It contained 10.1 pH and 12.0% HS (based on the colorimetric method, including moisture).

Lentil of the CDC Maxim variety was selected as the crop to be tested. The project was completed in Alberta on 28 test plots measuring 1.8 m × 6.0 m (or 10.8 m^2) each. The soil was loam with Solonetzic clay underneath. It had 6.0% OM, 5.4 pH, 0.2 mS/cm EC, and 9.3 meq/100 g CEC. Nutrients in the soil were as follows (all in kg/ha): 24 NO_3-N, 36 P, 671 K, 40 S, 2 Cu, 41 Mn, 13 Zn, 2 B, 523 Fe, and 4 Cl. The soil also contained 1180 Ca, 271 Mg, and 100 Na (all in ppm).

The crop was direct seeded on each plot in May 2016. MAP, rated at 11-52-0, was broadcasted during seeding. The rates were 0 (Control 0), 22 (Control 1), and 45 kg/ha (Control 2), representing 0, 50, and 100% application rates. These were reduced rates, which were lower than the recommended rate of around 90 kg/ha. Liquid HA was applied on the surface during seeding. Since this product contained 1.8% K_2O, the amount of K was compensated for each treatment (including Control 0, 1, and 2 as well as other treatments) in order to provide a fair comparison between treatments. The product rates were 0 (Control 0, 1, and 2), 9 (Treated 1-1 and 2-1), and 27 l/ha (Treated 1-2 and 2-2), labelled as "HA0," "HA1," and "HA2," respectively.

Each treatment was replicated four times (see Table 6.20 for the experimental protocol). Harvest was made in September 2016 and yields for each plot were recorded in t/ha. All data were analyzed using Minitab™ for means, standard deviations, and ANOVA ($P = 0.05$). Fisher's least significant difference ($\alpha = 0.05$) was completed for all treatments. Different alphabetical groupings (a, b, and ab) indicated that the values were statistically different.

As expected, Control 0 (0% MAP + HA0) resulted in the lowest yield at 1.54b t/ha (see Figure 6.13). Control 1 (50% MAP + HA0) had a slight increase at 1.64b t/ha (or 6.5% higher that Control 0). Treated 1-1 (50% MAP + HA1) showed a better yield at 2.07ab t/ha or 26.2% higher than Control 1. Treated 1-2 (50% MAP + HA2) did not significantly change the yields (2.21ab t/ha). This indicated that increasing liquid HA beyond a certain point would not help. Increasing nutrients to 100% (Control 2: 100% MAP + HA0) significantly enhanced the yields compared to Control 0 to 2.16ab t/ha (0% MAP + HA0) or 40.3% higher than Control 0, as well as compared to Control 1 (50% MAP + HA0) or 31.7% higher. Incorporating liquid

Table 6.20 Experimental protocol.

Treatment	Plot size m^2	MAP		Liquid HA		#
		Rate	kg/ha	Rate	l/ha	
Control 0	1.8 m × 6.0 m (= 10.8 m^2)	0%	0	HA0	0	4
Control 1		50%	22			
Control 2		100%	45			
Treated 1-1		50%	22	HA1	9	
Treated 2-1		100%	45			
Treated 1-2		50%	22	HA2	27	
Treated 2-2		100%	45			

Note: # = number of replications.

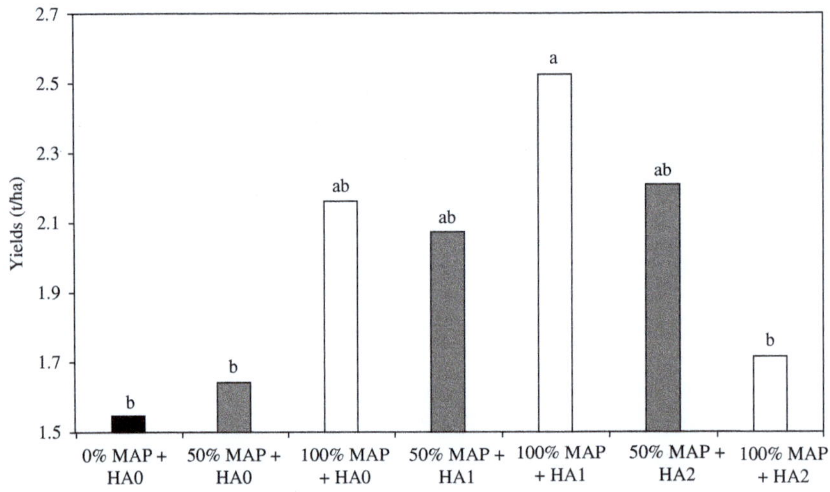

Figure 6.13 Treatment vs. yields.

HA as Treated 2-1 (100% MAP + HA1) significantly enhanced the yields to 2.53a t/ha (or 17.1% higher than Control 2). Increasing liquid HA in Treated 2-2 (100% MAP + HA2) significantly lowered the yields to 1.72ab t/ha (or 20.4% lower than Control 2).

These results showed that very high amounts of liquid HA and nutrients resulted in reduced responses. Under optimum conditions, the addition of this humic product enhanced crop yields or maintained them at reduced nutrient inputs. Treated 1-1 and 1-2 (50% MAP + HA1 and 50% MAP + HA2) performed similarly to Control 2 (100% + HA0). This indicated that 50% nutrients with the addition of liquid HA at any rate would result in similar yields to that of crops with 100% nutrients only.

These findings showed that the addition of HS improved yields best when the growing environment was unfavorable, i.e. when the soil had low amounts of nutrients. When the soil contained sufficient nutrients, the effects of HS were not obvious. When nutrients in soil were sufficient and a more than adequate amount of HS was incorporated, yields were negatively affected. The same trend was also observed when more than sufficient nutrients were present in soil and an adequate amount of HS was added. The mechanism behind these results was not investigated in this project, however similar findings were also observed in Sections 6.6.1 (rice) and 6.6.6 (canola). Statistically significant differences were observed in this project (lentil).

The economy of this project was evaluated based on the application of liquid HA (at 9 l/ha) and 100% NPK to enhance crop yields from 2.16 (Control 2) to 2.53 (Treated 2-1) t/ha. The price for liquid HA was $1.03/l (delivered to site, Canadian Humalite International 2020d) and at the specified rate of application, it increased production costs by $9.27/ha. Applying this product on the field using a high clearance sprayer increased operating costs by $12.29/ha (Manitoba Agriculture and Resource Development 2020b). The increase in total production costs was $21.56/ha. An increase in crop yields of 0.37 t/ha (at a target price for lentil of $504/t, Saskatchewan Crop Insurance Corporation 2020) would give $186.48/ha in gross revenue. This indicated that the application of liquid HA resulted in a net revenue increase of $164.92/ha.

The economy of this project was also evaluated based on the application of liquid HA (at 9 l/ha) to maintain crop yields between 2.07 and 2.16 t/ha (Treated 1-1 and Control 2; neither results were statistically different) with a 50% reduction in nutrients. The application of liquid HA increased production costs by $21.56/ha. The price for MAP was $0.72/kg (delivered to site, Canadian Humalite International Inc. 2020d). At 45 kg/ha of this product a 50% reduction in nutrients resulted in savings of $16.20/ha. This indicated that the application of liquid HA was not economically worthwhile. This option would only be worthwhile if liquid HA was applied together with other liquid nutrients or chemicals (if required) in one pass. Production costs for liquid HA would be reduced to $9.27/ha, resulting in a net revenue increase of $6.93/ha.

In summary, liquid HA optimized yields of lentil. This product at 9 l/ha applied together with 45 kg nutrients/ha resulted in the best yield. Liquid HA at 27 l/ha applied with 22 kg nutrients/ha resulted in the best yield improvement over the control of the same nutrients. The economy of this project was worthwhile when the humic product was applied to increase crop yields at the same nutrient input, or when applied together with reduced nutrients in one pass to maintain crop yields.

6.6.8 Soil Compaction

Soil compaction was defined as the compression of soil particles into a smaller volume, reducing the size of pore space available for water and air. Soil compaction might be caused by the combination of soil tillage and raindrop or irrigation water impact, hardpan tillage-induced compaction, wheel traffic-induced compaction, and topographical-influenced compaction. It affected both tilled and non-tilled crop productions (McKenzie 2010b). The soil parameters to be analyzed included density, porosity, structure, penetration resistance, biota, water infiltration, and percolation. Soil compaction could impair water infiltration, crop emergence, root penetration, and the uptake of water and nutrients by crops. Overall, soil compaction reduced soil productivity and crop yields (McKenzie 2010b and Duiker 2004).

The management of soil compaction included the reduction of trafficking, axle loads, contact pressures, and tillage. Additionally, using fibrous and tap-rooted crops in rotation and natural soil processes of wetting-drying and freeze-thaw helped to manage soil compaction. Another strategy was to maintain or increase OM contents in soil. This could be achieved by reducing soil disturbance, promoting biologically healthy soil practices, and adding OM in the form of crop residues, manure, or compost (McKenzie 2010b and Duiker 2004). The addition of OM to soil was found to enhance soil properties (Malkawi et al. 1999).

Two product efficacy projects were completed to evaluate the efficacy of dry HS and liquid HA for improving the physical properties of soil with respect to its compaction.

Dry HS was evaluated in the first project. It had a mean particulate sizing of 2.1 mm, less than 5% water solubility, 4.1 pH, 60.2% HS (including moisture), 1.1% Ca, 0.3% Mg, and 0.3% Na. This product contained negligible amounts of toxic metals, toxic organics, and pathogens. The project was completed on a 42-ha farm in Alberta from October 2015 until October 2016. Three test plots containing wheat crop residues of 0.6 ha each were allocated for Control, Treated 1, and Treated 2.

In October 2015, ten soil compaction readings at 15 cm deep were completed on each plot using a penetrometer. Ten soil samples were taken at 15 cm deep from each plot, and each was analyzed for moisture. For each plot, the samples were aggregated and analyzed for soil structure, pH, EC, and OM. All plots had practically the same soil properties, containing a high percentage of clay (see Table 6.21). Dry HS was then broadcasted on the Control, Treated 1, and Treated 2 plots at

Table 6.21 Soil parameters before application (October 2015).

	Control plot	Treated 1 plot	Treated 2 plot
Sand:silt:clay (%)	17:31:52	15:35:50	13:33:54
pH (no unit)	6.2	6.1	6.2
EC (mS/cm)	1.4	1.4	1.3
OM (%)	7.3	7.3	7.3

application rates of 0, 180, and 450 kg/ha, respectively. In April 2016, 30 soil compaction readings at 15 cm deep were made on each plot using a penetrometer. Ten soil samples were taken at 15 cm deep from each plot, and each was analyzed for moisture. In October 2016, 30 soil compaction readings at 15 cm deep were completed on each plot using a penetrometer. Ten soil samples were taken at 15 cm deep from each plot, and each was analyzed for moisture. All data were analyzed for means, standard deviations, ANOVA, and Fisher's least significant difference ($\alpha = .05$) using Matlab® v.17.

Before the application of dry HS in October 2015 (see Figure 6.14), each test plot had similar compaction readings (mean for all data = 1470a kPa) and moisture contents (mean for all data = 27.1A %). After six months of application (April 2016, see Figure 6.15), HS significantly reduced compaction from 1520a (mean for Control) down to 1240b kPa (mean for Treated 1 with 180 kg/ha dry HS) and 1160b kPa (mean for Treated 2 with 450 kg/ha dry HS). This product did not significantly increase moisture (mean for all data = 26.3A %). After 12 months of application (October 2016, see Figure 6.16), HS reduced compaction from 1060a (mean for Control) down to 980ab kPa (mean for Treated 1 with 180 kg/ha dry HS) and 830b kPa (mean for Treated 2 with 450 kg/ha dry HS). HS did not significantly increase moisture (mean for all data = 31.2A %). Soil compaction was affected by OM (HS enhanced crop residue breakdown) and soil moisture (October 2016 was the wetter month). The effect of dry HS at 180 kg/ha on soil compaction was reduced after 12 months, while its application at 450 kg/ha maintained its performance. Different alphabetical groupings (a, b, ab, A, B, and AB) indicated that the values were statistically different.

Liquid HA was evaluated in the second project. It had 9.7 pH, almost 100% water-soluble, 11.9% HS (based on the colorimetric method, including moisture), and 1.8% K_2O. This product contained negligible amounts of other nutrients, toxic metals, toxic organics, and microorganisms. The project was completed in grape yards located in Region V, Chile from October 2016 until March 2017. Table grapes of the Thompson Seedless variety were planted in

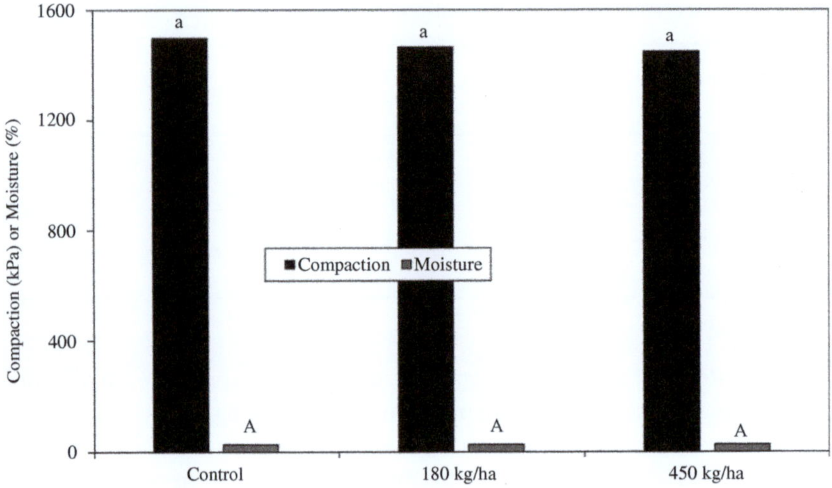

Figure 6.14 Treatment vs. compaction and moisture (October 2015).

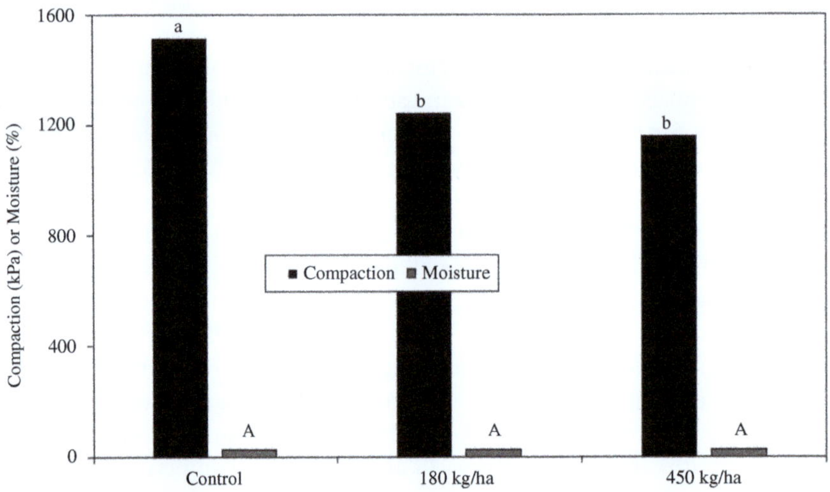

Figure 6.15 Treatment vs. compaction and moisture (April 2016).

1997, 3 m apart in sandy soil. Soil properties in the area were as follows: 7.3 pH, 1.5% OM, and 15 meq/100 g CEC.

In October 2016, 42 compaction readings were made at 15 cm soil depth using a penetrometer, followed by 41 readings at 30 cm depth, and 36 readings at 45 cm depth (labelled as Control or Time 0). The humic product was then applied to soil

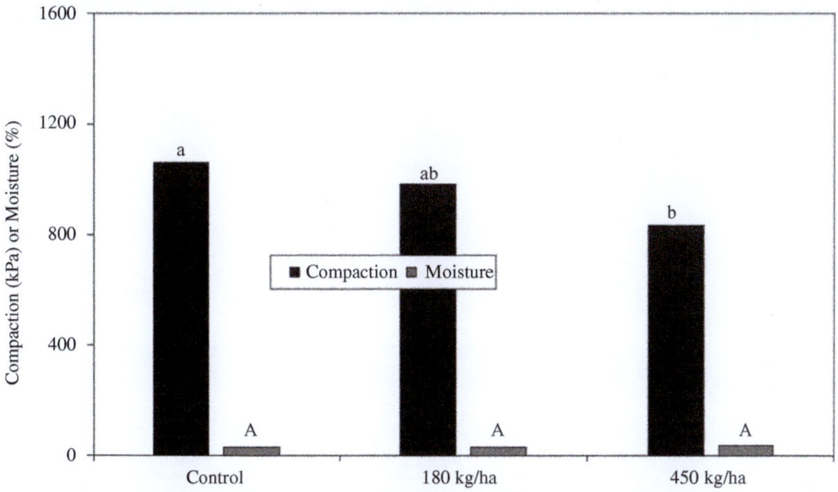

Figure 6.16 Treatment vs. compaction and moisture (October 2016).

for the whole yard using a drip irrigation system at a rate of 600 l/ha. Two months later in December 2016, 24 compaction readings were made at 15 cm soil depth using a penetrometer, followed by 24 readings at 30 cm depth, and 23 readings at 45 cm depth (labelled as Time 1). Five months later in March 2017, 41 compaction readings were made at 15 cm soil depth using a penetrometer, followed by 31 readings at 30 cm depth, and 30 readings at 45 cm depth (labelled as Time 2). Each reading was completed in between two plants, randomly chosen throughout the farms. All data were analyzed (for all treatments) using R version 3.5.1 for means, standard deviations, ANOVA ($P = 0.05$), and Duncan's test ($\alpha = 0.05$).

Control (Time 0) had a mean reading of 1360a kPa at 15 cm depth. Liquid HA reduced it to 450c kPa after 2 months (Time 1) and 890b kPa after 5 months (Time 2). At 30 cm soil depth, Time 0 had a mean reading of 1860A kPa. Liquid HA reduced it to 790B kPa after 2 months (Time 1) and 1110B kPa after 5 months (Time 2). At 45 cm soil depth, Time 0 had a mean reading of 2220p kPa. Liquid HA reduced it to 1470q kPa after 2 months (Time 1) and 1930p kPa after 5 months (Time 2). Different alphabetical groupings (a to c, A to B, and p to q) indicated that the values were statistically different (see Figure 6.17).

The effect of HS was more obvious closer to the soil surface where the product was applied, as it should be. The effect of HS was also reduced with time, which made sense due to the continuous addition of moisture throughout the season. The best responses were recorded during Time 1 (2 months after product application), especially at 15 cm soil depth. Readings at Time 2 (5 months after product

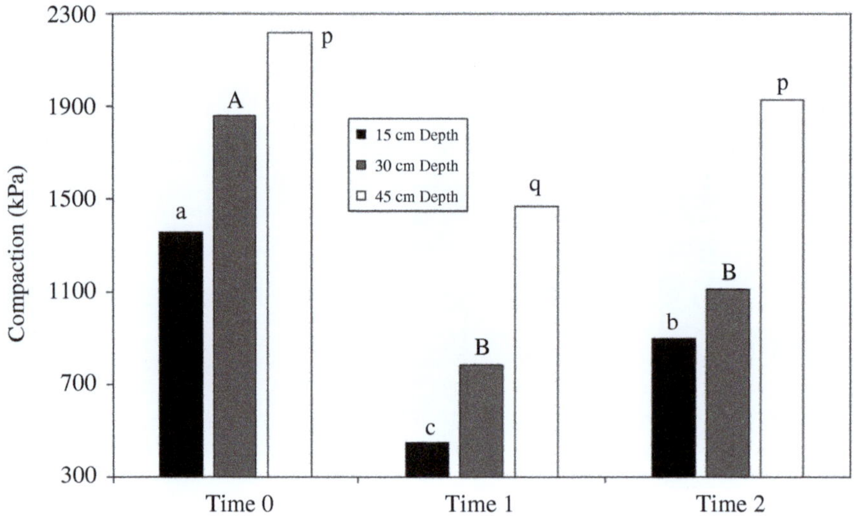

Figure 6.17 Treatment vs. compaction.

application) were higher than those of Time 1, but still lower than those of Time 0. Time 0 and Time 2 readings at 45 cm soil depth were not significantly different.

In summary, the application of dry HS significantly reduced compaction of high-clay soil by up to 18.4% over the control after six months of application. An application rate of 180 kg/ha dry HS was found to be optimum. The application of liquid HA significantly reduced compaction of sandy soil by up to 66.9% after two months of application at an application rate of 600 l/ha. While both results could not be directly compared due to each study using a different type of soil, these findings showed the difference between raw (low water solubility) and processed (high water solubility) products. Larger amounts of raw HS were required to be effective, but their effects would last longer due to a slower breakdown in soil.

6.6.9 Barley

In this project, the efficacy of liquid FA was evaluated on yields of barley. This product contained 0.3% HA and 0.7% FA (total HS = 1.0%, based on ISO 19822, including moisture) and was practically 100% soluble in water at any pH. Its pH was 10.0 and other than 0.8% K_2O, it contained very low amounts of other nutrients, toxic metals, toxic organics, and microorganisms.

The evaluation was completed in 2018 in an outdoor agricultural farm located in Alberta. The soil was Solonetzic clay with 4.5% OM, 6.1 pH, 0.4 mS/cm total salts, and 20.1 meq/100 g CEC. Its nutrient levels were 17 NO_3-N, 38 P, 348 K, 162 S,

1,002 Ca, and 419 Mg (all units in ppm). A fertilizer blend consisting of 80% urea (46-0-0), 11% MAP (11-52-0), and 9% KCl (0-0-60) was broadcasted at a rate of 184 kg/ha, providing 70N $10P_2O_5$ $10K_2O$ kg/ha. Barley of the CDC Copeland variety was seeded in May 2018 on test plots measuring 1.8 m × 6.7 m (or 12.1 m^2) each. Two weeks after seeding, a liquid fertilizer with a 4-16-5 rating was foliar applied at rates of 0 (Control 0), 33 (Control 1), and 100% (Control 2), together with liquid FA (also foliar applied) at rates of 0 (FA0), 70 (FA1), and 140 ml/ha (FA2), respectively. Note: 100% rate for 4-16-5 fertilizer = 7.0 l/ha. The amount of K within liquid FA (at 0.8% K_2O) was compensated for each treatment (including the control) to allow for a fair comparison between treatments.

Each treatment was replicated five times (see Table 6.22). Harvest was completed in September 2018. Yields were recorded in t/ha and all data were analyzed (for all treatments) using R version 3.5.1 for means, standard deviations, ANOVA ($P = 0.05$), and Duncan's test ($\alpha = 0.05$). Note: different alphabetical groupings (a to b) indicated that the values were statistically different.

As shown in Figure 6.18, Control 0 (0% nutrients and FA0) produced the lowest yield (1.31b t/ha), followed by Control 1 (33% nutrients and FA0) at 1.51ab t/ha (or 15.3% higher than Control 0). Increasing nutrients to 100% nutrients and FA0 (Control 2) did not significantly change the yields (1.49ab t/ha). Incorporating FA1 and FA2 into the low input of nutrients significantly increased yields to 1.59a (Treated 1-1) and 1.62a (Treated 1-2) t/ha, respectively, up to 7.3% higher than Control 1. Doubling liquid FA apparently did not significantly improve yields (Treated 1-1 vs. 1-2). Incorporating FA1 and FA2 into the high input of nutrients did not significantly improve yields: 1.55ab (for Treated 2-1) and 1.53ab (for Treated 2-2) t/ha, respectively, up to 4.0% higher than Control 2.

Table 6.22 Experimental protocol.

Treatment	Plot size m^2	Nutrients (4-16-5) Rate	Nutrients (4-16-5) l/ha	Liquid FA Rate	Liquid FA ml/ha	#
Control 0	1.8 m × 6.7 m (= 12.1 m^2)	0%	0.0	FA0	0	5
Control 1		33%	2.3			
Control 2		100%	7.0			
Treated 1-1		33%	2.3	FA1	70	
Treated 2-1		100%	7.0			
Treated 1-2		33%	2.3	FA2	140	
Treated 2-2		100%	7.0			

Note: # = number of replications.

Most interestingly, 100% nutrients with or without liquid FA (Control 2, Treated 2-1, and Treated 2-2) all performed similarly to Control 1 (33% nutrients without liquid FA). Only when liquid FA was applied to 33% nutrients (Treated 1-1 and 1-2) were significant yield improvements recorded. This indicated that 100% nutrients were already more than adequate for this application and so liquid FA did not help to increase crop yields. In fact, at its highest rate (FA2), the yield was slightly reduced. These findings were consistent with those of Sections 6.6.1 (rice), 6.6.6 (canola), and 6.6.7 (lentil), which showed that very high applications of HS and nutrients resulted in reduced yields.

The economy of this project was evaluated based on the foliar application of liquid FA (at 140 ml/ha) together with liquid 4-16-5 (at 2.3 l/ha) in one pass to enhance crop yields from 1.51 (Control 1) to 1.62 (Treated 1-2) t/ha. The price for liquid FA was $18.03/l (delivered to site, Canadian Humalite International Inc. 2020d), and at this rate of application, it increased production costs by $2.52/ha. An increase in crop yields of 0.11 t/ha (at a target price for barley of $190/t, Manitoba Agriculture and Resource Development 2020a) would bring in $20.9/ha in gross revenue. This indicated that the application of liquid FA resulted in a net revenue increase of $18.38/ha.

The economy of this project was also evaluated based on the foliar application of liquid FA (at 70 ml/ha) together with liquid 4-16-5 (at 2.3 l/ha) to enhance crops yields from 1.31 (Control 0) to 1.59 (Treated 1-1) t/ha. Prices for liquid FA and liquid 4-16-5 were $18.03 and $2.44/l (delivered to site, Canadian Humalite International Inc. 2020d). At application rates of 70 ml and 2.3 l/ha, respectively,

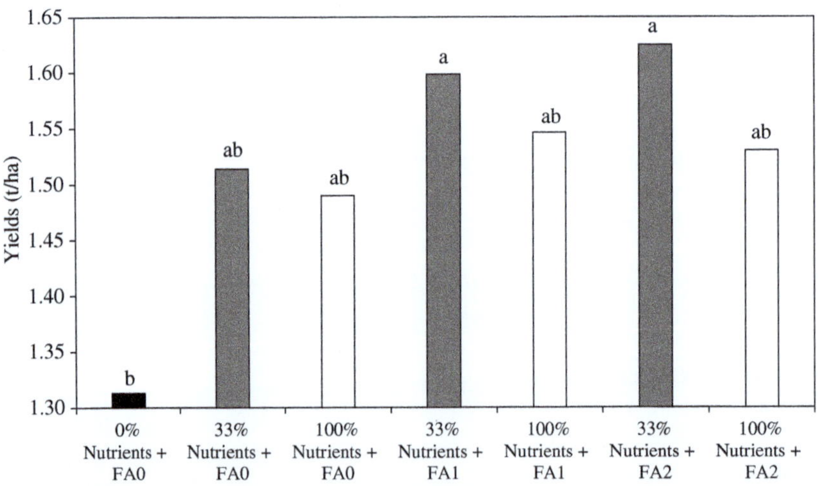

Figure 6.18 Treatment vs. yields.

using both products increased production costs by \$6.87/ha. Applying both products on the field in one pass using a high clearance sprayer increased production costs by \$12.29/ha (Manitoba Agriculture and Resource Development 2020b). In total, the application of liquid FA and liquid 4-16-5 increased production costs by \$19.16/ha. An increase in crop yields of 0.28 t/ha would bring in \$53.2/ha in gross income. The application of liquid FA and liquid 4-16-5 resulted in a net revenue increase of \$34.04/ha.

In summary, liquid FA applied foliar to seedlings at small application rates enhanced yields of barley. The best result was achieved when applying 2.3 l/ha liquid 4-16-5 and 140 ml/ha liquid FA. Higher rates of nutrients and FA were not recommended. The economy of this project was found to be worthwhile when liquid FA was used to increase crop yields.

6.6.10 Wheat (2)

The efficacy of pelletized HS was evaluated on yields of wheat. This product had a 2.1 mm mean particulate sizing, 3.9 pH, and 65.6% HS (based on the colorimetric method, including moisture). It contained very low amounts of nutrients, toxic metals, toxic organics, and pathogens. This product efficacy project was completed in 2019 on test plots located in Alberta measuring 1.8 m × 6.0 m (or 10.8 m^2) each. The soil had the following parameters: 5.7 pH, 5.2% OM, 15 ppm NO_3-N, 81 ppm P, 220 ppm K, 21 ppm S, 1130 ppm Ca, 243 ppm Mg, 0.6 ppm B, 0.7 ppm Cu, 95 ppm Fe, 37 ppm Mn, 4.3 ppm Zn, 33 ppm Na, and 24 meq/100 g CEC.

Wheat of the Brandon variety (developed by Semiarid Prairie Agricultural Research Centre Agriculture and Agri-Food Canada in Swift Current, Saskatchewan) was seeded in May 2019. A fertilizer blend containing 74% urea (46-0-0), 11% MAP (11-52-0), and 15% KCl (0-0-60) was applied at rates of 83 and 111 kg/ha to provide 29N 4.5P_2O_5 7.5K_2O (75% nutrients) and 39N 6P_2O_5 10K_2O (100%) kg/ha, respectively. Pelletized HS was applied at rates of 0 (PH0), 28 (PH1), 55 (PH2), and 110 (PH3) kg/ha. On each plot, both nutrients and pelletized HS were applied side-band during seeding, near where the future roots would develop.

There were eight different treatments completed with six replications each (see Table 6.23). Harvest was completed in September 2019 and yields were recorded in t/ha. All data were analyzed (for all treatments) using R version 3.5.1 for means, standard deviations, ANOVA ($P = 0.05$), and Duncan's test ($\alpha = 0.05$). Note: different alphabetical groupings (a, ab, and b) indicated that the values were statistically different.

As shown in Figure 6.19, Control 1 (75% nutrients + PH0) produced the lowest yield at 5.57a t/ha. Adding PH1 (Treated 1-1) increased its yields to 5.88ab t/ha (5.6% higher). Adding PH2 and PH3 (Treated 1-2 and 1-3, respectively) did not significantly change the yields (5.83ab and 5.77ab t/ha, respectively). These yields were in fact lower, although they were not statistically different.

Control 2 (100% nutrients + PH0) enhanced the yields over Control 1 (75% nutrients + PH0) to 5.88ab t/ha (5.6% higher), as expected. Adding PH1 (Treated 2-1) increased the yields to 6.03b t/kg. Adding PH2 and PH3 (Treated 2-2 and 2-3, respectively) significantly reduced yields (5.76ab and 5.87ab t/ha, respectively).

Table 6.23 Experimental protocol.

Treatment	Plot size m^2	NPK Rate	NPK kg/ha	Pelletized HS Rate	Pelletized HS kg/ha	#
Control 1	1.8 m × 6.0 m (= 10.8 m^2)	75%	29-4.5-7.5	PH0	0	6
Control 2		100%	39-6-10			
Treated 1-1		75%	29-4.5-7.5	PH1	28	
Treated 2-1		100%	39-6-10			
Treated 1-2		75%	29-4.5-7.5	PH2	55	
Treated 2-2		100%	39-6-10			
Treated 1-3		75%	29-4.5-7.5	PH3	110	
Treated 2-3		100%	39-6-10			

Note: # = number of replications.

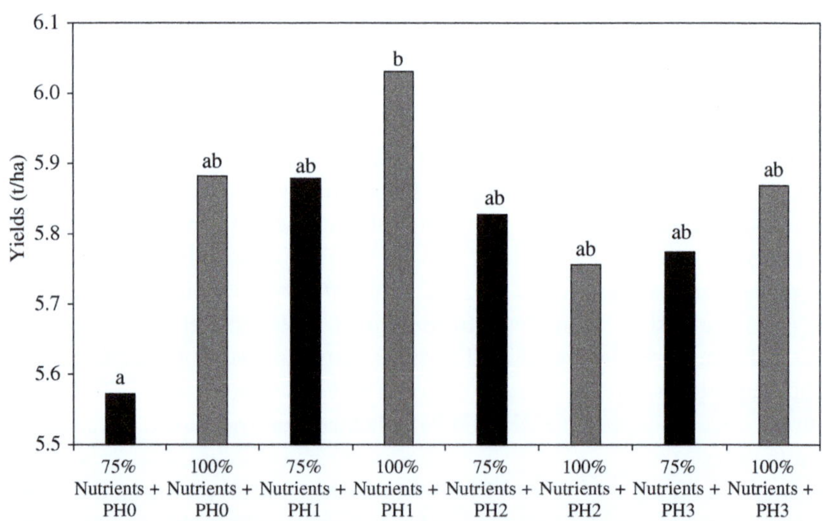

Figure 6.19 Treatment vs. yields.

These results showed that HS enhanced crop yields best when the growing condition was not optimum (i.e. when there were lower nutrients in the soil). From Control 1 (75% nutrients + PH0) to Treated 1-1 (75% nutrients + PH1), there was an improvement of 5.6%. From Control 2 (100% NPK + PH0) to Treated 2-1 (100% NPK + PH1), the increase was only 2.6%. Increasing pelletized HS rates did not help, especially at higher amounts of nutrients. This indicated that at very high applications of HS and nutrients, responses were not always better. These findings were consistent with those of Sections 6.6.1 (rice), 6.6.6 (canola), 6.6.7 (lentil), and 6.6.9 (barley).

The economy of this project was evaluated based on the application of pelletized HS (at 28 kg/ha) and 100% NPK to enhance crop yields from 5.88 (Control 2) to 6.03 (Treated 2-1) t/ha. The price for pelletized HS was \$0.51/kg (delivered to site, Canadian Humalite International 2020d), and at this rate of application, it increased production costs by \$14.28/ha. In this case, the humic product was applied together with the fertilizer blend in one pass. An increase in crop yields of 0.15 t/ha (at a target price for wheat at \$250/t, Manitoba Agriculture and Resource Development 2020a) would bring \$37.50/ha in gross revenue. This indicated that the application of pelletized HS resulted in a net revenue increase of \$23.22/ha.

The economy of this project was also evaluated based on the application of pelletized HS (at 28 kg/ha) to maintain crop yields between Control 2 and Treated 1-1 at a 25% reduction in nutrients. The application of pelletized HS increased operating costs by \$14.28/ha. The prices for urea, MAP, and KCl were \$0.62, \$0.72, and \$0.35/kg, respectively (delivered to site, Canadian Humalite International Inc. 2020d). The price for the fertilizer blend was \$0.59/kg. A reduction of 28 kg/ha of this product (111 – 83 kg/ha) resulted in a saving of \$16.52/ha. The application of pelletized HS resulted in a net revenue increase of \$2.24/ha.

In summary, pelletized HS at a rate of 28kg/ha applied side-band together with 39N-6P_2O_5-10K_2O kg/ha resulted in the best yield at 6.03 t/ha. The best yield improvement (at 5.6% higher than that of the control) was achieved at a pelletized HS rate of 28 kg/ha applied together with 29N-4.5P_2O_5-7.5K_2O.This project was found to be economically worthwhile when the product was applied to increase crop yields at the same nutrient input, or when applied to maintain them at reduced nutrients.

6.6.11 Other Soil and Crop Product Efficacy Projects

Fourteen other product efficacy projects were completed in Vietnam, Idaho, Alberta, and Quebec (see Table 6.24). Twelve projects on crop yields were completed in agricultural fields and two projects on the emergence of seedlings and the growth of a beneficial fungus were completed in a laboratory.

A product efficacy project was completed in an agricultural field in Lam Dong, Vietnam, from June to July 2011 to evaluate the efficacy of liquid HA (10.0 pH, 12.0% HS, and 1.7% K_2O) on lettuce. For Control, nutrients were irrigated into the soil at rates of 250N and 85P_2O_5 kg/ha. For Treated, the same nutrients at the same rates were irrigated into the soil with a foliar application of liquid HA at a rate of 1.25 l/ha four times after 5, 10, 15, and 20 days of planting (or a total of 5 l/

Table 6.24 Other soil and crop product efficacy projects.

No	Year	Target	Product	Location	Parameter
11	2011	Lettuce	Liquid HA	Field, Vietnam	Yield
12	2011	Chrysanthemum	Liquid HA	Field, Vietnam	Yield
13	2011	Tea	Liquid HA	Field, Vietnam	Yield
14	2011	Tomato	Liquid HA	Field, Vietnam	Yield
15	2011 to 2012	Coffee	Liquid HA, Dry HS	Field, Vietnam	Yield
16	2011 to 2012	Watermelon	Liquid HA, Dry HS	Field, Vietnam	Yield
17	2011	Corn	Dry HS	Field, Idaho	Yield
18	2012	Wheat	Liquid HA	Field, Alberta	Yield
19	2014	Soybean	Liquid HA	Field, Quebec	Yield
20	2014	Corn	Liquid HA	Field, Quebec	Yield
21	2014 to 2015	Shrub	Dry HS	Field, Alberta	Yield
22	2016	Lentil, pea	Liquid HA	Laboratory, Alberta	Emergence
23	2019	Wheat	Liquid HA	Field, Alberta	Yield
24	2020	Fungus	Liquid HA	Laboratory, Alberta	Growth

ha for the whole project). During harvest, the production of lettuce was enhanced with the application of liquid HA. The average number of leaves per plant increased from 20.4 (Control) to 23.1 (Treated); an increase of 13.2%. Crop yields increased from 30.46 (Control) to 37.48 (Treated) t/ha; an increase of 23.0%.

Another project was completed in an agricultural field of the same region from June to July in 2011 with the aim of evaluating the efficacy of the same liquid product on chrysanthemum. For Control, nutrients were applied to the soil at rates of 150N-75P_2O_5-135K_2O kg/ha. For Treated, the same nutrients at the same rates were applied to the soil with the foliar application of liquid HA at a rate of 1.9 l/ha three times after 5, 12, and 17 days of planting (or a total of 5.7 l/ha for the whole project). During harvest, crop yields of chrysanthemum were enhanced with the application of liquid HA from 22.40 (Control) to 26.81 (Treated) t/ha; an increase of 19.7%.

The third project was completed in an agricultural field of the same region from March to June in 2011 to evaluate the efficacy of the same liquid product on oolong tea. For Control, nutrients were applied to the soil at rates of 160N-70P_2O_5-70K_2O kg/ha. For Treated 1, 2, 3, 4, and 5, the same nutrients at the same rates were applied to the soil with liquid HA foliar applied at rates of 3.9, 5.4, 6.6, 7.9, and 9.3 l/ha, respectively. One third of the rate was equally foliar applied 3, 18, and 33 days from the previous harvest. The average number of buds per m^2 increased from 441.9 (Control) to 452.7, 462.6, 463.1, 474.5, and 477.6 (Treated 1, 2, 3, 4, and 5, respectively); an increase of up to 8.1% compared to that of the control. Average bud weights were enhanced from 115.8 (Control) to 123.9, 127.2, 128.2, 127.8, and 133.0 g/100 buds (Treated 1, 2, 3, 4, and 5, respectively); an increase of up to 14.9% compared to that of the control. Crop harvests increased from 2.05 (Control) to 2.19, 2.20, 2.33, 2.42, and 2.45 t/ha (Treated 1, 2, 3, 4, and 5, respectively); an increase of up to 19.5% compared to that of the control. The best result was observed at the highest rate of humic product application. The responses were almost linear so that the optimum condition could not be identified (see Figure 6.20).

The fourth project was completed in an agricultural field of the same region in 2011 to evaluate the efficacy of the same liquid product on cherry tomato. For Control, nutrients at 135N-170P_2O_5-220K_2O were applied to the soil during transplant. For Treated 1, 2, and 3, the same nutrients at the same rates were applied to the soil with the same liquid humic product applied foliar at rates of 1.1, 1.5, and 1.9 l/ha, respectively. At those rates, liquid HA was applied once 10 weeks after transplanting, and once every week until harvest (21 weeks). During harvest, the product improved the average weights per plant from 943 (Control) to 946, 1,052, and 1,026 g (Treated 1, 2, and 3, respectively); an increase of up to 11.6% compared to that of the control. The product enhanced the average number of fruits per plant from 134.3 (Control) to 135.0, 155.3, and 156.0 (Treated 1, 2, and 3,

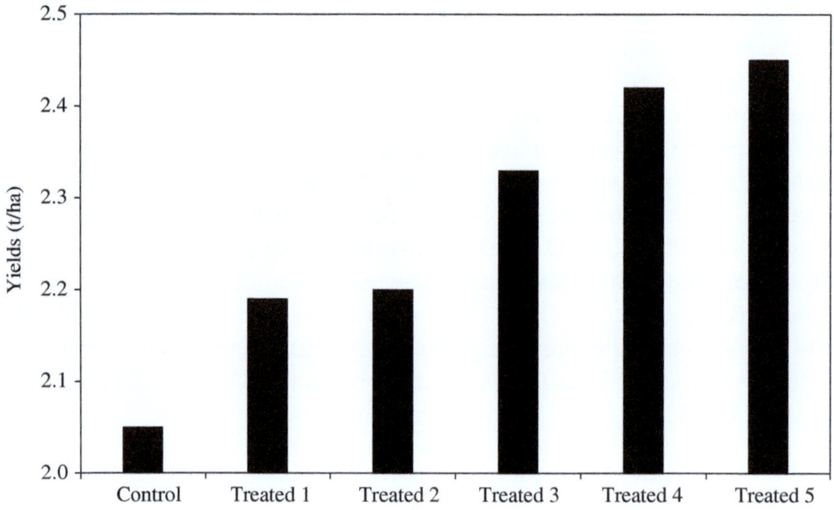

Figure 6.20 Treatment vs. yields.

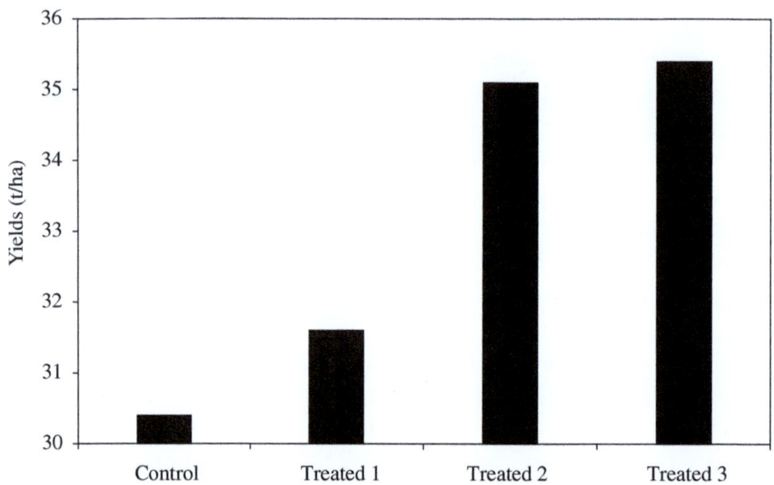

Figure 6.21 Treatment vs. yields.

respectively); an increase of up to 16.2% compared to that of the control). Most importantly, liquid HA enhanced fruit yields from 30.4 (Control) to 31.6, 35.1, and 35.4 t/ha (Treated 1, 2, and 3, respectively); an increase of up to 16.4% compared to that of the control. In this case, the best result was also observed at the highest product rate. However, the two highest rates resulted in similar responses, indicating that the positive effect of HA was already at its peak (see Figure 6.21).

A product efficacy project was completed in an agricultural farm in Dak Lak, Vietnam, from 2011 to 2012 to evaluate the efficacy of the same liquid product and dry HS (pH = 3.9 and HS = 55.9%, including moisture) on Robusta coffee. For Control, nutrients of $180N\text{-}55P_2O_5\text{-}180K_2O$ were applied to the soil twice. For Treated, the same nutrients at the same rates were applied to the soil with liquid HA foliar applied at a rate of 3 l/ha (applied three times throughout the growing season) and dry HS applied to the soil at a rate of 500 kg/ha. The application of both products enhanced the average development of branches per plant from 6.0 (Control) to 7.7 (Treated); an increase of 28.3%. The average fully developed coffee cherries were also improved from 78 (Control) to 92 (Treated); an increase of 17.9%. Treated also resulted in a higher average cherry weight of 180.2 g/1000 cherries compared to that of Control at 170.3 g/1000 cherries (5.8% higher). Average bean/cherry ratios were enhanced from 19.2 (Control) to 23.3% (Treated); an increase of 21.4%. Most importantly, liquid HA and dry HS enhanced fruit yields from 3.30 (Control) to 4.00 (Treated) t/ha; an increase of 21.2%.

A product efficacy project was completed in an agricultural farm in Kien Giang, Vietnam, on the efficacy of the same liquid and dry products on watermelon of the F1 TN522 variety. For Control, nutrients at rates of $175N\text{-}150P_2O_5\text{-}110\ K_2O$ kg/ha were applied with irrigation water. For Treated, the same nutrients at the same rates were applied to the soil with liquid HA foliar applied at a rate of 2.25 l/ha (applied four times throughout the growing season) and dry HS applied to the soil at a rate of 500 kg/ha. The application of both products enhanced the average vine lengths 60 days after transplanting from 327.4 (Control) to 347.3 cm (Treated); an increase of 6.1%. The average vine diameters 47 days after transplanting increased from 9.2 (Control) to 9.4 mm (Treated); an increase of 2.2%. The average plant weights increased from 2.1 (Control) to 2.5 kg (Treated); an increase of 19.0%. The average fruit lengths increased from 47.7 (Control) to 51.4 cm (Treated); an increase of 7.8%. The average fruit diameters increased from 16.5 (Control) to 17.7 cm (Treated); an increase of 7.3%. Most importantly, the average fruit yields increased from 25.41 (Control) to 28.02 t/ha (Treated); an increase of 10.3%.

A product efficacy project was completed in an agricultural farm in Idaho in 2011 to evaluate the efficacy of dry HS (pH = 3.9 and 60.1% HS, including moisture) on corn. This product was broadcasted at a rate of 340 kg/ha during the previous fall to allow the product to breakdown in the soil. Nutrients were broadcasted at rates of $168N\text{-}56P_2O_5\text{-}100K_2O$ kg/ha plus 34N kg/ha applied side-dress during seeding in 2011. The control plot consisted of nutrients only, while the treated plot consisted of both nutrients and dry HS. During harvest in 2011, it was found that crop yields were enhanced with the application of dry HS from 14.50 (Control) to 16.30 (Treated) t/ha; an increase of 12.4%.

Enhanced liquid HA (pH = 6.3, HS = 7.7%, with $9N\text{-}5P_2O_5\text{-}3K_2O$) was evaluated in a product efficacy project on wheat of the Harvest variety, grown in loamy soil with Solonetzic clay underneath (pH = 5.8 and OM = 6.0%) from May to September 2012 in Alberta. The project was completed on test plots, each measuring 3.3 m × 6.7 m (or 22.1 m^2), and was replicated four times. For Control, $34N\text{-}11P_2O_5$ kg/ha was applied to the soil during seeding. For Treated 1, Treated 2, and Treated 3, the same nutrients were applied to soil during seeding together with 9, 27, and 81 l/ha of enhanced liquid HA, respectively. During harvest, yields of wheat were recorded at 3.16, 3.37, 3.35, and 3.20 t/ha, respectively (see Figure 6.22). While the real efficacy of HS could not be evaluated due to the lack of a real control (representing HS only, without nutrients), the general pattern clearly showed that at higher rates of HS and nutrients, yields were not improved and were in fact reduced. The best result was observed at 9 l/ha of the enhanced liquid HA, showing an increase of 6.6%.

Liquid HA (pH = 10.1, HS = 12.0%, and K_2O = 1.8%) was evaluated on soybeans (DK 28-60 variety) grown in silty loam soil (pH = 6.3 and OM = 4.3%) from May to October 2014 in Quebec. For Control, the soil was fertilized with MAP (168 kg/ha) and KCl (168 kg/ha) applied side-banded. For Treated, the same nutrients were applied with 9 l/ha liquid HA applied in row seed place. The whole tested plot was 7.5 ha. During harvest, yields of soybean were recorded at 3.15 (Control) and 3.51 (Treated) t/ha, showing an increase of 11.4%.

The same product was evaluated on corn (DK 34-47 variety) grown in clayey soil (pH = 5.0 and OM = 6.8%) from May to October 2014 in Quebec. For the control plot, the soil was fertilized with $84N\text{-}90P_2O_5\text{-}90K_2O$ kg/ha (broadcasted) and 3.4 t

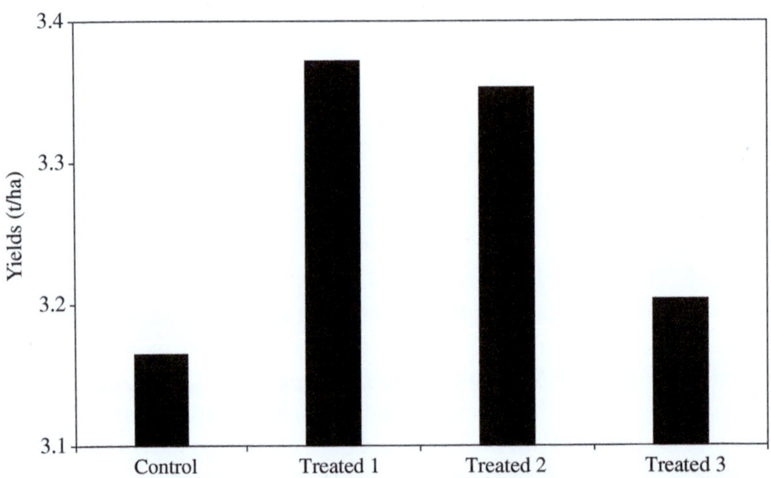

Figure 6.22 Treatment vs. yields.

of lime/ha (side-banded) before seeding. Then, 3N-11P_2O_5-1K_2O kg/ha was applied in row seed place. For the treated plot, the same nutrients were applied to the soil plus 14 l/ha of the liquid product applied broadcast. The whole tested plot was 1.5 ha. During harvest, yields of corn were recorded at 13.67 (control) and 14.49 (treated) t/ha, showing an increase of 6.0%. While it was not really fair to compare both results from Quebec head to head due to different target crops and soil quality, it was noticeable that the better result was achieved for the one with lower soil OM and applied nutrients, representing a less favorable growing environment.

Dry HS (pH = 4.0 and HS = 60.0%, including moisture) was evaluated on soil, located in Alberta, which had a 1380 kg/m^3 bulk density, 1.2% total organic C, 7.1 pH, 1.1 mS/cm EC, 6.3 μg/g NO_3-N, 10.0 μg/g total P, and 194.0 μg/g total K. Awned wheat and fringed brome grass were seeded on the soil in April 2014. The product was then applied on the surface of undisturbed topsoil at a rate of 0.5 t/ha. Woody species (green alder, dogwood, jack pine, balsam polar, golden aspen, gray willow, and soap berry) were planted in May 2015. Fireweed and goldenrod herbs were planted in June 2015. White spruce was then planted in early August 2015. This product efficacy project was completed on a 25 m × 25 m (or 625 m^2) test plot and was replicated four times. Plant cover analysis was conducted per plot in mid-August 2015, with five sub-samples each (see Table 6.25). It was found that the application of dry HS improved total plant covers from 37.0 (mean) ± 26.9 (SD) % to 44.0 (mean) ± 30.1 (SD) %, an increase of 18.9% over the control. The best improvement was achieved for woody species, which showed an increase of 9.5 times over the control.

The effect of liquid HA (pH = 10.1, HS = 12.0%, and K_2O = 1.8%) was evaluated alongside two commercial fertilizers on the growth of lentil and pea seeds grown in a laboratory. Fertilizer 1 contained 2% N, 3% K_2O, 0.3% Mo, and 0.2% Cobalt (Co),

Table 6.25 Plant cover analyses.

	0 t/ha (control)		0.5 t/ha (treated)	
	Mean (%)	**SD (%)**	**Mean (%)**	**SD (%)**
Native grass	5.0	5.2	5.1	6.6
Native herb	3.8	4.8	4.2	7.0
Woody	0.2	0.7	1.9	4.0
Non-native herb	28.0	30.5	32.8	29.3
Total	37.0	26.9	44.0	30.1

Note: SD = standard deviation.

representing the lower nutrient rating. Fertilizer 2 contained 4% N, 16% P_2O_5, 5% K_2O, 2.5% Cu, 2% Zn, and 1% Mn, representing the higher nutrient rating. The seeds were randomly selected (with approximately 190 seeds per lot) and grown on test paper. For Control, distilled water (pH = 6.0) was used. For Treated 1, distilled water plus 1.2 ml liquid HA per kg of seeds was used. For Treated 2, distilled water plus 1.2 ml Fertilizer 1 per kg of seeds was used. For Treated 3, distilled water plus 1.2 ml Fertilizer 2 per kg of seeds was used. Each treatment was replicated twice, and the results were totalled. Radicle emergence was recorded as a percentage of the initial number of seeds per day. For lentil seeds after ten days, Control showed 79% emergence, liquid HA produced 81% emergence, Fertilizer 1 produced 70% emergence, and Fertilizer 2 produced 70% emergence. For lentil seeds after 14 days, Control showed 84% emergence, liquid HA produced 84% emergence, Fertilizer 1 produced 79% emergence, and Fertilizer 2 produced 78% emergence. For pea seeds after ten days, Control showed 72% emergence, liquid HA produced 82% emergence, Fertilizer 1 produced 73% emergence, and Fertilizer 2 produced 78% emergence. For pea seeds after 14 days, Control showed 93% emergence, liquid HA produced 94% emergence, Fertilizer 1 produced 84% emergence, and Fertilizer 2 produced 91% emergence. These findings showed that after 10 days, liquid HA sped up the germination over Control by 2.5% (for lentil seeds) and 13.9% (for pea seeds). After 14 days, all results were practically identical. Fertilizer 1 and Fertilizer 2 did not perform that well, even when compared to Control.

Liquid HA (pH = 10.0, HS = 12.1%, and K_2O = 1.8%) was evaluated on hard red spring wheat grown outdoors in Alberta in Orthic dark brown Chernozem soil of a medium texture. Seeding was completed in May 2019, in which 135-0-11-11 kg/ha N-P_2O_5-K_2O-SO_4 were applied to the soil side-band. For the control plots, nutrients of 8N-29P_2O_5-4K_2O kg/ha were applied in furrows at a rate of 47 l/ha. For the treated plots, the same nutrients were applied in furrows at 37 l/ha (a reduction of 21.3%), together with liquid HA at a rate of 10 l/ha. This project was completed in a randomized block design with four replications for each treatment. Harvest was completed in October 2019. It was found that the mean crop yields for the control and treated plots were not statistically different from one another (at approximately 3.9 t/ha). These findings showed that HS maintained crop yields at reduced nutrients.

The same liquid product was evaluated in the laboratory on the growth of a beneficial native soil fungus. One drop (300 μL) of liquid HA was placed onto an individual 9 cm diameter plate filled with either water agar (no nutrients) or potato dextrose agar. The product strengths were 0 (control, i.e. deionized water), 25, 124, and 620 ppm. The product drop created a visible circular area on the surface of the agar medium. Each plate was then inoculated with *P. bilaiae* at two opposite locations on the edge of the area, identified as Colony 1 and Colony 2. Each plate was evaluated after 7, 14, and 21 days. On water agar plates, the growth

Table 6.26 Colony growth.

Product rate ppm	% Size change compared to day 7			
	Colony 1		Colony 2	
	Day 14	Day 21	Day 14	Day 21
0 (Control)	0	0	0	0
25	100	100	100	100
124	150	183	233	467
620	129	157	113	125

of bacterial colonies was visible within the area where the product was applied. The bacteria came from the product itself. A further analysis indicated that this product contained 430 cfu/g HPC bacteria, with zero or below detection limits of non-beneficial types including *Salmonella*, *E. coli*, fecal coliform, and *S. aureus* (Canadian Humalite International Inc. 2020d). These bacteria were found to not have either beneficial or detrimental effects on the growth of fungus, indicated by the fact that there were no significant changes observed in the fungus colony size. On potato dextrose agar, liquid HA was found to enhance the colony size (see Table 6.26). Compared to day 7, the size of Colony 1 and Colony 2 did not change on the control (0 ppm liquid HA). Compared to day 7, liquid HA at 124 ppm enhanced the size of Colony 1 and Colony 2 by up to 183 and 467%, respectively. While still better than the control, product rates of 25 and 620 ppm did not produce better results.

A subsequent laboratory study was completed using the same product on the growth of *P. bilaiae* on potato dextrose agar plates at a concentration of 10^{-7} g per plate. Liquid HA was applied to each plate at rates of 0 (control), 25, 125, and 620 ppm. Each treatment was replicated five times. All plates were incubated for six days, during which uniform growth of fungal colonies were observed. It was found that the product did not have any negative effects on the fungal growth. Based on the mean values of all replications, the highest humic product rate improved the colony counts by 7.5% over the control.

6.6.12 Beef Cattle

One important key in animal farming is the feed efficiency, which is important for good health, growth, and yields. It may also be associated with a reduction in the gaseous by-products released by animals, which may become a major issue during the operation. From the perspective of organic practice, the use of feed supplements would be preferable to medicinal and other non-natural approaches. The hypothesis here is that supplements improve feed digestion within the animal's digestive system, either chemically and/or biologically, thus optimizing its utilization. It is important to note that feed supplements contained little or no nutrients and therefore had to be added together with feeds.

Among other supplements, HA were found to be effective in promoting the growth of beneficial organisms (Visser 1985) and enhancing nutrients uptake within the digestive system (Visser 1973), thus improving the digestion of feed. HS optimized the growth and yields of poultry, including the production of eggs (Pistova et al. 2016 and Arpasova et al. 2016). It also optimized the growth and yields of swine (Fuchs et al. 1995 and Ji et al. 2006), as well as reducing the production of ammonia (NH_3) gas from swine operations (Ji et al. 2006). Limited data were available for cattle (McMurphy et al. 2011) and more product efficacy projects would be required to draw similar conclusions here.

In this project, dry HS was evaluated on beef cattle. This product had 3.6% water solubility, 3.9 pH, 52.3% OM, and 51.9% HS (based on the colorimetric method, including moisture). It contained 26.8% moisture, 800 kg/m^3 bulk density, and particulates that were 0.2 to 2.4 mm in size; making it suitable when blended with common feed.

The product contained some amounts of crude fiber, ash, carbohydrates, crude protein, aluminum (Al), and silica (Si). It had small to very small amounts of crude fat and nutrients, i.e. Ca, Mg, N, Fe, S, K, Na, P, Cl, B, Mn, Zn, Cu, and molybdenum (Mo). Dry HS had very small amounts of toxic metals, i.e. lead (Pb), arsenic (As), nickel (Ni), cadmium (Cd), and mercury (Hg). Toxic organics and pathogens were negligible (see Table 6.27, Canadian Humalite International Inc. 2020d). By eliminating these variables, a fair evaluation could therefore be made on the effects of HS on the target animals.

The main objective of this project was to study the effect on feed digestion and ruminal microbial activity, the improvement of which should lead to a reduction in gaseous by-product formation. The project was completed from September 2016 until April 2017 in Alberta. Angus × Hereford cross beef heifers were chosen as the animals to be tested.

The first part of the project consisted of in-vitro (laboratory) experiments, in which rumen fluid of 60% barley silage and 40% concentrate was collected from cannulated cattle. Dry HS (plus the addition of buffers) was added to the fluid at 0 (control), 0.9, 1.8, 2.7, and 3.6% of total feed by weight. The incubation times were 12 and 48 hours. Parameters to be analyzed included disappearances of N and starch, dry matter digestibility, neutral detergent fiber digestibility, volatile fatty acids, NH_3-N, and methane (CH_4) production. These were all monitored at 3, 6, 9, 12, 24, and 48 hours after incubation. Each treatment was replicated five times and statistical analyses were completed on a 95% confidence interval.

Table 6.27 Dry HS technical data.

Parameter	Value	Unit
As is analysis (including moisture)		
pH	3.9	–
Water solubility	3.6	%
Bulk density	800	kg/m^3
Particulates > 2.4 mm	< 1	%
Particulates 0.8–2.4 mm	78	%
Particulates 0.2–0.8 mm	21	%
Particulates < 0.2 mm	< 1	%
Moisture	26.8	%
OM	52.3	%
HA + FA (= HS)[a]	51.9	%
Fecal coliform	Below detection limit	#/g
E. coli	Below detection limit	#/g
Salmonella	Negative	–
Dry matter analysis		
Ash	22.8	%
Crude protein	6.4	%
Crude fat	0.7	%
Carbohydrates	17.2	%
Crude fiber	52.9	%
Ca	1.21	%
Mg	1.00	%
N	0.64	%
Fe	0.55	%
S	0.38	%
K	0.32	%
Na	0.28	%
P	0.01	%
Cl	0.01	%
B	364	ppm
Mn	195	ppm
Zn	14	ppm

(*Continued*)

Table 6.27 (Continued)

Parameter	Value	Unit
Dry matter analysis		
Cu	7	ppm
Mo	3	ppm
Al	3.30	%
Si	2.94	%
Pb	12	ppm
As	4	ppm
Ni	3	ppm
Cd	1	ppm
Hg	61	ppb
Dioxins and furans	0.3	ppt

Source: Canadian Humalite International Inc. (2020a);
note:
[a] colorimetric method.

The second part of the project consisted of in-vivo (live animal) experiments, in which 60% barley silage, 35% dry rolled barley, and 5% minerals and vitamins were fed to the animals. Dry HS was added to the feed at 0 (control), 0.01, 0.02, and 0.03% of animal body weight (approximately 780 kg each). Feces and urine were analyzed for five days for digestibility. Rumen fluid was analyzed for pH, volatile fatty acid, NH_3-N, and organisms before and six hours after feeding. Blood samples were analyzed for urea, N, and glucose. A three-day measurement on CH_4 production was completed in respiratory chambers. Each treatment was replicated four times, and statistical analyses were completed on a 95% confidence interval.

The in-vitro experiments found that dry HS significantly increased feed (as true dry matter) disappearances from 627.1A (control) up to 690.4B mg/g feed (10.0% higher) after 12 hours, and from 731.5a (control) up to 789.2b mg/g feed (7.9% higher) after 48 hours (see Figure 6.23). Microbial N increased from 25.8A (control) up to 31.0B mg/g true dry matter feed (20.2% higher) after 12 hours, and from 9.0a (control) up to 13.9b mg/g true dry matter feed (54.4% higher) after 48 hours (see Figure 6.24). CH_4 gas production was significantly reduced from 32.5A (control) down to 29.8B ml/g true dry matter feed (8.3% reduction) after 24 hours, and from 39.4a (control) down to 34.4b ml/g true dry matter feed (12.7% reduction) after 48 hours (see Figure 6.25). Some improvements were also observed on disappearances of N, starch, and neutral detergent fiber. There was also a reduction in NH_3-N and volatile fatty acid. Note: different alphabetical groupings (a, ab, and b, as well as A, AB, and B) indicated that the values were statistically different.

The in-vivo experiments found that dry HS significantly raised ruminal pH towards neutral, from 6.24b (control) to 6.54a (see Figure 6.26). Feed digestibility, blood parameters, NH_3-N, and ruminal organism activity and diversity were not significantly improved. No detrimental effects were observed. CH_4 gas production was reduced from 32.7 (control) down to 31.6 g/kg feed (as true dry matter), and from 47.6 down to 44.7 g/kg feed (as digestible dry matter), representing 3.4 and 6.1 % reductions, respectively (see Figure 6.27). Note: different alphabetical groupings (a, ab, and b) indicated that the values were statistically different.

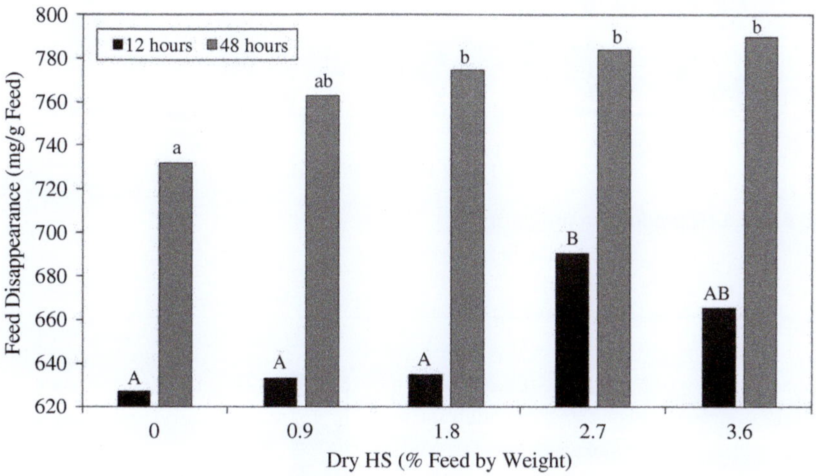

Figure 6.23 Feed disappearance (in-vitro).

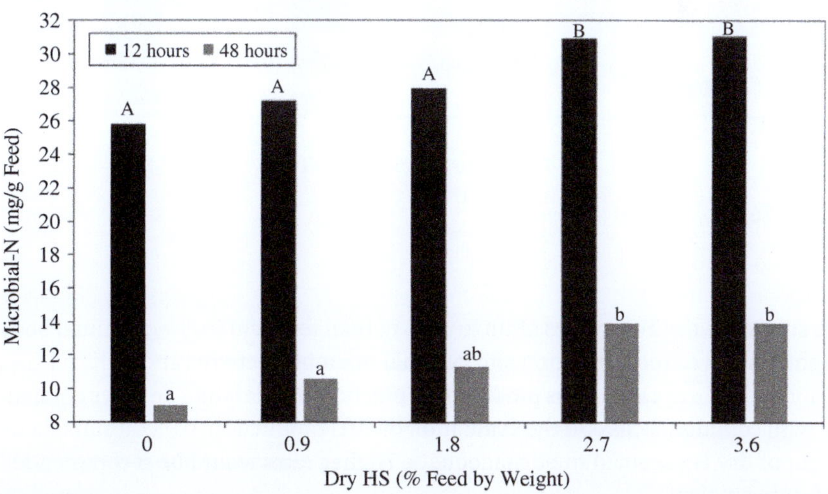

Figure 6.24 Microbial-N (in-vitro).

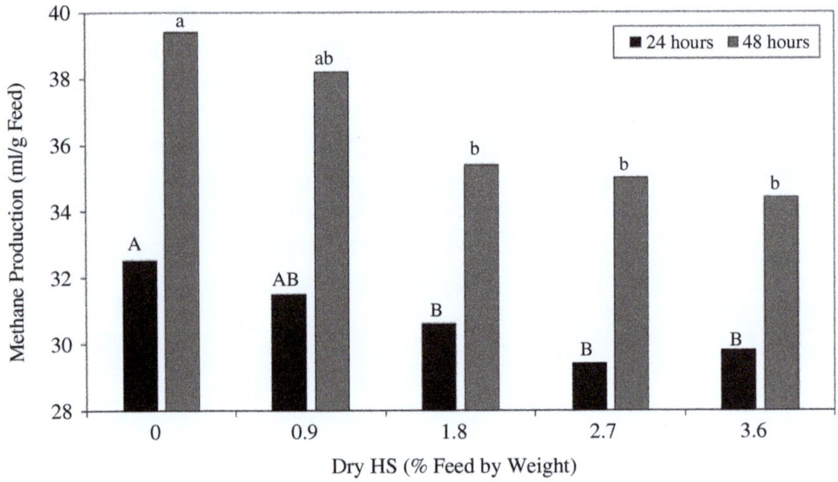

Figure 6.25 CH_4 production (in-vitro).

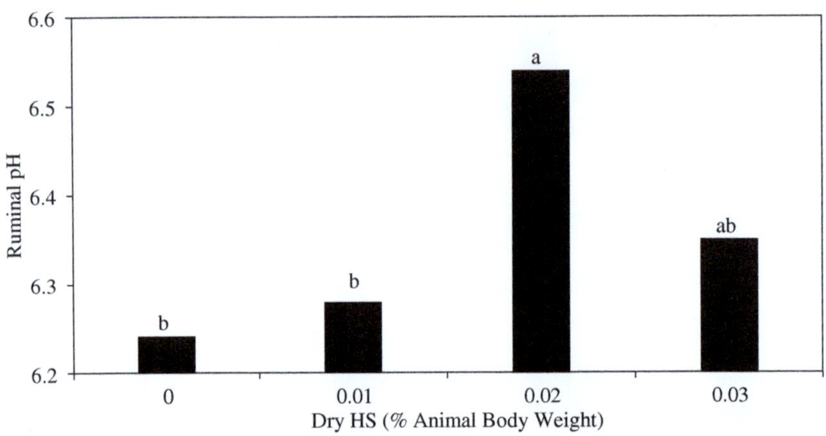

Figure 6.26 Ruminal pH (in-vivo).

In summary, dry HS at rates of up to 3.6% of total feed and 0.03% of animal body weight, improved feed digestion and ruminal microbial activity, and reduced CH_4 production in beef cattle. This product was practical for blending and was compatible with common feeds. On the evaluation of CH_4 production, 0.03% animal body weight of dry HS seemed to be inadequate. Higher rates would be recommended for future studies.

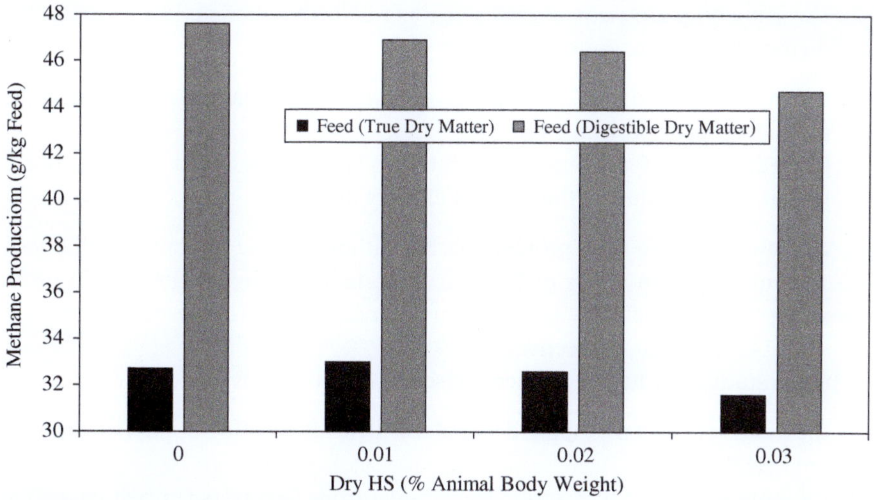

Figure 6.27 CH_4 production (in-vivo).

6.7 Cost–Benefit Study

A cost–benefit study was completed to evaluate the overall feasibility of the use of humic products in agriculture. Return on investment (ROI), defined as [(gross revenue – production costs) ÷ production costs] × 100%, was calculated for crop growing without humic products (control) and was compared with ROI for crop growing with humic products (treated). For this study, product efficacy projects completed on barley (2018), spring wheat (2019), canola (2014), and corn (2011 and 2014) were selected, representing different crops with low to high target prices. Dry HS, pelletized HS, liquid HA, and liquid FA were evaluated, representing different humic products with low to high product prices: $0.49/kg (in 1 t supersacks), 0.58/kg (in 1 t supersacks), 1.10/l (in 1000 l totes), and 18.10/l (in 1000 l totes), respectively. Note: all prices were distributor prices including $0.12/kg (or l) for one full truckload transported across North America (Canadian Humalite International Inc. 2020d).

End users (farmers) with commercial farms of 512 to 7168 ha were contacted to obtain detailed production costs and the expected gross revenue for each crop grown in recent years. Unfortunately, the information was either incomplete or inconsistent from one end user to another, which made its use in this study unreliable. Instead, guidelines developed by Manitoba Agriculture and Resource Development (2020a) were adopted and the information was deemed applicable for the same crop grown in a similar region of North America. In this study, the following assumptions were made:

1) the increase in crop yields would be proportional to that observed during the product efficacy project; and
2) the reduction of fertilizers that maintained crop yields would be proportional to that observed during the product efficacy project.

As shown in Table 6.28, growing barley without humic products for the year 2020 would result in a loss of 11.2% in ROI, mainly due to a low target price. Growing spring wheat, canola, and corn without humic products would give 4.6, 5.3, and 14.0% in ROI, respectively.

In this study, four humic application scenarios were considered for each crop:

1) the humic product was applied together with fertilizers (in one pass) to improve crop yields;
2) the humic product was applied separately from fertilizers (in two passes) to improve crop yields;
3) the humic product was applied together with fertilizers (in one pass) to reduce fertilizers input while maintaining crop yields; and
4) the humic product was applied separately from fertilizers (in two passes) to reduce fertilizers input while maintaining crop yields.

Scenario 1 would increase chemical costs (humic optimum rate × product price). Scenario 2 would increase chemical costs (humic optimum rate × product price) as well as fuel and machinery costs. Scenarios 1 and 2 would increase gross revenues. Scenario 3 would increase chemical costs (humic optimum rate × product price) but reduce fertilizer costs (% reduction × fertilizer costs). Scenario 4 would increase chemical costs (humic optimum rate × product price), as well as fuel and machinery costs, but reduce fertilizer costs (% reduction × fertilizer costs). Scenarios 3 and 4 would maintain gross revenues.

The 2018 product efficacy project on barley showed that 100% nutrients were already more than adequate and would not help crop yields. Therefore, Scenarios 3 and 4 were not evaluated in this study. At the reduced rate (33% nutrients), the crop yield was recorded at 1.51 t/ha. With the application of liquid FA broadcast to seedlings at 70 ml/ha, the yield was enhanced to 1.59 t/ha (or 5.3% over the control). Scenario 1 resulted in an increase in operating costs of 70 ml/ha × \$18.10/l = \$1.27/ha and an increase in the target yield to 4.15 × 1.053 = 4.37 t/ha. This scenario resulted in a 6.7% loss in ROI, which represented an improvement of 40.2% over the control (with no humic application). Scenario 2 resulted in an increase in the operating cost of \$1.27/ha (humic) + \$91.60/ha (fuel and machinery) = \$92.87/ha, and increased the target yield to 4.37 t/ha. This scenario resulted in a 15.4% loss in ROI, which was worse than that of the control. It was concluded

Table 6.28 Production costs and gross revenue without humic.

	Barley	Spring wheat	Canola	Corn
Operating costs				
Seed ($/ha)	37.03	59.26	154.32	212.96
Fertilizer ($/ha)	123.80	157.51	199.33	245.48
Chemical[a] ($/ha)	108.37	116.05	75.70	64.81
Humic ($/ha)	N/A	N/A	N/A	N/A
Fuel ($/ha)	59.80	56.25	52.07	69.85
Machinery[b] ($/ha)	31.80	31.80	31.80	31.80
Insurance[c] ($/ha)	39.62	34.74	38.15	69.70
Others ($/ha)	19.14	19.14	19.14	19.14
Land taxes ($/ha)	37.03	37.03	37.03	37.03
Drying ($/ha)	0.00	0.00	0.00	86.42
Interest ($/ha)	12.56	14.07	16.72	23.01
Total ($/ha)	469.15	525.85	624.26	860.20
Fixed costs				
Land ($/ha)	165.58	165.58	165.58	165.58
Machinery ($/ha)	166.19	166.19	166.19	186.12
Storage ($/ha)	22.44	16.54	11.78	37.85
Total ($/ha)	354.21	348.31	343.55	389.55
Labor costs				
Total ($/ha)	65.19	65.19	65.19	65.19
Total costs ($/ha)	888.55	939.35	1,033.00	1,314.94
Target price ($/t)	190	250	477	180
Target yield (t/ha)	4.15	3.93	2.28	8.33
Gross revenue ($/ha)	788.50	982.50	1,087.56	1,499.40
ROI (%)	–11.2	4.6	5.3	14.0

Note:
[a] herbicides, fungicides, and insecticides;
[b] operating and lease;
[c] crops and hails; N/A = not applicable.

Table 6.29 Production costs and gross revenue for barley.

		Scenario 1	Scenario 2
	No humic	Liquid FA	
Operating costs ($/ha)	469.15	470.42	562.02
Fixed costs ($/ha)	354.21	354.21	354.21
Labor costs ($/ha)	65.19	65.19	65.19
Total costs ($/ha)	888.55	889.82	981.42
Target price ($/t)	190	190	190
Target yield (t/ha)	4.15	4.37	4.37
Gross revenue ($/ha)	788.50	830.3	830.3
ROI (%)	−11.2	−6.7	−15.4

Note: no evaluation completed for Scenarios 3 and 4.

that liquid FA enhanced the ROI when applied together with fertilizers in one pass (see Table 6.29).

The 2019 product efficacy project on wheat showed that at 100% nutrients, applying pelletized HS side-band at a rate of 28 kg/ha enhanced the yield by 2.6% (up from 5.88 to 6.03 t/ha). Scenario 1 resulted in an increase in operating costs by 28 kg/ha × $0.58/kg = $16.24/ha and an increase in the target yield to 3.93 × 1.026 = 4.03 t/ha. This scenario resulted in a 5.4% ROI, which was 17.4% higher than that of the control (with no humic application). Scenario 2 resulted in an increase in operating costs by $16.24/ha (humic) + $88.05 (machinery and fuel), and an increase in the target yield to 4.03 t/ha. This scenario resulted in a 3.5% loss in ROI (see Table 6.30).

The 2019 product efficacy project on wheat showed that at 75% nutrients, applying pelletized HS side-band at a rate of 28 kg/ha maintained yields at 5.88 t/ha. Scenario 3 resulted in an increase in operating costs of $16.24/ha but reduced fertilizer costs by 25% × $157.51 = $39.38/ha. This reduced operating costs by $23.14/ha while maintaining crop yields. Scenario 3 resulted in a 7.2% ROI, which was 56.5% higher than that of the control. Scenario 4 maintained crop yields and resulted in an increase in operating costs of $16.24/ha (humic) + $88.05 (fuel and machinery) – $39.38/ha (fertilizer) = $64.91/ha. This scenario resulted in a 2.2% loss in ROI (see Table 6.30).

It was concluded that pelletized HS enhanced the ROI when it was intended for the increase of crop yields or to maintain crop yields at reduced nutrients, as long as it was applied together with fertilizers in one pass.

Table 6.30 Production costs and gross revenue for wheat.

		Scenario 1	Scenario 2
	No humic	Pelletized HS	
Operating costs ($/ha)	525.85	542.09	630.14
Fixed costs ($/ha)	348.31	348.31	348.31
Labor costs ($/ha)	65.19	65.19	65.19
Total	939.35	955.59	1,043.64
Target price ($/t)	250	250	250
Target yield (t/ha)	3.93	4.03	4.03
Gross revenue ($/ha)	982.50	1,007.50	1,007.50
ROI (%)	4.6	5.4	−3.5

		Scenario 3	Scenario 4
	No humic	Pelletized HS	
Operating costs ($/ha)	525.85	502.71	590.76
Fixed costs ($/ha)	348.31	348.31	348.31
Labor costs ($/ha)	65.19	65.19	65.19
Total	939.35	916.21	1,004.26
Target price ($/t)	250	250	250
Target yield (t/ha)	3.93	3.93	3.93
Gross revenue ($/ha)	982.50	982.50	982.5
ROI (%)	4.6	7.2	−2.2

The 2014 product efficacy project on canola showed that at 100% nutrients, applying liquid HA broadcast at a rate of 9 l/ha enhanced crop yields by 13.8% (at 2.47 t/ha) over the control (at 2.17 t/ha). Scenario 1 resulted in an increase in operating costs of 9 l/ha × $1.10/l = $9.90/ha and an increase in the target yield to 2.28 × 1.138 = 2.59 t/ha. This scenario resulted in an 18.5% ROI, which was 3.5 times higher than that of the control (with no humic application). Scenario 2 resulted in an increase in operating costs of $9.90/ha (humic) + $83.87/ha (fuel and machinery) = $93.77/ha, and an increase in the target yield to 2.59 t/ha. This scenario resulted in a 9.6% ROI, which was 81.1% higher than that of the control (see Table 6.31).

Table 6.31 Production costs and gross revenue for canola.

		Scenario 1	Scenario 2
	No humic	Liquid HA	
Operating costs ($/ha)	624.26	634.16	718.03
Fixed costs ($/ha)	343.55	343.55	343.55
Labor costs ($/ha)	65.19	65.19	65.19
Total costs ($/ha)	1,033.00	1,042.90	1,126.77
Target price ($/t)	477	477	477
Target yield (t/ha)	2.28	2.59	2.59
Gross revenue ($/ha)	1,087.56	1,235.43	1,235.43
ROI (%)	5.3	18.5	9.6

		Scenario 3	Scenario 4
	No humic	Liquid HA	
Operating costs ($/ha)	624.26	604.26	688.13
Fixed costs ($/ha)	343.55	343.55	343.55
Labor costs ($/ha)	65.19	65.19	65.19
Total costs ($/ha)	1,033.00	1,013.00	1096.87
Target price ($/t)	477	477	477
Target yield (t/ha)	2.28	2.28	2.28
Gross revenue ($/ha)	1,087.56	1,087.56	1,087.56
ROI (%)	5.3	7.4	-0.8

The 2014 product efficacy project on canola showed that at 85% nutrients, applying liquid HA broadcast at a rate of 9 l/ha maintained yields between 2.17 and 2.21 t/ha. Scenario 3 resulted in an increase in operating costs of $9.90/ha but reduced fertilizer costs by 15% × $199.33 = $29.90/ha. This reduced operating costs by $20.00/ha while maintaining crop yields. Scenario 3 resulted in a 7.4% ROI, which was 39.6% higher than that of the control. Scenario 4 maintained crop yields and resulted in an increase in operating costs by $9.90/ha (humic) + $83.87

(fuel and machinery) – \$29.90/ha (fertilizer) = \$63.87/ha. This scenario resulted in a 0.8% loss in ROI (see Table 6.31).

It was concluded that liquid HA enhanced the ROI when it was intended for the increase of crop yields or when it was applied together with reduced fertilizers in one pass to maintain yields.

As shown in Table 6.32, the 2011 product efficacy project on corn showed that at 100% nutrients, applying dry HS broadcast at a rate of 340 kg/ha enhanced crop

Table 6.32 Production costs and gross revenue for corn.

		Scenario 1	Scenario 2
	No humic	Dry HS	
Operating costs (\$/ha)	860.20	1,026.80	1,128.45
Fixed costs (\$/ha)	389.55	389.55	389.55
Labor costs (\$/ha)	65.19	65.19	65.19
Total costs (\$/ha)	1,314.94	1,481.54	1,583.19
Target price (\$/t)	180	180	180
Target yield (t/ha)	8.33	9.36	9.36
Gross revenue (\$/ha)	1,499.40	1,684.80	1,684.80
ROI (%)	14.0	13.7	6.4

		Scenario 1	Scenario 2
	No humic	Liquid HA	
Operating costs (\$/ha)	860.20	875.60	977.25
Fixed costs (\$/ha)	389.55	389.55	389.55
Labor costs (\$/ha)	65.19	65.19	65.19
Total costs (\$/ha)	1,314.94	1,330.34	1,431.99
Target price (\$/t)	180	180	180
Target yield (t/ha)	8.33	8.82	8.82
Gross revenue (\$/ha)	1,499.40	1,587.60	1,587.60
ROI (%)	14.0	19.3	10.9

Note: no evaluation completed for Scenarios 3 and 4.

yields by 12.4% (at 16.3 t/ha) over the control (at 14.5 t/ha). Scenario 1 resulted in an increase in operating costs of 340 kg/ha × \$0.49/kg = \$166.60/ha and an increase in the target yield to 8.33 × 1.124 = 9.36 t/ha. This scenario resulted in a 13.7% ROI, which was slightly worse than that of the control (with no humic application). Scenario 2 resulted in an increase in operating costs of \$166.60/ha (humic) + \$101.65/ha (fuel and machinery) = \$268.25/ha, and an increase in the target yield to 9.36 t/ha. This scenario resulted in a 6.4% ROI, which was worse than that of the control. It was concluded that dry HS was not economically worthwhile in this regard. However, dry HS might still benefit the soil over the longer term by improving its quality. Scenarios 3 and 4 were not evaluated due to the lack of data.

As also shown in Table 6.32, the 2014 product efficacy project on corn showed that at 100% nutrients, applying liquid HA broadcast at a rate of 14 l/ha enhanced crop yields by 6.0% (at 14.49 t/ha) over the control (at 13.67 t/ha). Scenario 1 resulted in an increase in operating costs of 14 l/ha × \$1.10/l = \$15.40/ha and an increase in the target yield to 8.33 × 1.06 = 8.82 t/ha. This scenario resulted in a 19.3% ROI, which was 37.9% higher than that of the control (with no humic application). Scenario 2 resulted in an increase in operating costs of \$15.40/ha (humic) + \$101.65/ha (fuel and machinery) = \$117.05/ha, and an increase in the target yield to 8.82 t/ha. This scenario resulted in a 10.9% ROI, which was worse than the control. It was concluded that liquid HA enhanced the ROI when applied together with fertilizers in one pass. Scenarios 3 and 4 were not evaluated due to the lack of data.

The cost–benefit study presented here provided estimates on how ROI could be improved with the incorporation of humic products within the fertilization program. While dollar values and target yields might change, similar exercises could still be performed based upon the same principle, meaning that this study would be applicable for other cases. In general, liquid products performed better because of high-water solubility. Much smaller application rates would be required since humic components were instantaneously available to crops. Dry humic products would prove to be more beneficial to soil, improving its quality over the longer term. This study also emphasized the importance of product compatibility between humic products and common fertilizers in order for both to be blended and applied on the field in one pass. Another critical factor to consider was the target yield and price of the target crop, as well as operating costs, including fertilizer and fuel costs.

6.8 Closure

The product performance evaluation is to show the practicality and efficacy of commercial humic products for soil, crops, and livestock. Most product efficacy projects have been completed on soil and crops. Practicality means that the products can be applied on the field together with common nutrients using typical field applicators without creating any problems. Efficacy means that the products show clear benefits when applied on the field, with the main focus of improving the physical properties of the soil and enhancing crop yields, or maintaining yields at reduced nutrient inputs.

The basic properties of low-rank coal and humic products have to be taken into consideration as follows:

1) Low-rank coal contains soft and porous particulates with a high affinity for water. It contains both external moisture (moisture on the surface) and internal moisture (moisture within the particulates). Mixing dry HS with certain nutrients (urea 46-0-0, for example) at certain ratios may result in mushy blends due to the release of internal moisture to the surface. Within a field applicator, dry HS is best stored in a separate container from other nutrients and applied in a dedicated line (using the double shoot system), unless a compatibility test has been completed in the laboratory with satisfactory results (this is always recommended prior to application). Since dry HS is relatively soft, field applicators containing augers or other abrasive elements have to be avoided in order to maintain particulate integrity. Pelletized humic products are harder than common dry products, which may help with particulate integrity.
2) Liquid humic products are not a true solution. The mixture is more like a suspension and/or colloidal, so that it may separate in water over time. HA are components of HS soluble in water at high pH, FA are soluble at any pH, and humin is not soluble at any pH. Ideally, liquid HA does not contain humin, and liquid FA contains only FA. However, small amounts of humin may be found in liquid HA and some HA may be found in liquid FA. This results in the possible presence of settlement within the product, especially at boundary pH levels (i.e. at 6 to 7 pH for liquid HA and 4 to 5 pH for liquid FA). When blending liquid HA or FA with N, P, Ca, and Mg at certain ratios, gels and precipitates may form due to the reaction between HS and these nutrients. Metal salts (such as Zn, Fe, Mn) significantly lower the solution's pH, affecting its solubility. In most cases, humic products, common nutrients, and a large amount of water are blended in a tank before being applied on the field. Still, caution has to be taken, most importantly, with respect to the possibility of a significant drop in pH of the whole blend. The best practice is to keep the blend of the liquid humic product and a large amount of water in a separate tank from the one filled with a blend of common nutrients. Always ensure that screens and nozzles do not plug. A compatibility test is always recommended prior to field application.

In general, product compatibility evaluation emphasizes the importance of developing humic products that are compatible with common nutrients. The information presented in this chapter can be used as a starting point. Emulsifiers or anti-gelling/precipitating agents can be added to the products to improve their solubility. Careful consideration has to be taken with their status with respect to the organic registration (see Section 4.6.2).

Field projects have been completed for different commercial humic products (i.e. dry HS, liquid HA, liquid FA, and pelletized HS), which can be grouped as raw and processed products, containing low and high water solubility values, respectively. Raw products require larger application rates before positive responses can be observed than do processed products. Raw products, however, are retained better in soil, in which their effects last longer than those of processed products. In most cases, these products are applied in the fall for the following year's growing season.

Good projects are the ones with adequate information on the relevant growing media and environment. Each treatment is replicated and completed randomly on site, so that all data can be properly analyzed for the relevant statistical parameters. Not only is a control needed for each project, most importantly, nutrient corrections have to be made for all treatments. Humic products (especially processed products) may contain significant amounts of nutrients originating from extracting chemicals used during their manufacture, or intentionally added to boost their nutrient values. These nutrient amounts have to be compensated for in each treatment, in order to allow for a fair comparison between treatments.

Different types of soil and crops most likely show different responses. The focus should firstly be on the improvement of soil quality (particularly on mitigating the effect of salinity and on reduction of soil compaction) and secondly on the enhancement of crop productivity (producing higher yields, or maintaining them at reduced nutrients with the incorporation of humic products within the fertilization program). The effect of HS has been found to be best for the less favorable growing environment, i.e. soil with high salinity contents, crops grown in soil with low OM levels and nutrients, and crops grown with reduced nutrient inputs. Most interestingly, several product efficacy projects have found that higher rates of humic products and/or nutrients do not always produce better responses. In fact, negative responses have been observed at very high rates of humic products and/or nutrients. This indicates that there is an optimum rate for a specific application that has to be evaluated carefully in a dedicated project. Fortunately, optimum rates in reasonable ranges are usually applicable for similar types of soil and crops.

Only one product efficacy project has been completed for livestock on the reduction of gaseous by-product formation. More projects are required on livestock productivity, i.e. on increasing yields or maintaining yields at reduced feeds. Yields in this case can be animal body weight, egg production, or milk production.

For each product being evaluated, a unique application label can be generated. It should include the following information:

1) manufacturer identification;
2) product name and HS and/or nutrient claims (if any);
3) a short summary of its purpose and target soil, crops, or livestock;
4) recommended product rates in reasonable ranges;
5) a short summary of its safety and compatibility with selected nutrients or chemicals;
6) a short summary of its registration under a certain jurisdiction;
7) basic product parameters and packaging forms; and
8) instructions on product handling, including during field application.

Application labels are to be presented together with safety and registration documents.

Typically, a product application label is generated for a specific jurisdiction. Claims concerning HS can be tricky, since a given jurisdiction may recognize some analytical methods for HA or FA differently, resulting in different product claims for exactly the same product. The same product may have different functions, like improving the physical properties of soil, enhancing crop yields, or reducing the gaseous by-products generated by livestock. A separate application label is required for each application, which is generated from a specific product efficacy project. In this regard, numerous projects are constantly needed to provide extensive information on the use of commercial humic products for various functions.

In general, the application of liquid humic products broadcast to improve crop yields has been found to be economically worthwhile due to their high water solubility. Small application rates are required that positive responses can be observed immediately. When applied to maintain crop yields at reduced nutrient inputs, this application becomes advantageous if the nutrient costs are high. Reducing nutrient inputs, however, benefits soil quality in the long run.

Due to their low water solubility, dry humic products, when applied broadcast, require high application rates in order for positive results to be observed. It may take a few months for the products to break down in soil before crops can benefit from HS. This indicates that dry humic products are more beneficial to soil and crops over the long term. A cost–benefit analysis completed based on multiple years of crop data is therefore necessary to evaluate the true economy of this application. Dry humic products can be applied side-band close to the seeds at smaller rates. This application has generally been found to be economically worthwhile.

The field evaluation also shows the importance of product compatibility so that humic products and common fertilizers can be blended together and applied on the field in one pass. An additional pass results in an increase in operating costs, which should be avoided when possible.

Reference

Agriculture and Agri-Food Canada (2017). *What is hydroponics?* 1–2. Ottawa, ON.

APHA 2510 (1992). Conductivity. In: *Standard Methods for the Examination of Water and Wastewater*, 18th ed. (eds. A.E. Greenberg, L. S. Clesceri, and A.D. Eaton), 2-43 to 2-46. Washington, DC: American Public Health Association.

APHA 3120B (1992). Metals by plasma emission spectroscopy: Inductively Coupled Plasma (ICP) method. In: *Standard Methods for the Examination of Water and Wastewater*, 18th ed. (eds. A.E. Greenberg, L. S. Clesceri, and A.D. Eaton), 3-34 to 3-40. Washington, DC: American Public Health Association.

APHA 4110B (1992). Determination of anions by ion chromatography: Ion Chromatography with chemical suppression of eluent conductivity. In: *Standard Methods for the Examination of Water and Wastewater*, 18th ed. (eds. A.E. Greenberg, L.S. Clesceri, and A.D. Eaton), 4-1 to 4-5. Washington, DC: American Public Health Association.

APHA 4500-H^+ (1992). pH value. In: *Standard Methods for the Examination of Water and Wastewater*, 18th ed. (eds. A.E. Greenberg, L.S., Clesceri, and A.D. Eaton), 4-65 to 4-69. Washington, DC: American Public Health Association.

APHA 5310B (1992). Total organic carbon (TOC): high-temperature combustion method. In: *Standard Methods for the Examination of Water and Wastewater*, 18th ed. (eds. A.E. Greenberg, L. S. Clesceri, and A.D. Eaton), 5-21 to 5-23. Washington, DC: American Public Health Association.

APHA 9215D (1992). Heterotrophic plate count: membrane filter method. In: *Standard Methods for the Examination of Water and Wastewater*, 18th ed. (eds. A.E. Greenberg, L.S. Clesceri, and A. D., Eaton), 9-38 to 9-39. Washington, DC: American Public Health Association.

Arpasova, H., Kacaniova, M., Pistova, V., et al. (2016). Effect of probiotic and humic acid on egg production and quality parameters of laying hens eggs. *Animal Science and Biotechnologies* 49(2): 1–9.

Asik, B.B., Turan, M.A., Celik, H., and Katkat, A.V. (2009). Effects of humic substances on plant growth and mineral nutrients uptake of wheat (*Triticum durum* cv. Salihli) under conditions of salinity. *Asian Journal of Crop Science* 1(2): 87–95.

Beckman Coulter (2011). Instruction for use. *LS 13 320 Laser Diffraction Particle Size Analyzer* 1–246. Brea, CA.

Brechner, M. and Both, A.J. (2013). *Hydroponic Lettuce Handbook* 6–18. Ithaca, NY: Cornell University Controlled Environment Agriculture Program, Cornell University.

Canadian Humalite International Inc. (2020a). Raw material and products. *Laboratory Analyses and Technical Data* 1–656. Edmonton, AB.

Canadian Humalite International Inc. (2020b). Commercial products. *Field Applicability* 1–44. Edmonton, AB.

Canadian Humalite International Inc. (2020c). Commercial products. *Registration and Application Labels* 1–286. Edmonton, AB.

Canadian Humalite International Inc. (2020d). Commercial products. *Efficacy Projects* 1-179. Edmonton, AB.

Cardon, G.E., Davis, J.G., Bauder, T.A., and Waskom, R.M. (2007). Managing saline soils. *Fact Sheet No. 0.503* 1–5. Fort Collins, CO: Colorado State University.

Chen, Y. and Aviad, T. (1990). Effects of humic substances on plant growth. In: *Humic Substances in Soil and Crop Sciences: Selected Readings* (eds. P. MacCarthy, C.E. Clapp, R.L. Malcolm, and P.R. Bloom), 161–186. Madison, WI: American Society of Agronomy/Soil Science Society of America.

Cornell University (2010). Competency area 4: nutrient sources, analyses, application methods. *Northeast Region Certified Crop Adviser (NRCCA) Study Resources* 1–4. Ithaca, NY.

Dudley, J.B., Pertuit Jr., A.J., and Toler, J.E. (2004). Leonardite influences zinnia and marigold growth. *HortScience* 39(2): 251–255.

Duiker, S.W. (2004). *Effects of Soil Compaction* 1–12. University Park, PA: College of Agricultural Sciences, Pennsylvania State University.

EEC A.6 (1992). Water solubility. Dir 92/69/EEC (O.J. L383 A). In: *Official Journal of the European Communities No. L 383/113* (member K. Van Miert), 1–9. Aberdeen, UK: Commission Directive 92/69/EEC.

Evans, M.R. and Li, G. (2003). Effect of humic acids on growth of annual ornamental seedling plugs. *HortTechnology* 13(4): 661–665.

Evans, R.G. (2001). *Center Pivot Irrigation* 1–22. Sidney, MT: USDA-Agricultural Research Service.

Follett, R.H., Murphy, L.S., and Donahue, R.L. (1981). *Fertilizers and Soil Amendments* 302–392 and 424–457. Englewood Cliffs, NJ: Prentice-Hall.

Fuchs, B., Orda, J., Pres, J., and Muchowicz, M. (1995). The effect of feeding piglets up to the 100th day of their life with peat preparation on their growth and physiological and biochemical indices. *Archivum Veterinarium Polonicum* 35(1-2): 97–107.

Garvey, S. (2017). Ag-market sees new interest in dry-product spreaders. http://www.country-guide.ca/columns/machinery-guide/ag-market-sees-new-interest-in-dry-product-spreaders/ (accessed 12 February 2020).

Goyal, M.R. 2015. Principles of micro irrigation. In: *Sustainable Micro Irrigation: Principles and Practices Vol. 1, Ch. 11* (ed. M. R. Goyal), 198–236. Oakville, ON: Apple Academic Press.

Grattan, S.R. and Grieve, C.M. (1999). Salinity-mineral nutrient relations in horticultural crops. *Scientia Horticulturae* 78: 127–157.

Green, M. (1999a). Direct seeding systems: terms, definitions and explanations. *Agdex 570-7* 1–6. Edmonton, AB: Alberta Agriculture, Food, and Rural Development.

Green, M. (1999b). Ground opener systems. *Agdex 519-24* 1–7. Edmonton, AB: Alberta Agriculture, Food, and Rural Development.

Green, M. and Eliason, M. (1999). Equipment issues in crop residue management for direct seeding. *Agdex 519-4* 1–3. Edmonton, AB: Alberta Agriculture, Food, and Rural Development.

Grisso, R., Hipkins, P., Askew, S.D., et al. (2013). Nozzles: selection and sizing. *Publication 442-032* 1–12. Petersburg, VA: Virginia Polytechnic Institute and State University.

Hartwigsen, J.A. and Evans, M.R. (2000). Humic acid seed and substrate treatments promote seedling root development. *HortScience* 35(7): 1231–1233.

ISO 19822:2018(E) (2018). *Fertilizers and soil conditioners-determination of humic and hydrophobic fulvic acids concentrations in fertilizer materials*. Geneva, Switzerland: International Organization for Standardization.

Ji, F., McGlone, J.J., and Kim, S.W. (2006). Effects of dietary humic substances on pig growth performance, carcass characteristics, and ammonia emission. *Journal of Animal Science* 84: 2482–2490.

Khan, M.A., Ungar, I.A., and Showalter, A.M. (2000). Effects of salinity on growth, water relations and ion accumulation of the subtropical perennial halophyte *Atriplex griffthii* var. *Stocksii*. *Annals of Botany* 85(2): 225–232.

Krogmeier, M.J. and Bremner, J.M. (1989). Effects of phenolic acids on seed germination and seedling growth in soil. *Biology and Fertility of Soils* 8: 116–122.

Lamar, R., Olk, D.C., Mayhew, L., and Bloom, P.R. (2014). A new standardized method for quantification of humic and fulvic acids in humic ores and commercial products. *Journal of AOAC (Association of Official Analytical Chemists) International* 97(3): 721–730.

Lethbridge College (2020). Aquaculture centre of excellence. www.lethbridgecollege.ca/departments/aquaculture-centre-of-excellence (accessed 6 March 2020).

Malkawi, A.I.H., Alawneh, A.S., and Abu-Safaqah, O.T. (1999). Effects of organic matter on the physical and the physicochemical properties of an Illitic soil. *Applied Clay Science* 14: 257–278.

Manitoba Agriculture and Resource Development (2020a). 2020 cost of production. *Crops* 1–12. Winnipeg, MB.

Manitoba Agriculture and Resource Development (2020b). 2020/2021 cost of production. *Farm Machinery* 1–44. Winnipeg, MB.

McKenzie, R.H. (2010a). Management of sodic soil in Alberta. *Agdex 518-20* 1–6. Edmonton, AB: Alberta Agriculture.

McKenzie, R.H. (2010b). Agricultural soil compaction: causes and management. *Agdex 510-1* 1–10. Lethbridge, AB: Alberta Agriculture and Rural Development.

McMurphy, C.P., Duff, G.C., Sanders, S.R., et al. (2011). Effects of supplementing humates on rumen fermentation in Holstein steers. *South African Journal of Animal Science* 41(2): 134–140.

Morrison, R.I. (1963). Products of the alkaline nitrobenzene oxidation of soil organic matter. *Journal of Soil Science* 14(2): 201–216.

O'Donnell, R.W. 1973. The auxin-like effects of humic preparations from leonardite. *Soil Science* 116(2): 106–112.

Pistova, V., Arpasova, H., Hrncar, C., et al. (2016). The effect of the humic acid and herbal additive supplement on production parameters of broiler chicken. *Animal Science and Biotechnologies* 49(2): 166–169.

Rauthan, B.S. and Schnitzer, M. (1981). Effects of a soil fulvic acid on the growth and nutrient content of cucumber (*Cucumis satimus*) plants. *Plant and Soil* 63(3): 491–495.

Reynolds, A.G., Wardle, D.A., Drought, B., and Cantwell, R. (1995). Gro-Mate soil amendment improves growth of greenhouse-grown 'Chardonnay' grapevines. *HortScience* 30(3): 539–542.

Saskatchewan Crop Insurance Corporation (2020). Prices. http://www.scic.ca/?/ci/prices (accessed 26 February 2020).

Shafeek, M.R, Helmy, Y.I., Nadia, M.O., and Rizk, F.A. (2013). Effect of foliar fertilizers with nutritional compound and humic acid on growth and yield of broad bean plants under sandy soil conditions. *Journal of Applied Science Research* 9(6): 3674–3680.

Stevenson, F.J. (1965). Gross chemical fractionation of organic matter. In: *Methods of Soil Analysis. Part 2. Chemical and Microbiological Properties, Agronomy Monograph 9.2* (ed. C.A. Black), 1409–1421. Madison, WI: Soil Science Society of America.

Stevenson, F.J. (1994). *Humus Chemistry: Genesis, Composition, Reactions*, 2nd ed., 24–58. New York, NY: Wiley-Interscience.

Swift, R.S. (1996). Organic matter characterization. In: *Methods of Soil Analysis, Part 3 Chemical Method, Ch. 35* (eds. D.L. Sparks, A.L. Page, P.A. Helmke, R.H. Loeppert, P.N. Soltanpour, M.A. Tabatabai, C.T. Johnston, and M. E. Sumner), 1011–1069. Madison, WI: Soil Science Society of America/American Society of Agronomy.

Thurman, E.M., Wershaw, R.L., Malcolm, R.L., and Pinckney, D.J. (1982). Molecular size of aquatic humic substances. *Organic Geochemistry* 4:27–35.

Valley (2018). Valley irrigation. http://www.valleyirrigation.com (accessed 25 April 2019).

Verdi, P., Cano, M., and Liem, E. (2015). Low concentrations of humic substances enhanced the growth of *Lactuca sativa* L. *Journal of Experimental Agriculture International (formerly American Journal of Experimental Agriculture)* 5(6): 555–561.

Visser, S.A. (1973). Some biological effects of humic acids in the rat. *Acta Biologica et Medica Germanica* 31: 569–581.

Visser, S.A. (1985). Effect of humic acids on numbers and activities of micro-organisms within physiological groups. *Organic Geochemistry* 8(1): 81–85.

Walker, D.J. and Bernal, M.P. (2005). Plant mineral nutrition and growth in a saline Mediterranean soil amended with organic wastes. *Communications in Soil Science and Plant Analysis* 35(17-18): 2495–2514.

Walker, D.J. and Bernal, M.P. (2008). The effect of olive mill waste compost and poultry manure on the availability and plant uptake of nutrients in a highly saline soil. *Bioresource Technology* 99: 396–403.

Wang, H., Wu, Z., Zhou, Y., et al. (2012). Effects of salt stress on ion balance and nitrogen metabolism in rice. *Plant, Soil, and Environment* 58(2): 62–67.

Wang, X.J., Wang, Z.Q., and Li, S.G. (1995). The effect of humic acids on the availability of phosphorous fertilizers in alkaline soils. *Soil Use and Management* 11: 99–102.

Water Quality (2003). *Determination of nitrogen: determination of bound nitrogen (TN$db), following oxidation to nitrogen oxides*. London, UK: British Standards.

Whitehead, D.C. (1964). Identification of p-hydroxybenzoic, vanillic, p-coumaric and ferulic acids in soils. *Nature* 202: 417–418.

Wolf, T. (2018). Sprayers 101. http://www.sprayers101.com (accessed 1 August 2019).

Yan, N., Marschner, P., Cao, W., et al. (2015). Influence of salinity and water content on soil microorganisms. *International Soil and Water Conservation Research* 3: 316–323.

7
Conclusion

It is important to emphasize that low-rank coal is not the same as other, more well-known, coal materials on the market, i.e. anthracite, bituminous, and sub-bituminous. Low-rank coal has lower calorific values than those of other materials. It can still be combusted like any other organic matter (OM)-containing materials. In fact, according to Alberta Energy (2017), its main recommended use has been for the generation of electricity, while other coal materials are primarily used for metallurgical applications, household cooking, and heating fuel. With respect to its application in agriculture, as discussed in this book, low-rank coal is limited to lignite and leonardite, which includes Alberta low-rank coal (humalite), with calorific values of 19 300 kJ/kg or less (see Chapter 2).

While its use as a source of calories is justified due to the abundance of low-rank coal in the continent and the high demand of electricity, this practice results in the emission of carbon dioxide (CO_2) to the atmosphere. The question is: why not utilize low-rank coal for something with a much lower negative impact on the environment like agriculture, which is still beneficial to our needs? This question is salient due to the unique richness of this material as a source of humic substances (HS, or "humic"). However, it is important to note that this new application is not suggested as an immediate replacement for its current usage as a source of calories for the generation of electricity. Instead, the suggestion of a new application provides an alternative should an opportunity arrive for this new market sector to be generated and expanded.

From the perspective of scientific research, it has been known for many years that HS enhance the productivity of agriculture (i.e. soil, crops, and livestock). This knowledge can be found as early as the 1900s in Europe, followed by America, Asia, and other parts of the world (see Chapter 4). One of the most important societies in the world concerning HS has been the International Humic Substances Society (IHSS). Founded in 1981 in the US, this association

Low-Rank Coal Applications in Agriculture: Humic Analyses, Products, and Performance, First Edition. L. Edwin Liem.

establishes and maintains standard samples of humic acids (HA) and fulvic acids (FA) from lignite, fresh water, mineral soil, and organic soil. It also assembles their characterization data. IHSS promotes public understanding of HS by having international gatherings involving leading experts in collaboration with the IHSS committee. While some practical work has been completed and presented or published by its members, its main objective is to further scientific education and research concerning HS (International Humic Substances Society 2020).

On the practical side of things, the justification for a new use for low-rank coal begins with the raw material itself. Other coal materials are rich in OM, but they contain very little HS. There are materials other than low-rank coal that also contain HS. These include carbonaceous shale, natural soil or water, peat or peat moss, manure, compost, and municipal or industrial waste. The challenge here is whether any of this material is feasible for manufacture into commercial products. Factors to be considered here are the material's strength (or concentration) of HS, texture, moisture, and impurities (ash, heavy metals, pathogens, and toxic organics). Low-rank coal contains a very concentrated amount of HS, very small amounts of heavy metals, and negligible amounts of pathogens and toxic organics. With respect to handling, processing, and transportation, low-rank coal is classed as a non-hazardous material, an attribute many do not realize it has (see Chapter 2). While low-rank coal has unique properties, the collection of this material from the mine site, and its handling and processing into dry, liquid, pelletized, and dry soluble products is still reasonably manageable. The fact that it can practically and safely be manufactured into good-quality humic products of various types means that low-rank coal has an advantage over other OM-containing materials. Still, product handling at the client or sub-contractor's site must be done properly in order to ensure the integrity of the product as well as the practicality and safety of the work (see Chapter 5).

While it is indeed feasible to manufacture low-rank coal into commercial products, challenges to this practice have been raised by the agricultural community. As Lyons and Genc (2016) note, inconsistent results, exaggerated claims, and questionable application rates have been the main concerns regarding the use of commercial humic products. It has been recommended that future work be done to further explore product characterization, mechanisms and interactions with plants (including nutrient cycling), wider ranges of application rates, and different application methods.

The main objective of this book is to present the efforts and outcomes from the use of low-rank coal in agriculture. While most of the work has been focused on its practicality, the scientific and technical components of this topic have also been considered, and valid answers to the aforementioned challenges can be now be presented without any partiality or bias.

The first challenge concerns the consistency of commercial humic products. It is important to remember that for soil and crop applications, humic products function as soil amendments, not fertilizers. Numerous efforts have been made to apply information attained from laboratory studies (greenhouses and growth chambers) or test plots to commercial agricultural fields. Unfortunately, this has often led to mixed results. HS have been found to be effective in certain cases but not in others. Also, similar products have been found to perform differently when applied in similar growing conditions. These inconsistencies create major confusion and wariness of drawbacks with respect to the use of humic products in agriculture, where reliable profit is the determining factor.

A common assumption is that HS are no different than soil OM. Their recommended rates have to be high enough to build up significant amounts of OM on the fields to be effective. There is some truth to this statement in that dry humic products (non-water-soluble in nature) require high application rates on the fields to show positive results. A second common assumption is that low-rank coal is not beneficial for agriculture, like in the case of other coal materials. Good results obtained in the field, it is thought, are rather caused by other factors, independent of humic commercial products. This extreme view is not correct at all. However, there is an important insight that needs to be understood: higher rates of product application do not always translate into better results. In fact, in some cases, the responses are negative. A third assumption is that HS only work under certain conditions in the laboratory, but not in real-life applications. There is also some truth to this. HS have been found to work best in certain growing conditions, but this is regardless of whether they are in laboratories or in agricultural fields. In order to address these concerns, a review of the literature available in public domains and actual work with original data on commercial humic products has been completed, the results of which can be summarized as follows.

Commercial humic products may contain different strengths of active ingredients, which affect their effective rates of application. Product claims have to be interpreted to verify the actual contents of HA and FA, as products that claim to have similar functions can in fact be totally different from one another. Some products contain little or no humic components, but have similar properties and claim to have these components (also known as adulterants), resulting in confusion with respect to their application and efficacy. Amongst the various analytical methods, it is recommended that the colorimetric method and ISO 19822 be used as the standard methods for analysis. They are to be used side by side so that an accurate quantification of humic products can be made. Still, this method cannot differentiate certain adulterants that may skew the results. Under certain circumstances, more comprehensive analytical methods are required (see Chapter 3).

Commercial humic products can either be water-soluble or non-water-soluble. Non-water-soluble products require a much higher application rate than water-soluble products in order to provide the same response. However, this does not mean that water-soluble products are better. Each product has a different usage and purpose. Non-water-soluble products break down in soil slowly and are good for improving its quality over a longer period of time. It is common to apply these products in fall or winter for the next year's growing season. Water-soluble products provide readily available HS to soil or crops, and can be applied before or during seeding, or at the seedling stage. Application methods on the field also make a difference. For example, broadcast application always requires larger product amounts to have the same or similar results as that of side-band application (see Chapters 4 and 6).

Commercial humic products may contain components with different molecular weights. HA have a larger mean molecular weight than that of FA. HA products are suitable for both soil and foliar applications. FA products are best for foliar (not for soil) applications, as they may leach through the media. Processed products contain chemicals added during their manufacturing processes. These chemicals may contain nutrients within them, affecting product efficacy. Or, nutrients may be deliberately added to the finished products to boost their performance. When evaluating the efficacy of commercial products for the market, any effects due to the presence of nutrients must be negated in order to allow for a fair comparison (see Chapters 4 and 6).

Commercial humic products may not be compatible with agricultural chemicals. While their interactions are complex, pH (or acidity) is one of the most important parameters to consider. HA are not soluble at low pH, while FA are soluble at any pH. Fertilizers-containing phosphorus (P) may have neutral to low pH. Micronutrient-based fertilizers-containing copper (Cu), iron (Fe), manganese (Mn), and zinc (Zn) may have low to extremely low pH. Some herbicides, pesticides, and surfactants are effective at certain pH ranges. In addition, HS tend to be insoluble and form mushy blends when mixed with certain nutrients, including nitrogen (N), resulting in a reduced efficiency when applied on the field. Preliminary compatibility tests of commercial humic products and common nutrients in the laboratory are always recommended (see Chapter 6).

Higher rates of product application do not always lead to better results. Excessive amounts of commercial humic products in combination with nutrients have been shown not to improve yields, and even sometimes to reduce them. As is evident from the literature, HS, among other things, enhance living cells' metabolism and nutrients uptake. This may explain the adverse effects of excessive amounts of HS on the cell and organism in general. The toxic effects of common nutrients in excessive amounts have also been identified as a possible cause. Under an optimum growing condition, HS either have no effect or only a minor effect. Humic products show a more dramatic performance when used alongside unfavorable growing conditions (i.e. soil with low OM, high salinity, heavy sand, heavy clay, low moisture, and reduced nutrients). Field growing conditions vary with location, weather, climate, soil moisture, and soil quality, which are very difficult to replicate. Different crops react differently to HS and even different varieties of the same crop may perform differently. All of these factors underline the importance of product efficacy projects, completed under specific growing conditions, for identifying an optimum rate for a given commercial humic product in combination with nutrients (see Chapters 4 and 6).

The second challenge concerns the exact mechanism for how HS enhance the productivity of soil and crops. Laboratory studies, including growing crops in pots inside a greenhouse with a controllable growing environment, are able to explain the detailed mechanism. HS increase nutrient availability due to a larger uptake of nutrients by crops, which is most likely due to a better crop growth. The most accepted understanding of this mechanism is that HS promote the growth of roots and increase soil carbon (C) inputs (see Chapter 4). However, what actually happens on the field is more complicated, as it involves interactions between many variables. These variables include the types and characteristics of the growing media, the amounts of water and nutrient being applied, the crop types and varieties, the product rates and application methods, and the weather. Unless the information is intended only for crops grown in pots inside a greenhouse, it may not scale up well for those crops grown in agricultural fields. In many cases, laboratory results are not even recognized by end users and regulators as valid information that can be transferred to agricultural field work.

Most information in Chapter 6 is based upon crops grown in agricultural fields. There are other growing practices, including hydroponic and aquaponic systems, which are gaining more popularity in recent years. These practices use water instead of soil as the growing medium, resulting in a different path and mechanism by which HS and nutrients are taken up by crops. Most crops cultivated in these systems are also different from the ones grown in agricultural fields. The applied nutrients may not be the same either. For example, nitrates are the preferred source of N instead of urea, and so on. Specific projects need to be completed in order for the exact mechanism for how HS benefit crops grown in hydroponic and aquaponic systems to be identified.

The third challenge concerns the long-term benefits of commercial humic products. It is not easy to maintain the same or similar growing conditions for more than one growing season, even with only one variable to be substituted. In actual fields (or even test plots), many variables come into play, such as the weather and soil characteristics. It is often difficult to control for what and how to grow, since everything is based on market demand. Commercial humic products are often inconsistently applied from one season to another. This is especially true when no positive responses have been observed during the first try of the products and this discourages applications for subsequent years. Further efforts are recommended beyond the work completed in this book. It is ideal to select several commercial fields of at least one-quarter section (64 ha) each, located in different regions, dedicated only to product efficacy projects.

Humic products are to be applied continuously from one growing season to the next for several years. Crop rotations are to be carefully documented. Parameters on soil and crops should be consistently monitored before and after humic product applications. Actual nutrients taken up by crops and/or their accumulation in soil are then to be monitored and documented for many years of application. Most importantly, a long-term study with many variables with repeated measurements, which include adequate controls, is the only way to identify the exact mechanism for how HS benefit the productivity of soil and crops.

In addition, one of the most important parameters that can only be measured under a long-term condition concerns the microbiology aspects of the mechanism, on which this book has very limited information. HS do enhance the growth of plants and soil microorganisms, but it is also very important to note that beneficial soil organisms and plants live in symbiosis. Plant roots release exudates that are beneficial for the growth of soil organisms. In return, beneficial organisms provide mineral nutrients, suppress the growth of detrimental organisms, and produce plant hormones (see Chapter 4). The study of the effects of HS on microbiology may always be started from a laboratory under a controllable environment. When the results are promising, this preliminary work should be scaled up for an actual field accordingly.

While the main focus of this book has been on soil and crops, similar challenges have also been encountered in the case of livestock. HS have been proven in the laboratory to enhance the metabolic parameters of living cells and nutrients bioavailability, translating to improved feed efficiency as well as animal health, growth, and yields. Still, real-life applications are subject to interactions among variables, such as animal types, varieties, health, and ages, as well as applied nutrient amounts, product rates, application methods, and growing conditions (see Chapters 4 and 6). An exact mechanism for how HS enhance livestock productivity can only be identified through a long-term study on living animals, which will involve many variables with repeated measurements and adequate controls.

The fourth challenge concerns the economy of commercial humic products. The confusion around this issue arises because mixed results have been reported from different cost–benefit studies. Some have tended to claim that the products are not worth their dollar value. While there are many factors to be considered, this conclusion is not correct.

When improving crop yields, applying liquid (processed) products is generally a better choice than applying raw (dry) products, especially when they are broadcasted over the field. Raw (dry) products are economical when applied side-band close to the seeds, while their broadcast application is more beneficial to the soil. The improvement of soil quality, however, can only be observed over a long-term field evaluation. When maintaining crop yields at reduced nutrient inputs, humic products are generally economical when the nutrient costs are very high (see Chapter 6). Still, the reduction of nutrients is actually beneficial for the quality of soil, mainly on the growth and health of beneficial soil organisms. The improvement is, again, only observable over a long-term field evaluation. Cost–benefit studies based on multiple growing seasons can be completed to show that the improvement of soil quality, in the long run, results in improved crop yields. The economy of the use of humic products can therefore be justified.

The fifth challenge concerns the status of low-rank coal, HS, and commercial humic products within regulatory bodies. Not all of them recognize HS, that a commercial product has to be registered under a specific category, such as being a source of OM (see Chapter 4). It is true that OM is beneficial for agriculture, but HS have more useful properties that need to be addressed and recognized. Also, when registered as a source of OM, the recommended effective application rate tends to be inflated. This might be acceptable in the case of dry (raw) products (which are low in water solubility), since their recommended application rates are high in the first place. Liquid (processed) products require much lower application rates, due to their high water solubility. There is also a movement that may in the end lead to humic products being considered pesticides or plant biostimulants, which is not in line with the original purpose of these products for crop application.

A similar challenge has been observed in the livestock sector, where commercial humic products may have to be registered under categories such as coloring agents or a source of Fe (see Chapter 4). These challenges contribute to the confusion concerning efficacy, since they mean that the products are not being used as intended and, most importantly, the recommended rates given do not reflect those that actually provide optimum results.

In this context, the Humic Products Trade Association (HPTA) has been vocal about the needs of commercial humic manufacturers, not only with respect to regulatory issues, but also regarding standardization of analytical methods and the provision of technical information about HS to the public (see Chapters 3 and 4). HPTA has been in constant negotiations with various regulatory bodies and has shared the results of these with its members and the general public.

Despite all of these challenges, the good news is that HS are more and more understood and recognized by major regulatory bodies. In recent years, end users have become aware of the practical and economical benefits of humic products, and are incorporating them as an integral component of the whole fertilization program. More manufacturers are including HS as an ingredient in their specialty nutrient packages. Looking at the current trend on the use of HS in agriculture, the global demand for humic products is expected to significantly increase in the coming years. It has been claimed that the growing demand for HA and FA products in the world market will increase from $351 million in 2011 to $1.0 billion in 2020 (P&S Market Research 2015).

From a practical perspective, and considering all perceived limitations, it can be concluded that commercial humic products manufactured from low-rank coal (when applied at reasonable rates) have been proven to be beneficial for agriculture. HS enhance the growth and health of soil, crops, and livestock. Nutrients uptake is improved when humic products are incorporated within the fertilizing or feeding program. Yields are increased at the same nutrient input or maintained at a reduced input. Reducing the amount of added nutrients benefits soil in general. HS lower gaseous by-products in livestock production. Using low-rank coal in agriculture instead of as a source of calories also promotes a healthier environment since it effectively reduces the amount of CO_2 released into the atmosphere. This positions the coal industry in a more environmentally friendly direction, which is a beneficial scenario for both coal and agricultural sectors. In terms of economy, the cost–benefit analysis shows that there is a significant advantage to the use of commercial humic products in crop application. Additionally, the success of this alternative use of low-rank coal is expected to promote greater interest in the manufacture of other secondary or inferior-quality materials, by-products, or residuals into commercial products for the benefit of the agricultural sector and environment.

Information presented in this book is limited. Further work is encouraged on the consistency, mechanism, long-term effects, soil microbiology aspects, and current regulatory status of low-rank coal and humic applications in agriculture. Detailed work on hydroponic, aquaponic, and aquaculture systems, as well as livestock production, is greatly recommended. Another application of low-rank coal outside the agricultural sector that is worth mentioning concerns its pharmaceutical benefits for human health. And more on the environmental side, the implications of low-rank coal in sustainable agriculture for climate change and its long-term environmental impact also need to be explored.

In closing, this book has been written from a practical point of view and is intended for various readers, including field practitioners, end users, marketers, operators, regulators, researchers, those in academia, and others who are interested in this subject matter. Its main objective is to show that low-rank coal and HS are beneficial for agriculture, hopefully minimizing any confusion and misunderstanding about this topic. The science behind the work presented here is well substantiated and it is fair to conclude that humic products are effective and safe. Commercial humic products manufactured from low-rank coal at their reasonable and optimum rates are practical and economical for real-life applications on the field.

Reference

Alberta Energy (2017). Coal types. http://www.alberta.ca/coal-types.asp (accessed 13 August 2020).

International Humic Substances Society (2020). Introduction to the International Humic Substances Society (IHSS). www.humic-substances.org (accessed 4 February 2020).

Lyons, G. and Genc, Y. (2016). Commercial humates in agriculture: real substance or smoke and mirrors? *Agronomy (Basel)* 6(4) Article 50: 1–8.

P&S Market Research (2015). Industry insights by application (foliar, soil, seed) by crop type (row crops, turfs & ornamentals, fruits & vegetables, others). *Global Humic Acid Market Size, Share, Development, Growth and Demand Forecast to 2020* 26–33. Delhi, India.

Index

Low-Rank Coal Applications in Agriculture: Humic Analyses, Products, and Performance, First Edition. L. Edwin Liem.

m

Biography

Edwin completed his BSc (civil engineering) from Parahyangan Catholic University, Indonesia in 1990, where he was in the top 10% of his graduating class. He obtained his PhD (civil-environmental engineering) from the University of Alberta, Canada in 1998, specializing in physical, chemical, and biological processes for the treatment of water, waste, and soil.

Edwin worked on many local government- and corporate-sponsored projects during his MSc and PhD studies, and was also a recipient of a Canadian government research grant. During his employment with engineering companies, he was involved in various industrial projects, including work related to low-rank coal from 2001 to 2004. Since 2004, he has been in the product manufacturing business utilizing low-rank coal as the input material for oil/gas drilling, mining remediation, and agricultural applications. He is responsible for all technical aspects of the work, including securing government financial support for technology development.

Edwin is the first author of five articles published in major scientific journals. He has also published in other journals and magazines, and presented his work at conferences and seminars. He is a member of the Humic Products Trade Association, International Humic Substances Society, and Association of Professional Engineers and Geoscientists of Alberta. He lives in Edmonton, Alberta, Canada, and is accessible at www.linkedin.com/in/edwin-liem-44750234/.

Low-Rank Coal Applications in Agriculture: Humic Analyses, Products, and Performance, First Edition. L. Edwin Liem.

Figure 2.2 Warning label element.

Low-Rank Coal Applications in Agriculture: Humic Analyses, Products, and Performance, First Edition. L. Edwin Liem.

Figure 5.1 Dry HS

Figure 5.2 Liquid HA and FA

Figure 5.3 Pelletized and dry soluble HS

Figure 6.6 Early plant emergences.